Soil Hydrology
in a Changing Climate

Editors: Humberto Blanco, Sandeep Kumar and Stephen H. Anderson

CSIRO PUBLISHING

CRC Press
Taylor & Francis Group
Boca Raton London New York

CRC Press is an imprint of the
Taylor & Francis Group, an **informa** business

A catalogue record for this book is available from the National Library of Australia.

ISBN: 9781486313778 (hbk)
ISBN: 9781486313785 (epdf)
ISBN: 9781486313792 (epub)

Published in print in Australia and New Zealand, and in all other formats throughout the world, by CSIRO Publishing.

CSIRO Publishing
Private Bag 10
Clayton South VIC 3169
Australia

Telephone: +61 3 9545 8400
Email: publishing.sales@csiro.au
Website: www.publish.csiro.au
Sign up to our email alerts: publish.csiro.au/earlyalert

Published in print only, throughout the world (except in Australia and New Zealand), by CRC Press, with ISBN 9781032372068.

CRC Press
6000 Broken Sound Parkway NW, Suite 300, Boca Raton, FL 33487-2742
and
2 Park Square, Milton Park, Abingdon, Oxon, OX14 4RN
Website: www.routledge.com

CRC Press is an imprint of Taylor & Francis Group, LLC

Front cover: (top) A flooded switchgrass field near Mead, eastern Nebraska (photo by Humberto Blanco); (bottom left to right) A field after corn silage harvest in the western US Corn Belt; A corn field intercropped with summer cover crops near Sao Paulo, Brazil; Cropland east of Clay Center, Nebraska in spring (photos by Humberto Blanco)

Edited by Natalie Korszniak
Cover design by Cath Pirret
Typeset by Envisage Information Technology
Index by Max McMaster
Printed in China by 1010 Printing International Ltd

CSIRO acknowledges the Traditional Owners of the lands that we live and work on across Australia and pays its respect to Elders past and present. CSIRO recognises that Aboriginal and Torres Strait Islander peoples have made and will continue to make extraordinary contributions to all aspects of Australian life including culture, economy and science. CSIRO is committed to reconciliation and demonstrating respect for Indigenous knowledge and science. The use of Western science in this publication should not be interpreted as diminishing the knowledge of plants, animals and environment from Indigenous ecological knowledge systems.

The paper this book is printed on is in accordance with the standards of the Forest Stewardship Council® and other controlled material. The FSC® promotes environmentally responsible, socially beneficial and economically viable management of the world's forests.

MIX
Paper from responsible sources
FSC www.fsc.org FSC® C016973

Jul22_01

Contents

Preface

Fluctuating climates with frequent droughts, flooding episodes, heat waves and other weather extremes are causing major challenges for soil and water management in many parts of the world. Developing management practices that maintain or enhance soil resilience within such challenges is a high priority. A discussion of how different soil management practices impact soil hydrological characteristics can improve our understanding of soil and water management under changing conditions. This book is designed for land managers, soil conservationists, researchers and others involved in soil and water conservation and management.

Soil hydrology includes water infiltration, soil water storage and drainage, all of which are critical processes for agricultural production. Efforts to modify or redesign current soil management practices to improve soil hydrological characteristics need evaluation to sustain agricultural production and meet the demands of the growing human population. For example, improved soil management practices should store more soil water under hotter and drier conditions and release water under wetter conditions resulting from fewer but more intense rainstorms and flooding events.

Soil management determines soil hydrological properties, which, in turn, influence how agricultural systems will respond to changing conditions. Managing soils is about adjusting current soil management practices to improve soil hydrological characteristics. Current soil management practices were designed to address past climate. Redesigning or developing improved soil management practices must be a goal to address current and future fluctuating climates. This book was designed to discuss these concerns.

Chapters 1 and 2 of the book examine the role of land use dynamics in the climate system and interactions between climate and managed terrestrial ecosystems. They also discuss the role of climate changes in soil hydrology and provide a broad overview of the implications of fluctuating climates on soil water dynamics. Chapter 3 deals with how different soil management practices impact soil health and its relationships with soil water dynamics. It synthesises information and assesses the relationship between soil health and soil water under different agricultural systems.

Chapters 4–8 discuss relationships of conservation tillage systems, diversified cropping systems, crop residue management, conservation buffers and forest land use with soil hydrology. They specifically discuss the impacts of these management practices on soil hydraulic processes, including water infiltration, water storage and water use efficiency. The implications of improved management practices, including no-till farming, the use of cover crops, crop residue baling and grazing and the use of annual and perennial vegetation, on soil water are discussed. Chapter 9 covers a growing topic in soil water management, which is how urban land management affects stormwater run-off and water dynamics. It also discusses how the rapidly changing urban environments influence overall soil hydrology and the resilience of urban systems against fluctuating climates.

Chapter 10 focuses on wetland ecosystems and soil hydrological processes and how these systems are affected by changing climates. It provides insights into wetland management for managing soil water. Chapter 11 discusses how hydrological modelling can assist with soil water management decisions under fluctuating climates. It provides case studies of modelling of soil water management.

The editors and authors hope this book will be a valuable resource for land managers, soil conservationists and others to better understand how different management practices affect soil and water dynamics and how such management practices can be adjusted to enhance agricultural sustainability and environmental quality under fluctuating climates.

List of contributors

Rebecca Abney
Forest and Disturbed Soils, Warnell School of Forestry and Natural Resources, University of Georgia, Athens, Georgia, USA

Salah M. Alagele
Department of Desertification Combat, College of Agricultural Engineering Sciences, University of Baghdad, Baghdad, Iraq

Stephen H. Anderson
Soil and Environmental Sciences, School of Natural Resources, University of Missouri, Columbia, Missouri, USA

Nicholas J. Balster
Department of Soil Science, University of Wisconsin – Madison, Madison, Wisconsin, USA

Claire Baffaut
USDA-Agricultural Research Service, Cropping Systems and Water Quality Research Unit, Columbia, Missouri, USA

R. Louis Baumhardt
USDA-Agricultural Research Service, Conservation and Production Research Laboratory, Bushland, Texas, USA

Humberto Blanco
Department of Agronomy and Horticulture, University of Nebraska – Lincoln, Lincoln, Nebraska, USA

Edward P. Boswell
Department of Planning and Landscape Architecture, University of Wisconsin – Madison, Madison, Wisconsin, USA

Christine Costello
Department of Agricultural and Biological Engineering, Pennsylvania State University, University Park, Pennsylvania, USA

Aaron Lee M. Daigh
Department of Soil Science, School of Natural Resource Sciences, North Dakota State University, Fargo, North Dakota, USA

Nandita Gaur
Soil Physics, Crop and Soil Science Department, University of Georgia, Athens, Georgia, USA

Sagar Gautam
Sandia National Laboratories, Pleasanton, California, USA

Glenn R. Guntenspergen
USA Geological Survey, Eastern Ecological Science Center, Laurel, Maryland, USA

Jerry L. Hatfield
USDA-Agricultural Research Service, National Laboratory for Agriculture and the Environment, Ames, Iowa, USA (retired)

Gary L. Hawkins
Water Resource Management, Crop and Soil Science Department, University of Georgia, Athens, Georgia, USA

Jane M. F. Johnson
USDA-Agricultural Research Service, North Central Soil Conservation Research Laboratory, Morris, Minnesota, USA

Carter W. Johnson
Department of Natural Resource Management, South Dakota State University, Brookings, South Dakota, USA

Sandeep Kumar
Department of Agronomy, Horticulture, and Plant Science, South Dakota State University, Brookings, South Dakota, 57007, USA

Rattan Lal
Center for Carbon Management and Sequestration, The Ohio State University, Columbus, Ohio, USA

Robert J. Lascano
USDA-Agricultural Research Service, Cropping Systems Research Laboratory, Lubbock, Texas, USA

Matthew R. Levi
Soil Pedology, Crop and Soil Science Department, University of Georgia, Athens, Georgia, USA

Ali Mehmandoost
Earth and Life Institute Environmental Science, Universite Catholique de Louvain, Louvain-la-Neuve, Belgium

Peter L. O'Brien
USDA-Agricultural Research Service, National Laboratory for Agriculture and the Environment, Ames, Iowa, USA

Quang Phung
Department of Biomedical, Biological and Chemical Engineering, University of Missouri – Columbia, Columbia, Missouri, USA

Sabrina J. Ruis
Department of Agronomy and Horticulture, University of Nebraska – Lincoln, Lincoln, Nebraska, USA

Udayakumar Sekaran
Department of Agricultural Sciences, Clemson University, Edisto Research and Education Center, Blackville, South Carolina, USA

Allen Thompson
Biomedical, Biological and Chemical Engineering, University of Missouri – Columbia, Columbia, Missouri, USA (retired)

Anita M. Thompson
Department of Biological Systems Engineering, University of Wisconsin – Madison, Madison, Wisconsin, USA

Ranjith P. Udawatta
School of Natural Resources and Center for Agroforestry, University of Missouri, Columbia, Missouri, USA

Kenneth M. Wacha
USDA-Agricultural Research Service, National Soil Erosion Laboratory, West Lafayette, Indiana, USA

1

Land use changes and fluctuating climate

Rattan Lal

Introduction

The Anthropocene (Crutzen 2006) began with the onset of agriculture ~10 000 years ago (Ruddiman 2003), but has intensified since the Industrial Revolution *c.* 1750. Settled agriculture was triggered by the so-called 'long summer' after the last Ice Age, when the climate warmed and stabilised. In turn, settled agriculture, and the attendant increased access to food, was the principal driver of the increase in the human population from 5–10 million approximately ten millennia ago to 7.8 billion in 2020, with projections of 9.8 billion by 2050 and 11.4 billion by 2100 (United Nations 2019). Because of the expansion of agriculture, ~40% of the ice-free land has been converted from natural ecosystems (i.e. forests, prairies, wetlands, alluvial floodplains) to agroecosystems. Agroecosystems occupy ~5 billion ha of land, of which 1.5 billion ha is under cropland and 3.5 billion ha under grazing lands. The expansion of agriculture, along with urbanisation, industrialisation and the use of fossil fuel, has changed the atmospheric chemistry (Table 1.1; World Meteorological Organization (WMO) 2020). Historically, land use and land use change (LULUC) are the major contributors to emissions of carbon dioxide into the atmosphere (Table 1.2; Friedlingstein *et al.* 2020).

The data presented in Table 1.1 show that between 1750 and 2019, atmospheric concentrations of carbon dioxide (CO_2) increased by 148% (from 278 to 410 ppm), those of methane (CH_4) increased by 260% (from 722 to 1877 ppb) and those of nitrous oxide (N_2O) increased by 123% (from 270 to 332 ppb; WMO 2020). These marked changes in atmospheric chemistry were brought about by LULUC since the onset of settled

Table 1.1. Changes in atmospheric concentrations of greenhouse gases since c. 1750 (adapted from World Meteorological Organization 2020)

Pre-Industrial concentrations of CO_2, CH_4 and N_2O were 278 ppm, 722 ppb and 270 ppb, respectively

	CO_2 (ppm)	CH_4 (ppb)	N_2O (ppb)
Mean (±s.d.) 2019 data	410.5 ± 0.2	1877 ± 2	332.0 ± 0.1
Increase since 1750	148	260	123
Absolute increase (2018–19)	2.6	8	0.9
% Increase (2018–19)	0.64	0.43	0.27
Mean annual increase (year^{-1})	2.37	7.3	0.36

Table 1.2. Estimated emissions of carbon from land use and land use change and fossil fuel combustion (adapted from Friedlingstein *et al.* 2020)

Unless indicated otherwise, data are given as the mean±SD

	Cumulative flux (Gt C)	
	1750–2019	1850–2020
Source		
Fossil fuel	445 ± 20	455 ± 20
Land use change	255 ± 70	210 ± 60
Total emissions	700 ± 75	665 ± 65
Sink		
Atmosphere	285 ± 5	270 ± 5
Ocean	170 ± 20	165 ± 20
Terrestrial	230 ± 60	215 ± 55
Residual	20	20
Terrestrial sink (% total emissions)	32.9	32.3

agriculture. Ruddiman (2003) estimated that cumulative emissions from LULUC since the onset of agriculture to *c.* 1750 were ~320 Pg C. Between 1750 and 2019, cumulative C emissions from LULUC are estimated at 255 ± 70 Pg, compared with 445 ± 20 Pg from fossil fuel combustion (Table 1.2). Of the 700 Pg of anthropogenic emissions since *c.* 1750, 36.4% were contributed by LULUC. During 2019, emissions from fossil fuel combustions were 10 Pg, compared with 1.5 Pg (13%) from LULUC (Friedlingstein *et al.* 2020). Thus, LULUC has strongly perturbed the global carbon (C) budget and has caused marked changes in biogeochemical and biogeophysical cycles (Pielke *et al.* 2002; Feddema *et al.* 2005; Krinner *et al.* 2005; Sitch *et al.* 2005; Fig. 1.1). In combination with emissions from fossil fuel combustion, changes in atmospheric chemistry due to LULUC have increased global temperatures by almost 1.2°C since the Industrial Revolution, and the global mean temperature is currently increasing at a rate of 0.2°C per decade.

Therefore, the objective of this chapter is to deliberate on the LULUC–climate change nexus. The specific goal is to describe the positive feedback from LULUC on climate change, and how the trend can be reversed to a negative feedback through re-carbonisation of the terrestrial biosphere. The last row of data in Table 1.2 shows that land-based sinks have absorbed ~32% of the anthropogenic emissions since *c.* 1750. Therefore, another objective of this chapter is to identify policy interventions that can promote: (1) the adoption of LULUC to lead to negative emissions (i.e. withdrawal of CO_2 from the atmosphere and transferral into vegetation and soil); (2) management systems that adapt and mitigate anthropogenic climate change; and (3) the adoption of policies that restore degraded soils, depleted ecosystems and denuded landscapes.

LULUC and global warming

Natural ecosystems have large C reserves, with diverse above- and below-ground components (Fig. 1.2). Two primary components are biomass-C and soil-C. Biomass-C comprises above-ground biomass (AGB), below-ground biomass (BGB) and living biomass or biota in soil. The live biomass in a healthy soil weighs ~5 Mg ha^{-1}. Soil-C comprises two separate

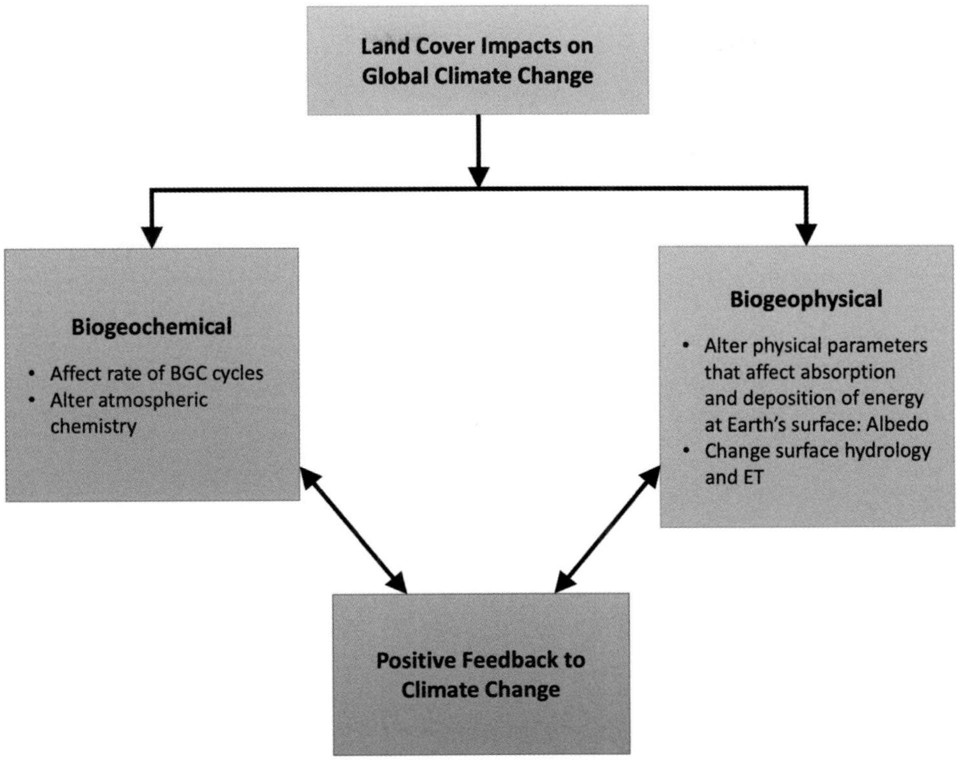

Fig. 1.1. Effects of land use and land use conversion on global climate change. BGC, biogeochemical; ET, evapotranspiration.

Table 1.3. Effect of land use and land use change on the decadal scale global carbon budget (recalculated from Friedlingstein *et al.* 2020)

Total anthropogenic emissions are given as the mean ± s.d. over the decade specified. LULUC, land use and land use change

	1960–69	1970–79	1980–89	1990–99	2000–09	2010–19	2019
Total anthropogenic emissions (Pg C $year^{-1}$)	4.5 ± 0.7	5.9 ± 0.7	6.7 ± 0.8	6.7 ± 0.8	6.7 ± 0.8	10.9 ± 0.9	11.5 ± 0.9
Emissions from LULUC (% of total emissions)	33.3	22.0	19.4	18.4	15.4	14.7	15.7
Terrestrial sink (% of total emissions)	28.8	35.6	29.9	34.2	31.9	31.2	27.0

but related components: soil organic carbon (SOC) and soil inorganic carbon (SIC). In addition to live biomass, which is labile and thus highly dynamic, another two components of SOC are partially and completely decomposed organic materials. The latter is called humus, which is relatively recalcitrant and has a long mean residence time (MRT) in the soil. The SIC consists of pedogenic and lithogenic carbonates. Pedogenic carbonates are

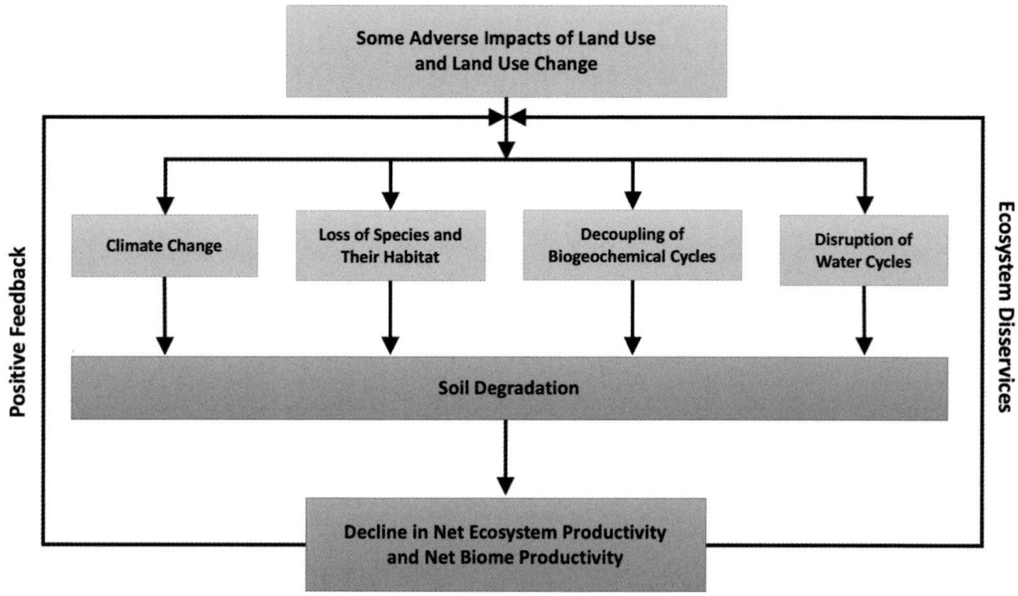

Fig. 1.2. Schematic diagram showing the adverse effects of land use and land use change on global warming and aggravation of reductions in ecosystem services.

formed within the soil (Fig. 1.2, arrow), linking living biomass to pedogenic carbonates through the release of CO_2, its dissolution to form a dilute carbonic acid and reprecipitation upon reaction with cations brought in from outside the system. Lithogenic carbonates are derived from the weathering of the parent material. The combined total magnitude of the SOC and SIC pools in world soils to a depth of 1 m is estimated at 2500 Pg C (Lal *et al.* 2021); this is an important component of the global carbon budget and carbon dynamics and has a strong effect on the global carbon cycle. Specifically, large-scale LULUC leads to transfer of C (and other elements, such as N, P and S) from the terrestrial biosphere into the atmosphere and hydrosphere. Therefore, LULUC creates a C deficit through the loss of C hitherto contained in both the biomass (above and below ground) and soil. In addition to adverse effects on the quality and quantity of ecosystem services (e.g. food, feed, fibre, fuel), the loss of C from the terrestrial biosphere creates a positive feedback and accelerates anthropogenic global warming.

Under extreme conditions, decoupling of some critical biogeochemical cycles (e.g. C, water, N) could lead to the depletion of terrestrial C stocks contained in the soil and vegetation, the emission of greenhouse gases (GHGs), acceleration of global warming and intensification of drought/flood cycles. For example, under an extreme climate change scenario, temperature is projected to increase by 3.8°C and precipitation is projected to decrease by 30% in the Amazon region (Lapola *et al.* 2011). Under these conditions, the projected climate change may adversely affect crop yields, such as reducing the yield of soybeans by 44%, leading to additional deforestation of 181 000 km^2 (+20%) in the Amazon and 240 000 km^2 (+273%) in the Cerrado compared with a moderate climate change scenario (Lapola *et al.* 2011). Therefore, reducing C emissions by protecting natural ecosystems and averting new deforestation are critical for the sustainable management of natural resources and to minimise the risk of accelerated global warming. Over and above the increased risks of global warming, deforestation and LULUC also lead to loss of biodiversity (Feeley and

Rehm 2012; Manchego *et al.* 2017). Gomes *et al.* (2019) reported that deforestation could decrease Amazon tree species richness by up to 58%. Furthermore, deforestation could lead to a more intense dry season (Staal *et al.* 2020) because of the loss of water through surface run-off and evaporation.

Although industrial emissions from fossil fuel combustion contribute as much as 87% to total anthropogenic emissions, for example accounting for 10 Pg of the total emissions of 11.5 Pg C in 2019 (Friedlingstein *et al.* 2020), there is a need for an objective consideration of a broader paradigm of addressing global warming by adopting a holistic approach to the management of the terrestrial biosphere and its C reserves (soil, vegetation, wetlands). Careful management of global LULUC is critical to mitigating climate change and avoiding ecosystem collapse (McAlpine *et al.* 2010). Furthermore, protection of ecologically sensitive ecoregions (e.g. tropical rainforest, wetlands, peatlands, steep lands) is also essential to save critical habitat for wildlife and maintain habitat connectivity to reduce species extinctions (Tabor *et al.* 2018).

Effects of LULUC on the global carbon cycle

Historic and current LULUC have strongly affected the global C cycle. Table 1.3 presents the global C budget on a decadal scale from 1960 to 2019, and highlights the importance of LULUC as both a source and a sink of atmospheric CO_2. A careful review of the data shows that LULUC as a source (expressed as a percentage of total anthropogenic emissions) is much less than LULUC as a land-based C sink (also expressed as a percentage of total emissions). Therefore, conversion to a restorative land use (by sparing land and returning it to nature) can have a negative effect on anthropogenic global warming. This concept of sparing land for nature (Lal 2021) is an important policy consideration whose significance cannot be overstated (Fig. 1.3). The strong nexus, or interaction, between vegetation cover, climate change and anthropogenic perturbations must be clearly understood for biome-specific situations.

The permanence, or MRT, of SOC/biomass C also depends on the current state and future trends of the terrestrial biosphere in relation to historic climate variability and anthropogenic land use. Kaplan *et al.* (2012) studied European terrestrial ecosystems over 500 years from 1500 to 2000 and concluded that, despite the effect of land use change on the C inventory, the current state of the terrestrial biosphere in Europe is controlled mostly by LULUC during the 20th century. Sequestration of C in the terrestrial biosphere between 1500 and 2000 was higher than that during the 20th century because the increase in photosynthesis by the CO_2 fertilisation effect is limited by leaf transpiration and drought. However, anthropogenic land use change caused emissions of ~25 Pg C during the 20th century. Kaplan *et al.* (2012) concluded that, with judicious management, European forests have the capacity to sequester 5–12 Pg C.

Similar trends in soil/biomass C dynamics have been surmised from studies of the prehistoric land use of ancient civilisations. A pertinent example is that of the Maya Lowlands of Mexico and Guatemala. Records indicate that deforestation by the ancient Maya began ~4000 years ago (Douglas *et al.* 2018). Mean soil transit time (MTT) was assessed by Douglas *et al.* (2018) on the basis of plant waxes (MTT_{wax}) in sediment cores from three lakes in the Maya Lowlands. The sediment core data showed a decrease in MTT_{wax} for the period of 2300 to 800 BCE over the past 3500 years, indicating a relatively shorter storage time for C in lake catchment soil. The shorter storage time was attributed to deforestation. Despite subsequent reforestation, Kaplan *et al.* (2012) argued that

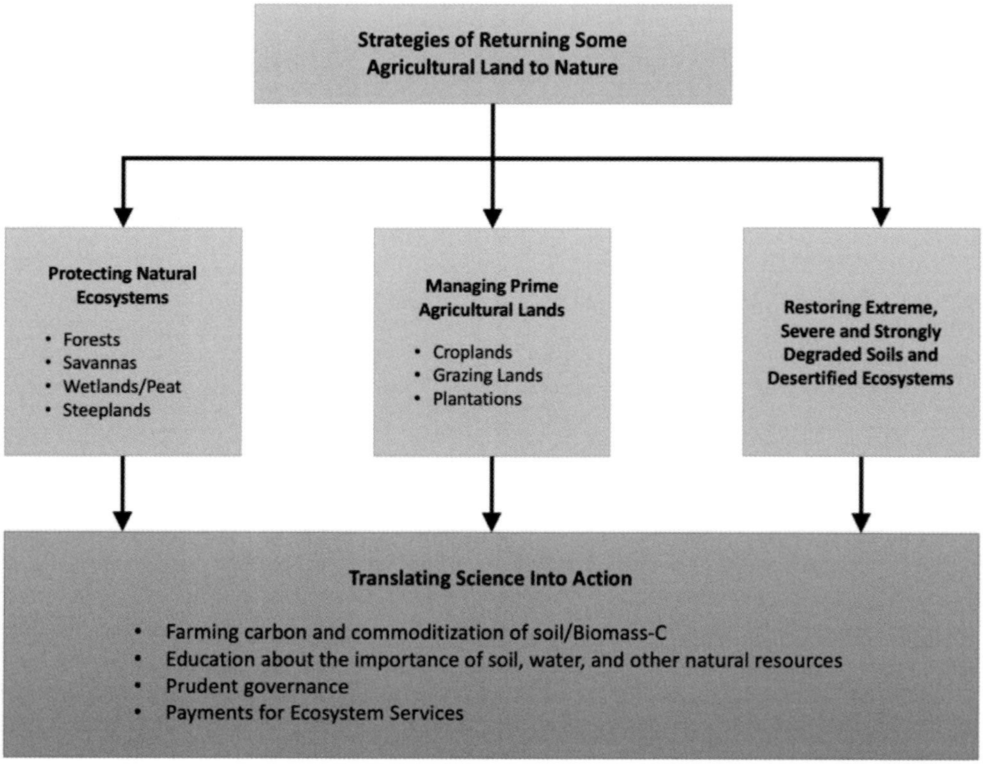

Fig. 1.3. Strategies for returning agriculturally marginal lands to nature in ecologically sensitive ecoregions.

MTT_{wax} never recovered to pre-deforestation levels. Therefore, the present large-scale deforestation in the tropics (Amazon, Indonesia, Malaysia, central Africa) may have a long-lasting effect on soil C sinks, which may not recover even a long time after reforestation.

Stinchcomb *et al.* (2011) studied pre-colonial (1100–1600 AD) sedimentation related to prehistoric maize agriculture and climate change in eastern North America. They reported geoarchaeological evidence of increasing floodplain sedimentation and prehistoric land use intensification in the Delaware River Valley during the Medieval Climate Anomaly–Little Ice Age transition. Stinchcomb *et al.* (2011) documented that the combined effect of prehistoric land use and climate change strongly influenced eastern North American floodplain development several centuries before the onset of European migration. Gierga *et al.* (2016) reported similar findings based on compound-specific radiocarbon analysis of Soppensee Lake sediments, whereby early prehistoric land use led to remobilisation of pre-aged SOC.

Together, these studies document the following: (1) there is a strong interaction between land use, climate change and human perturbation on vegetation and terrestrial C stocks and fluxes; (2) deforestation has a long-lasting effect on SOC stocks, which may not recover to pre-disturbance levels for centuries; and (3) deforestation markedly reduces the MRT of soil and overall terrestrial C stocks and aggravates global warming. These concepts have also been demonstrated by others, as outlined in Table 1.4.

Table 1.4. Effects of deforestation and land use change (LUC) on climate change

GHG, greenhouse gasses

Parameter	Effects	References
Temperature	Deforestation increases land surface and near-surface temperatures	Stoy (2018)
Local climate and weather	Increased temperatures (~2°C) in deforested compared with centres of forested areas Decreased latent heat flux, increased sensible heat flux and changes in the surface energy budget lead to changes in precipitation patterns Deforestation reduces evapotranspiration and increases rainfall in some areas, decreases rainfall in other areas, and changes regional climate Deforestation reduces moisture cycling, as well as regional rainfall (by up to 40%) There is a notable meridional (equator to pole) gradient in the climate response to LUC	Llopart *et al.* (2018), Spracklen *et al.* (2018), Perugini *et al.* (2017), Pires and Costa (2013), van der Molen *et al.* (2011)
Carbon cycle	LUC affects terrestrial C stocks Climate change can alter the C storage capacity of the terrestrial biosphere and the land-based C sink capacity	Humpenöder *et al.* (2015)
Drought	Fire and drought as a result of deforestation and increased GHGs may aggravate forest degradation via positive feedback	Coe *et al.* (2013)

Effects of vegetation on the hydrological cycle under changing climate

Hydrological processes are sensitive to climate change (Rasouli *et al.* 2019). Climate and land use changes are expected to intensify the water cycle, increase surface run-off, reduce soil water storage and increase the intensity and duration of pedological and agronomic drought. Therefore, it is pertinent to identify and implement site-specific management strategies to minimise land use-induced changes to the hydrological cycle. Furthermore, the hydrological regime is affected not only by land use, but also by climate change and land use changes, especially deforestation and the conversion of natural to managed ecosystems, which may lead to aridisation of the landscape (Fig. 1.4).

Based on a study in the Xinanjiang basin, east China, Guo *et al.* (2020) reported that both land use and land cover change, alone and in combination, increased stream flow. The direction of stream flow also depends on the magnitude of the impermeable area. Guo *et al.* (2020) observed that, for a lower impermeable area, the effects of climate change are stronger than those induced by land use/land cover change, but that land use/land cover change has a greater effect in a more impermeable area. Regardless, changes in annual mean stream flow are primarily driven by land use/land cover change, and climate change may decrease the effects of land use/land cover change.

Some modelling studies indicate strong warming and a strong decline in precipitation over southern Europe and the Mediterranean basin (Alo and Anagnostou 2017). However, most models consider a static vegetation cover, or vegetation fixed at present-day levels. Taking into consideration changes in the structure of vegetation over time can significantly alter the

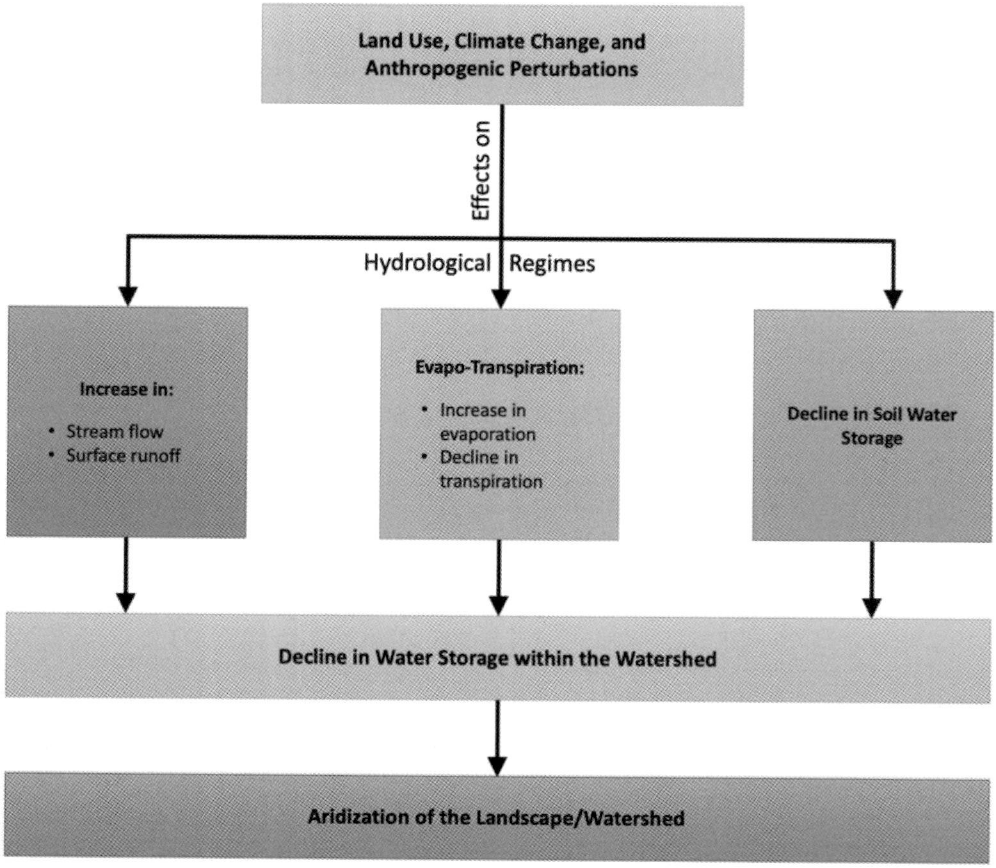

Fig. 1.4. Deforestation, global warming and anthropogenic perturbations may aggravate aridisation and depletion of the moisture regime with long-term consequences on net ecosystem productivity, net biome productivity and terrestrial (soil, vegetation) C stocks. SIC, soil inorganic carbon; SOC, soil organic carbon.

projected results (Alo and Anagnostou 2017). Similar observations were reported for future climate changes over West Africa taking into consideration the climate–vegetation interaction: the effect of dynamic vegetation feedback was more pronounced in semi-arid regions, with little effect in the humid regions of West Africa (Yu *et al.* 2016). Global vegetation distribution is also influenced by climate change and anthropogenic activities. However, the relative impact of anthropogenic perturbations versus climate change is neither widely studied nor adequately understood. In this regard, Zhang and Huang (2019) compared the existing vegetation under human influence with the potential vegetation without human interference, observing that human perturbations caused stronger damage to global vegetation change than those induced by climate change. Zhang and Huang (2019) defined 'eco-fragile' regions as those where vegetation experienced both human disturbance and climate change.

Tsarouchi and Buytaert (2018) observed that land use change may exacerbate the effects of climate change on water resources in the Ganges basin, which, at present, experiences monsoon flooding every year. It is feared that the future water demands in the upper Ganges region for the winter months may not be met. For a tropical catchment in Indonesia, Marhaento *et al.* (2018) observed that, individually, land use change and climate change will cause changes in water balance components, and that more drastic changes may occur if these factors are combined. For the Sudbury, Assabet and Concord rivers watershed in Massachusetts in the USA, Talib and Randhir (2017) observed that changes in climate and land cover would significantly increase run-off by 2100, as well as stream discharge.

Based on a modelling study of the Owabi catchment using the Soil-Water-Assessment-Tool (SWAT), Osei *et al.* (2019) observed that forest cover and topography play important roles in water loss, whereas evapotranspiration and surface run-off are the principal modulating processes. However, the effects of vegetation cover and physiology on the hydrological cycle are also influenced by the CO_2 fertilisation effect (refer above to the work of Kaplan *et al.* (2012) for Europe described above). Lemordant *et al.* (2018) also documented that the CO_2 fertilisation effect has a dominant role in evapotranspiration and evaporation fraction changes, and may have a major role in the long-term run-off component of the hydrological cycle. Therefore, predicting how increased concentrations of CO_2 will affect the hydrological cycle depends on the availability of N (and P, K, S), plant-available water capacity and the prevailing climate. Changes in the hydrological regime are driven by changes in precipitation and temperature regimes because of climate change, and that the land surface cover adjusts accordingly because the C and hydrological cycles are strongly coupled over land. These are priority researchable questions that need to be addressed.

Conclusions drawn from several studies show that LULUC and climate are two major controls that affect water fluxes and the hydrological balance across a broad range of spatiotemporal scales, and the strongly interconnected effects of these two controls are moderated by the CO_2 fertilisation effect and its determinants. Understanding these effects and their interactions are important for the identification and implementation of appropriate strategies for the sustainable management of water resources. It is precisely in this context that Wang and Van Oost (2019) documented the effects of these factors on an arboreal headwater catchment in Scotland and documented that land cover had stronger effects than the projected climate change. Wang and Van Oost (2019) reported that complete replacement of shrubs with trees under future climate change may increase evapotranspiration by 39% and evaporation by 21%.

Land use effects on vegetation and plant communities

LULUC affects the vegetation cover, and changes the dominant species in managed compared with natural ecosystems (Table 1.5). Based on the information for historical land use and climate change on tree–climate relationships in the Midwestern United States, Goring and Williams (2017) reported that the tree–climate relationships for 15 tree genera have undergone significant changes over the last two centuries (since *c.* 1800).

Human-induced drivers of global warming include: (1) the emission of GHGs from fossil fuel combustion; (2) deforestation and land use conversion; (3) increased transport of C and N, and their redistribution over the landscape; (4) changes in the disturbance regime; (5) biotic invasion and a shift in plant communities; and (6) changes in ground cover (Franklin *et al.* 2016). Therefore, being able to predict the effects of global warming on terrestrial plant communities can be important because the latter are a source of numerous and essential ecosystem services, including ecosystem productivity. Franklin *et al.* (2016) observed that persistent effects of land use change and human-induced fire regimes on vegetation can overshadow or interact with climate change. Site-specific experiments and monitoring studies are needed to be able to predict global vegetation changes in the

Table 1.5. Some examples of the effects of tropical deforestation on climate change and species diversity

Study	Region/process	Principle finding
Vargas Zeppetello *et al.* (2020)	Tropical deforestation	Local warming is larger than predicted from more than a century of climate change, with most extreme warming under large patches of deforestation
Senior *et al.* (2019)	Tropical forests	Extensive deforestation can impede range shift, increasing vulnerability to climate change
Spracklen *et al.* (2018)	Tropical deforestation	Deforestation reduces evapotranspiration, increases surface temperatures, alters rainfall patterns and reduces moisture recycling
Feeley and Rehm (2012)	Amazon forest	Deforestation can aggravate barriers to species migration and decrease interconnectivity
Franchito *et al.* (2012)	Tropical land, savannisation and climate change	The savanna biome may increase by 7.5% in the future, leading to the savannisation of tropical forests
Fearnside (2010)	Tropical rainforest	Deforestation leads, via positive feedback, to more global warming
ter Steege (2010)	Tropical regions	Massive conversion and degradation of natural habitat has detrimental effects on biodiversity and humanity
Poulter *et al.* (2010)	Amazon forest	Model study showed that by 2050 deforestation and fire would result in C loss of 7.4–20.3 Pg Deforestation and climate change may contribute 26–36 Pg C by 2100

21st century because different components of global change can have interacting effects on biodiversity and its community-level indicators (Clavero *et al.* 2011).

Urbanisation, especially in developing countries, is an important ramification of the Anthropocene in the 21st century, and its effect on land use change and the attendant consequences with regard to global warming must be understood. In the middle and lower reaches of the Yangtze River, Yao *et al.* (2015) assessed the effects of urbanisation on regional temperature and precipitation by using spatial analysis and numeric methods for the 20-year period from 1988 to 2008. Yao *et al.* (2015) reported that land use conversion by urbanisation significantly affected the regional temperature, but the effects on precipitation were not significant. Significant effects of urbanisation on food production, supply chains and future agricultural activities must also be addressed in the context of climate change and globalisation (Juliá and Duchin 2013).

Living in harmony with nature through afforestation and re-carbonisation of the terrestrial biosphere

There is a broad-based realisation that innovative management of the terrestrial biosphere has numerous co-benefits to humanity and nature (Table 1.6; Fig. 1.5). Thus, protecting natural ecosystems and avoiding deforestation are critical to limiting global warming (Chow *et al.* 2013), and there is a strong interest among policy makers in the sustainable management of land-based C sinks. This interest translated into the global call at the COP 21 declaration in Paris in 2015 regarding the need to limit global warming to 2°C, and preferably to 1.5°C (United Nations Framework Convention on Climate Change 2016;

Table 1.6. Examples of co-benefits of C sequestration in terrestrial ecosystems

Benefit	Principal findings	References
Increased productivity	Climate-resilient crop production by reducing risk of soil degradation	Droste *et al.* (2020)
Enhanced soil protection against erosion by creating healthy soils	Afforestation and forest management may reduce the risk of erosion by extreme events through C storage	Rodrigues *et al.* (2020), Op de Hipt *et al.* (2018), Averett (2016), Lacoste *et al.* (2015), Kostadinov (2010), DeLong *et al.* (2015)
Global warming limited to 2°C	Adverse effects of global warming on soil degradation and the decline in crop yield decline in Europe must be addressed Soil degradation may lead to caloric vulnerability up to 50 GCal ha^{-1}	Balkovič *et al.* (2018)
Soil characteristics as tools	Soil characteristics can be used to manage anthropogenic environmental changes and soil degradation that have adverse effects on food security; thus, soil management can preserve valuable ecosystem services from forested lands and improved cropping systems	Peterman and Bachelet (2012)

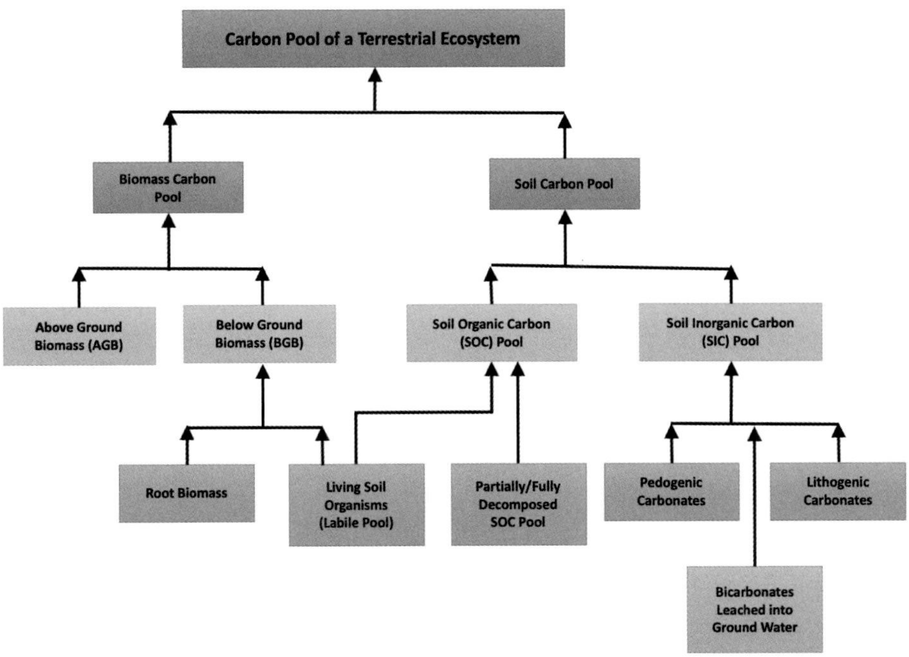

Fig. 1.5. Components of the C pool in an undisturbed terrestrial ecosystem.

Intergovernment Panel on Climate Change 2021). One of the global issues launched at COP 21 was the '4 per Thousand' or sequestration of C in world soils to 40-cm depth at a rate of 0.4% per year. It is also widely recognised that LULUC can strongly influence the terrestrial C and hydrological cycles, and that adverse impacts are exacerbated because of the negative impacts of climate change on net primary productivity (NPP) and terrestrial C stocks. In addition to the effects of deforestation on terrestrial C stocks, climate change can alter the C storage capacity of the terrestrial biosphere, which can influence the land-based mitigation potential. In this context, innovative management of tropical forests is critical to limiting global warming (Chow *et al.* 2013). For example, Humpenöder *et al.* (2015) and others have argued that climate policy must immediately stop deforestation and strongly encourage afforestation. It is critical that international agreements be established that commit to banning deforestation and reducing degradation (Mitchard 2018). However, the current governance structures and institutions are ineffective (Adelman 2015) and may not be able to bring about the much needed paradigm shift.

The global call to discourage any new deforestation is supported by data on the effects of deforestation on terrestrial C stocks and its positive feedback on climate change; some examples are briefly outlined below.

Poulter *et al.* (2010) reported that deforestation and fire in the Amazon may release 26–36 Pg C by 2100. Conversely, the global mitigation potential of afforestation (biomass C) between 2020 and 2100 was estimated at 155 Pg C by Lal *et al.* (2018) and at 191 Pg C by Humpenöder *et al.* (2015). In addition, Humpenöder *et al.* (2015) estimated an increase in NPP by 16 Pg C by 2100. Levy *et al.* (2004) modelled the effects of land use change on global scale fluxes of C from terrestrial ecosystems. In the absence of land use change between 1700 and 1990, the land-based sink stored 145 Pg C, but the terrestrial ecosystem became a net source of 97 Pg C when the land use change was represented for the same period. Rerunning the model between 1990 and 2100 without land use change showed a net C sink of 257–358 Pg C; however, land use change scaled with population change resulted in a net zero C sink for the period 1990–2100 (Levy *et al.* 2004).

Past changes in land use in the UK, characterised by decreasing agricultural use areas and increasing forested land areas, have made UK land use a net sink for GHGs (Rounsevell and Reay 2009). Bebber and Butt (2017) reported that protection of tropical areas reduced emissions by 4.88 Pg C between 2000 and 2018. Therefore, any future decline in agricultural land area would also increase the land-based C sink capacity. Herein lies the basic philosophy of returning some agriculturally marginal lands (30% or 50% of 5 billion ha under agriculture now) back to nature to sequester C: doing so would increase biodiversity, improve water quality and renewability and enable us to live in harmony with nature.

Thus, policy interventions are needed to increase the size of land-based C sinks, such as strategies to reduce C and N emissions through increased eco-efficiency, afforestation and a reduction in agricultural land areas. An increase in the area of younger forests would also strengthen the C sink (Mendelsohn and Dinar 2009). Policy interventions are needed to adapt rather than maladapt to climate change.

The need for returning some land to nature

Agroecosystems occupy ~40% of the ice-free land and contribute as much as 30% of all anthropogenic emissions (Lal 2021). Although global food production since 1960s has been among the greatest global success stories, 30–40% of all food produced is wasted and taken to landfills, where it leads to the pollution of surface and ground waters and the emission of GHGs (especially CH_4 and N_2O) into the atmosphere. Food and nutritional security have been achieved by the conversion of natural ecosystems to agroecosystems, the depletion and eutrophication of natural waters, the extinction of biodiversity and air pollution. There is a growing shift towards a preference for animal-based diets among growing and increasingly rich populations of emerging economies, with dire consequences regarding the use of land and water, as well as GHG emissions. Both undernutrition and malnutrition persist; there are an estimated 695 million undernourished people worldwide, with this number likely to have increased by 80–132 million by the end of 2020 because of the COVID-19 pandemic, and 2 billion malnourished. The problem lies with limited access to food because of poverty, poor distribution and unfair social and political environments. Any additional agronomic production, such as in Sub-Saharan Africa, South Asia, Central America or the Caribbean, must be achieved by narrowing the yield gap from the land already allocated rather than by bringing new land under agriculture. Indeed, humanity must reconcile the need to meet the food and nutritional demands of the growing population with the absolute necessity of improving the environment and mitigating climate change. Thus, the land-based C sink capacity must be enhanced by returning some agriculturally marginal lands to nature, by the afforestation of denuded landscapes and eroded ecosystems and the reflooding of drained wetlands. With the widespread adoption of recommended soil, crop, animal and water management practices, a large amount of agroecosystem land can be returned to nature. Living in harmony with nature is a win–win–win option for humanity.

Conclusions

LULUC, the conversion of natural to agricultural ecosystems, has been a source of GHGs and an important determinant of global warming ever since the dawn of settled agriculture ~10 millennia ago. Prehistoric land use for agriculture emitted as much as 320 Pg C to the atmosphere, with 255 Pg C having been emitted between 1750 and 2019. In comparison with the 575 Pg C emitted by land use change, 445 Pg C has been emitted from fossil fuel combustion since the onset of the Industrial Revolution.

Depletion of the C stock in the terrestrial biosphere (soil, vegetation, wetlands) has weakened and degraded several ecosystem services and created some disservices, such as global warming, the eutrophication and depletion of water, air pollution and species extinctions.

There is also a strong interaction between vegetation cover and climate change, and the effects of CO_2 fertilisation on NPP depend on the availability of essential nutrients (especially N) and water.

Restoration of degraded soils and depleted ecosystems, and the re-carbonisation of the terrestrial biosphere (storing atmospheric CO_2 in the soil and vegetation), are good strategies to reverse the degradation trends, mitigate climate change and improve water quality and renewability.

An innovative strategy to restore degraded soils and depleted ecosystems is to reduce land area allocated to agriculture by adopting the best management practices and narrowing the yield gap. Of the 5 billion ha of land allocated to agriculture, as much as one-third (1.5 billion ha) or one-half (2.5 billion ha) can be returned to nature through the adoption of a properly planned program of 'producing more from less' and 'using the best and saving the rest' for nature. Priorities to accomplish this goal include the following:

- site-specific experiments and monitoring studies to predict vegetation changes in the 21st century, because different components of global change can have interacting effects on biodiversity and its community-level indicators
- climate policy to immediately stop deforestation and strongly encourage afforestation

- policy interventions to increase the size of land-based C sinks, such as strategies to reduce C and N emissions through increased eco-efficiency, afforestation and a reduction in agricultural land areas
- research to predict how increased CO_2 concentrations will affect the hydrological cycle depending on the availability of N (and P, K, S), plant-available water capacity and the prevailing climate
- research to determine how future declines in agricultural land area would increase the land-based C sink capacity and mitigate climate change
- research to determine how land use change and climate change individually and in combination will affect water balance components.

References

Adelman S (2015) Tropical forests and climate change: a critique of green governmentality. *International Journal of Law in Context* **11**, 195–212. doi:10.1017/S1744552315000075

Alo CA, Anagnostou EN (2017) A sensitivity study of the impact of dynamic vegetation on simulated future climate change over southern Europe and the Mediterranean. *International Journal of Climatology* **37**, 2037–2050. doi:10.1002/joc.4833

Averett N (2016) Healthy ground, healthy atmosphere: recarbonizing the Earth's soils. *Environmental Health Perspectives* **124**, A31–A35. doi:10.1289/ehp.124-A30

Balkovič J, Skalský R, Folberth C, Khabarov N, Schmid E, Madaras M, Obersteiner M, van der Velde M (2018) Impacts and uncertainties of +2°C of climate change and soil degradation on European crop calorie supply. *Earth's Future* **6**, 373–395. doi:10.1002/2017EF000629

Bebber DP, Butt N (2017) Tropical protected areas reduced deforestation carbon emissions by one-third from 2000 to 2012. *Scientific Reports* **7**, 14005. doi.org/10.1038/s41598-017-14467-w

Chow J, Doria G, Kramer R, Schneider T, Stoike J (2013) Tropical forests under a changing climate and innovations in tropical forest management. *Tropical Conservation Science* **6**, 315–324. doi:10.1177/194008291300600302

Clavero M, Villero D, Brotons L (2011) Climate change or land use dynamics: do we know what climate change indicators indicate? *PLoS One* **6**, e18581. doi:10.1371/journal.pone.0018581

Coe MT, Marthews TR, Costa MH, Galbraith DR, Greenglass NL, Imbuzeiro HMA, Levine NM, Malhi Y, Moorcroft PR, Muza MN, *et al.* (2013) Deforestation and climate feedbacks threaten the ecological integrity of south-southeastern Amazonia. *Philosophical Transactions of the Royal Society of London. Series B, Biological Sciences* **368**, 20120155. doi:10.1098/rstb.2012.0155

Crutzen PJ (2006) The 'Anthropocene'. In *Earth System Science in the Anthropocene.* (Eds E Ehlers and T Krafft) pp. 13–18. Springer Berlin Heidelberg, Berlin, Heidelberg. doi:10.1007/3-540-26590-2_3

DeLong C, Cruse R, Wiener J (2015) The soil degradation paradox: compromising our resources when we need them the most. *Sustainability* **7**, 866–879. doi:10.3390/su7010866

Douglas PMJ, Pagani M, Eglinton TI, Brenner M, Curtis JH, Breckenridge A, Johnston K (2018) A long-term decrease in the persistence of soil carbon caused by ancient Maya land use. *Nature Geosicence* **11**, 645–649. doi:10.1038/s41561-018-0192-7

Droste N, May W, Clough Y, Börjesson G, Brady M, Hedlund K (2020) Soil carbon insures arable crop production against increasing adverse weather due to climate change. *Environmental Research Letters* **15**, 124034. doi:10.1088/1748-9326/abc5e3

Fearnside PM (2010) Global warming: how much of a threat to tropical forests? In *Survival and Sustainability. Environmental Earth Sciences.* (Eds H Gökçekus, U Türker and J LaMoreaux) pp. 1283–1292. Springer, Berlin, Heidelberg. doi:10.1007/978-3-540-95991-5_120

Feddema JJ, Oleson KW, Bonan GB, Mearns LO, Buja LE, Meehl GA, Washington WM (2005) The importance of land-cover change in simulating future climates. *Science* **310**, 1674–1678. doi:10.1126/science.1118160

Feeley KJ, Rehm EM (2012) Amazon's vulnerability to climate change heightened by deforestation and man-made dispersal barriers. *Global Change Biology* **18**, 3606–3614. doi:10.1111/gcb.12012

Franchito SH, Rao VB, Fernandez JPR (2012) Tropical land savannization: impact of global warming. *Theoretical and Applied Climatology* **109**, 73–79. doi:10.1007/s00704-011-0560-3

Franklin J, Serra-Diaz JM, Syphard AD, Regan HM (2016) Global change and terrestrial plant community dynamics. *Proceedings of the National Academy of Sciences of the United States of America* **113**, 3725–3734. doi:10.1073/pnas.1519911113

Friedlingstein P, O'Sullivan M, Jones MW, Andrew RM, Hauck J, Olsen A, Peters GP, Peters W, Pongratz J, Sitch S, *et al.* (2020) Global carbon budget 2020. *Earth System Science Data* **12**, 3269–3340. doi:10.5194/essd-12-3269-2020

Gierga M, Hajdas I, van Raden UJ, Gilli A, Wacker L, Sturm M, Bernasconi SM, Smittenberg RH (2016) Long-stored soil carbon released by prehistoric land use: evidence from compound-specific radiocarbon analysis on Soppensee Lake sediments. *Quaternary Science Reviews* **144**, 123–131. doi:10.1016/j.quascirev.2016.05.011

Gomes VHF, Vieira ICG, Salomão RP, ter Steege H (2019) Amazonian tree species threatened by deforestation and climate change. *Nature Climate Change* **9**, 547–553. doi:10.1038/s41558-019-0500-2

Goring SJ, Williams JW (2017) Effect of historical land-use and climate change on tree–climate relationships in the upper Midwestern United States. *Ecology Letters* **20**, 461–470. doi:10.1111/ele.12747

Guo Y, Fang G, Xu YP, Tian X, Xie J (2020) Identifying how future climate and land use/cover changes impact streamflow in Xinanjiang Basin, east China. *The Science of the Total Environment* **710**, 136275. doi:10.1016/j.scitotenv.2019.136275

Humpenöder F, Popp A, Stevanovic M, Müller C, Bodirsky BL, Bonsch M, Dietrich JP, Lotze-Campen H, Weindl I, Biewald A, *et al.* (2015) Land-use and carbon cycle responses to moderate climate change: implications for land-based mitigation? *Environmental Science & Technology* **49**, 6731–6739. doi:10.1021/es506201r

Intergovernment Panel on Climate Change (IPCC) (2021) *Sixth Assessment Report. Summary for Policymakers.* IPCC, Geneva, <https://www.ipcc.ch/report/sr15/summary-for-policymakers/>.

Juliá R, Duchin F (2013) Land use change and global adaptations to climate change. *Sustainability* **5**, 5442–5459. doi:10.3390/su5125442

Kaplan JO, Krumhardt KM, Zimmermann NE (2012) The effects of land use and climate change on the carbon cycle of Europe over the past 500 years. *Global Change Biology* **18**, 902–914. doi:10.1111/j.1365-2486.2011.02580.x

Kostadinov S (2010) Forests in Serbia as factor of soil and water protection against degradation in the conditions of global climate changes. In *Global Environmental Change: Challenges to Science and Society in Southeastern Europe.* (Eds V Alexandrov, M Gajdusek, C Knight and A Yotova) pp. 177–190. Springer Netherlands, Dordrecht. doi:10.1007/978-90-481-8695-2_14

Krinner G, Viovy N, de Noblet-Ducoudré N, Ogée J, Polcher J, Friedlingstein P, Ciais P, Sitch S, Prentice IC (2005) A dynamic global vegetation model for studies of the coupled atmosphere–biosphere system. *Global Biogeochemical Cycles* **19**, GB1015. doi:10.1029/2003GB002199

Lacoste M, Viaud V, Michot D, Walter C (2015) Landscape-scale modelling of erosion processes and soil carbon dynamics under land-use and climate change in agroecosystems. *European Journal of Soil Science* **66**, 780–791. doi:10.1111/ejss.12267

Lal R (2021) Feeding the world and returning half of the land back to nature. *Journal of Soil and Water Conservation* **76**, 75A–78A. doi:10.2489/jswc.2021.0607A

Lal R, Smith P, Jungkunst HF, Mitsch WJ, Lehmann J, Ramachandran Nair PK, McBratney AB, Sá JCM, Schneider J, Zinn YL, *et al.* (2018) The carbon sequestration potential of terrestrial ecosystems. *Journal of Soil and Water Conservation* **73**, 143A–152A. doi:10.2489/jswc.73.6.145A

Lal R, Bouma J, Brevik E, Dawson L, Field DJ, Glaser B, Hatano R, Hartemink AE, Kosaki T, Lascelles B, *et al.* (2021) Soils and sustainable development goals of the United Nations: an International Union of Soil Sciences perspective. *Geoderma Regional* **25**, e00398. doi:10.1016/j.geodrs.2021.e00398

Lapola DM, Schaldach R, Alcamo J, Bondeau A, Msangi S, Priess JA, Silvestrini R, Soares-Filho BS (2011) Impacts of climate change and the end of deforestation on land use in the Brazilian Legal Amazon. *Earth Interactions* **15**, 1–29. doi:10.1175/2010EI333.1

Lemordant L, Gentine P, Swann AS, Cook BI, Scheff J (2018) Critical impact of vegetation physiology on the continental hydrologic cycle in response to increasing CO_2. *Proceedings of the National Academy of Sciences of the United States of America* **115**, 4093–4098. doi:10.1073/pnas.1720712115

Levy PE, Friend AD, White A, Cannell MGR (2004) The influence of land use change on global-scale fluxes of carbon from terrestrial ecosystems. *Climatic Change* **67**, 185–209. doi:10.1007/s10584-004-2849-z

Llopart M, Reboita M, Coppola E, Giorgi F, da Rocha R, de Souza D (2018) Land use change over the Amazon Forest and its impact on the local climate. *Water (Basel)* **10**, 149. doi:10.3390/w10020149

Manchego CE, Hildebrandt P, Cueva J, Espinosa CI, Stimm B, Günter S (2017) Climate change versus deforestation: implications for tree species distribution in the dry forests of southern Ecuador. *PLoS One* **12**, e0190092. doi:10.1371/journal.pone.0190092

Marhaento H, Booij MJ, Hoekstra AY (2018) Hydrological response to future land-use change and climate change in a tropical catchment. *Hydrological Sciences Journal* **63**, 1368–1385. doi:10.1080/02626667.2018.1511054

McAlpine CA, Ryan JG, Seabrook L, Thomas S, Dargusch PJ, Syktus JI, Pielke RA, Etter AE, Fearnside PM, Laurance WF (2010) More than CO_2: a broader paradigm for managing climate change and variability to avoid ecosystem collapse. *Current Opinion in Environmental Sustainability* **2**, 334–346. doi:10.1016/j.cosust.2010.10.001

Mendelsohn R, Dinar A (2009) Land use and climate change interactions. *Annual Review of Resource Economics* **1**, 309–332. doi:10.1146/annurev.resource.050708.144246

Mitchard ETA (2018) The tropical forest carbon cycle and climate change. *Nature* **559**, 527–534. doi:10.1038/s41586-018-0300-2

Op de Hipt F, Diekkrüger B, Steup G, Yira Y, Hoffmann T, Rode M (2018) Modeling the impact of climate change on water resources and soil erosion in a tropical catchment in Burkina Faso, west Africa. *Catena* **163**, 63–77. doi:10.1016/j.catena.2017.11.023

Osei MA, Amekudzi LK, Wemegah DD, Preko K, Gyawu ES, Obiri-Danso K (2019) The impact of climate and land-use changes on the hydrological processes of Owabi catchment from SWAT analysis. *Journal of Hydrology. Regional Studies* **25**, 100620. doi:10.1016/j.ejrh.2019.100620

Perugini L, Caporaso L, Marconi S, Cescatti A, Quesada B, De Noblet-Ducoudré N, House JI, Arneth A (2017) Biophysical effects on temperature and precipitation due to land cover change. *Environmental Research Letters* **12**, 053002. doi:10.1088/1748-9326/aa6b3f

Peterman W, Bachelet D (2012) Climate change and forest dynamics: a soils perspective. In *Soils and Food Security.* (Eds RE Hester and RM Harrison) pp. 158–182. Royal Society of Chemistry, Cambridge.

Pielke RA, Marland G, Betts RA, Chase TN, Eastman JL, Niles JO, Niyogi DD, Running SW (2002) The influence of land-use change and landscape dynamics on the climate system: relevance to climate-change policy beyond the radiative effect of greenhouse gases. *Philosophical Transactions of the Royal Society A – Mathematical, Physical and Engineering Sciences* **360**, 1705–1719. doi:10.1098/rsta.2002.1027

Pires GF, Costa MH (2013) Deforestation causes different subregional effects on the Amazon bioclimatic equilibrium. *Geophysical Research Letters* **40**, 3618–3623. doi:10.1002/grl.50570

Poulter B, Aragão L, Heyder U, Gumpenberger M, Heinke J, Langerwisch F, Rammig A, Thonicke K, Cramer W (2010) Net biome production of the Amazon Basin in the 21st century. *Global Change Biology* **16**, 2062–2075. doi:10.1111/j.1365-2486.2009.02064.x

Rasouli K, Pomeroy JW, Whitfield PH (2019) Are the effects of vegetation and soil changes as important as climate change impacts on hydrological processes? *Hydrology and Earth System Sciences* **23**, 4933–4954. doi:10.5194/hess-23-4933-2019

Rodrigues AR, Botequim B, Tavares C, Pécurto P, Borges JG (2020) Addressing soil protection concerns in forest ecosystem management under climate change. *Forest Ecosystems* **7**, 34. doi:10.1186/s40663-020-00247-y

Rounsevell MDA, Reay DS (2009) Land use and climate change in the UK. *Land Use Policy* **26**(Suppl 1), S160–S169. doi:10.1016/j.landusepol.2009.09.007

Ruddiman WF (2003) The Anthropogenic greenhouse era began thousands of years ago. *Climatic Change* **61**, 261–293. doi:10.1023/B:CLIM.0000004577.17928.fa

Senior RA, Hill JK, Edwards DP (2019) Global loss of climate connectivity in tropical forests. *Nature Climate Change* **9**, 623–626. doi:10.1038/s41558-019-0529-2

Sitch S, Brovkin V, von Bloh W, van Vuuren D, Eickhout B, Ganopolski A (2005) Impacts of future land cover changes on atmospheric CO2 and climate. *Global Biogeochemical Cycles* **19**, GB2013. doi:10.1029/2004GB002311

Spracklen DV, Baker JCA, Garcia-Carreras L, Marsham JH (2018) The effects of tropical vegetation on rainfall. *Annual Review of Environment and Resources* **43**, 193–218. doi:10.1146/annurev-environ-102017-030136

Staal A, Flores BM, Aguiar APD, Bosmans JHC, Fetzer I, Tuinenburg OA (2020) Feedback between drought and deforestation in the Amazon. *Environmental Research Letters* **15**, 044024. doi:10.1088/1748-9326/ab738e

Stinchcomb GE, Messner TC, Driese SG, Nordt LC, Stewart RM (2011) Pre-colonial (A.D. 1100–1600) sedimentation related to prehistoric maize agriculture and climate change in eastern North America. *Geology* **39**, 363–366. doi:10.1130/G31596.1

Stoy PC (2018) Deforestation intensifies hot days. *Nature Climate Change* **8**, 366–368. doi:10.1038/s41558-018-0153-6

Tabor K, Hewson J, Tien H, González-Roglich M, Hole D, Williams J (2018) Tropical protected areas under increasing threats from climate change and deforestation. *Land (Basel)* **7**, 90. doi:10.3390/land7030090

Talib A, Randhir TO (2017) Climate change and land use impacts on hydrologic processes of watershed systems. *Journal of Water and Climate Change* **8**, 363–374. doi:10.2166/wcc.2017.064

ter Steege H (2010) Will tropical biodiversity survive our approach to global change? *Biotropica* **42**, 561–562. doi:10.1111/j.1744-7429.2010.00676.x

Tsarouchi G, Buytaert W (2018) Land-use change may exacerbate climate change impacts on water resources in the Ganges Basin. *Hydrology and Earth System Sciences* **22**, 1411–1435. doi:10.5194/hess-22-1411-2018

United Nations (UN) (2019) *World Population Prospects 2019. Highlights*. UN, New York, <https://population.un.org/wpp/Publications/Files/WPP2019_Highlights.pdf>.

United Nations Framework Convention on Climate Change (UNFCC) (2016) *Report of the Conference of the Parties on its twenty-first session, held in Paris from 30 November to 13 December 2015*. UNFCC, New York, <https://unfccc.int/resource/docs/2015/cop21/eng/10.pdf>.

van der Molen MK, van den Hurk BJJM, Hazeleger W (2011) A dampened land use change climate response towards the tropics. *Climate Dynamics* **37**, 2035–2043. doi:10.1007/s00382-011-1018-0

Vargas Zeppetello LRV, Parsons LA, Spector JT, Naylor RL, Battisti DS, Masuda YJ, Wolff NH (2020) Large scale tropical deforestation drives extreme warming. *Environmental Research Letters* **15**, 084012. doi:10.1088/1748-9326/ab96d2

Wang Z, Van Oost K (2019) Modeling global anthropogenic erosion in the Holocene. *The Holocene* **29**, 367–379. doi:10.1177/0959683618816499

WMO (2020) *WMO Greenhouse Gas Bulletin – No. 16: The State of Greenhouse Gases in the Atmosphere Based on Global Observations Through 2019*. WMO, Geneva, <https://library.wmo.int/index.php?id=21795&lvl=notice_display#.YZ8Fvy1L1UM>.

Yao X, Wang Z, Wang H (2015) Impact of urbanization and land-use change on surface climate in middle and lower reaches of the Yangtze River, 1988–2008. *Advances in Meteorology* **2015**, 395094. doi:10.1155/2015/395094

Yu M, Wang G, Pal JS (2016) Effects of vegetation feedback on future climate change over West Africa. *Climate Dynamics* **46**, 3669–3688. doi:10.1007/s00382-015-2795-7

Zhang X, Huang X (2019) Human disturbance caused stronger influences on global vegetation change than climate change. *PeerJ* **7**, e7763. doi:10.7717/peerj.7763

2

Climate fluctuations and soil hydrology

Jerry L. Hatfield, Peter L. O'Brien, Kenneth M. Wacha

Introduction

Agriculture, along with all other life on Earth, depends on water to maintain metabolic functions. In agricultural systems, water availability represents a balance between the ability of the soil to supply water and the climate as the ultimate source of water. The dynamics of water affect every segment of agriculture around the world. If we are to continue to generate the food, feed, fuel and fibre required for human existence, then we need to understand how climate change affects soil hydrology. There are basic principles that exist between climate fluctuations and soil hydrology applicable to all ecosystems around the world. Efforts by the Intergovernmental Panel on Climate Change (IPCC) document the changes in climate around the world, and these are valuable resources to help us understand how the hydrological cycle will be affected. Our goal in this chapter is to show how climate change will continue to affect soil hydrology and how soil management may increase the potential supply of water to agricultural systems. These principles apply to all agricultural systems; although the examples used focus on North America, the concepts are transferrable around the world.

Climate versus weather

Climate is the long-term average of weather at a given location, and we often report climate normals as the average of 30 years of weather. Climate describes what we can expect in terms of temperature or precipitation for a given year, month or day, but does not describe what is actually happening at a particular time. We consider climate normals as our planning tools for how we manage agricultural systems, whereas weather is the reality we need to incorporate into our short-term management decisions. An example of this is shown in the climate normal for monthly precipitation totals for Ames, Iowa, Macon, Georgia, and Davis, California (Fig. 2.1). These patterns describe the differences we would use for planning agricultural management around the precipitation totals across the months of the year.

Climate change and soil degradation

Climate change is a reality, and we tend to think that this is a new phenomenon; however, climate has always been changing. The difference with the current state is that the degree

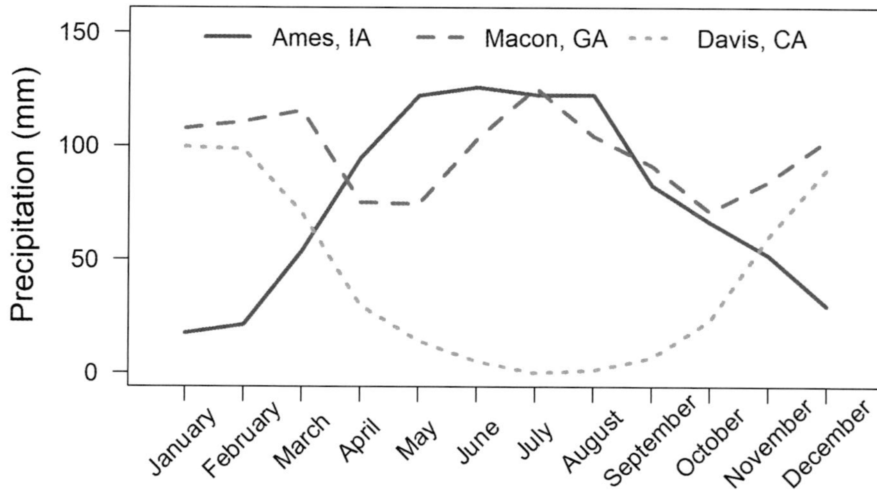

Fig. 2.1. Monthly normal precipitation for Ames, Iowa (IA), Macon, Georgia (GA) and Davis, California (CA), for the period 1980–2010.

of change is unlike what we have experienced in the recent past, and this has implications for our ability to efficiently produce the food required to sustain the ever-increasing population while protecting our natural resource base. The effects of climate change are exacerbated in agroecosystems because they are occurring in conjunction with, and contributing to, soil degradation. Hatfield and Walthall (2015) detailed a framework to assess our capabilities of meeting global food needs and found soil degradation was one of the major threats to secure food supply for future generations because degraded soils often reduce crop water availability. The recent IPCC report on soil degradation defined land degradation as a '*negative trend in land condition, caused by direct or indirect human-induced processes including anthropogenic climate change, expressed as long-term reduction or loss of at least one of the following: biological productivity, ecological integrity or value to humans*' (Olsson *et al.* 2019). Olsson *et al.* (2019) found that climate change would exacerbate land degradation because of an increased frequency, intensity and/or amount of heavy precipitation, and increased heat stress. That report from the IPCC provides a warning that our land resource is more fragile than we assume, and care must be taken to adopt strategies that will protect the land resource from further degradation.

Climate patterns

The changes in global climate are not uniform in space or time, but they are demonstrated in consistent trends across a wide range of regions over several decades. Climate change encompasses a wide range of metrics, but we focus on a series of temperature and precipitation trends that are directly relevant to soil hydrology. These metrics are: (1) overall annual temperatures; (2) seasonal temperature patterns; (3) overall precipitation; (4) seasonal precipitation patterns; and (5) extreme precipitation events.

Annual temperatures

Temperatures around the globe have shown an increasing trend, as global average temperatures have increased ~1°C from 1901 through 2016 (Hayhoe *et al.* 2018). Temperature projections at a global scale suggest an additional increase for the remainder of this century of between 0.2°C and 4.7°C, depending upon the assumptions used for the CO_2 increase in the atmosphere (Hayhoe *et al.* 2018). However, global temperature changes are not uniform across regions because land surface temperatures are warming more than temperatures over the oceans (Collins *et al.* 2013). Similarly, not all land surfaces are warming at a uniform rate due to differences in topography, urban density and natural climate variations. There has been a greater increase in temperatures in the northern latitudes than in the southern latitudes. For example, North Dakota has had the highest increase in average temperatures in the US, at 0.15°C per decade (National Oceanic and Atmospheric Administration (NOAA) 2013).

In the US, specifically, the average temperature has increased by 0.7°C (1.3°F) to 1.0°C (1.9°F) since record keeping began in 1895; most of this increase has occurred since *c.* 1970 (Walsh *et al.* 2014). The most recent decade was the nation's warmest on record. The changes in annual average temperatures are punctuated by changes in extreme temperature events. Heat waves have become more frequent and intense, especially in the West, whereas cold waves have become less frequent and intense across the nation. These trends are expected to be observed across the globe, as increasing global temperatures will result in more hot extremes and fewer cold extremes at both daily and seasonal time scales (Collins *et al.* 2013).

Seasonal temperatures

Just as temperatures are not increasing uniformly across the globe, nor are they increasing uniformly throughout the year. The warming average temperatures observed in the climate record show that the winter temperatures are warming more than the spring or summer periods, although these seasonal patterns also depend on location. For example, annual temperatures in Michigan since 1895 have increased at a rate of 0.02°C year^{-1}, but the increases have only been 0.005°C year^{-1} in the spring and summer seasons. Alternatively, Iowa shows an increasing temperature trend in the spring and a decreasing temperature trend for the summer months. Overall, these increasing temperatures are contributing to lengthening frost-free seasons (and corresponding growing seasons) across the US since 1980, with the largest increases occurring in the western states (Walsh *et al.* 2014). The extension of the growing season earlier in the spring in these regions is important for land managers of both natural areas and agroecosystems, because longer growing seasons may provide different opportunities (O'Brien *et al.* 2020). However, regions that are seeing warmer and earlier springs may also be more susceptible to 'false spring' events (Allstadt *et al.* 2015), where a late spring frost may have the potential for deleterious environmental effects (Kral-O'Brien *et al.* 2019).

In addition to changes at the seasonal level, temperature changes are evident at the daily scale. Generally, minimum temperatures are increasing at a faster rate than maximum temperatures. Just as temperatures during the coldest part of the year, winter, are warming faster than temperatures during the warmer parts of the year, temperatures during the coldest part of the day (most often night-time) are warming faster than temperatures during the warmest parts of the day. In fact, minimum temperatures are increasing while maximum temperatures are decreasing in some regions, including the Corn Belt of the US.

Total annual precipitation

Total annual precipitation is increasing worldwide, with precipitation expected to increase 1–3% for each 1°C increase in temperature (Collins *et al.* 2013). In the US, mean precipitation has increased by ~3% per year since the widespread drought in the 1930s. The increases in precipitation are linked with increasing temperatures because the water-holding capacity of air increases by 7% per 1°C (Trenberth 2011); thus, with increasing temperatures, the air can take up more water, and water vapour increases in the atmosphere. Like increasing temperatures, the overall increase in precipitation will not be uniform. Much of the northern US may be experiencing a steady increase in 30-year normal (e.g. Iowa; Plate 1), but overall precipitation is decreasing in the south-west and extreme south-east (Walsh *et al.* 2014). Notably, the number of years with low precipitation has decreased since 1895, although the high spatial variation in precipitation patterns still allows for some regions to be experiencing extended drought periods (Collins *et al.* 2013).

Seasonal precipitation

The seasonality of precipitation has also shifted in many regions worldwide. Across the upper Midwestern US, a greater proportion of annual precipitation is occurring in the spring months, whereas the proportion of precipitation occurring during the summer months is growing more variable, as projected by Walsh *et al.* (2014). An example of this shift is shown for Illinois, where spring and summer precipitation trends are shown since 1895 as totals for each season in Plate 2 and as fractions of the annual precipitation in Plate 3. Spring precipitation shows a positive upward trend, whereas summer trends are non-significant. Notably, there has been more variation in the combinations of rainfall in these seasons in recent years compared with the total data record; both the highest and lowest summer precipitation has occurred in recent years (Plate 2).

Precipitation intensity

Extreme weather events accompanied by precipitation, such as severe thunderstorms, hurricanes and winter storms, have increased in frequency and intensity since the 1950s (Walsh *et al.* 2014). These high-intensity precipitation events are a consequence of rising temperatures that allow for more water vapour to be held in the atmosphere (Trenberth 2011). Although extreme precipitation events are expected to increase across the US, many regional differences are expected (Kunkel *et al.* 1999). The increases have been observed more in the upper Midwest and north-east US, and they are concentrated more in the summer months than in other seasons of the year. Fig. 2.2 demonstrates this trend over the past four decades across the Midwest, where the number of days with precipitation >50 mm day^{-1} is increasing across this region. This trend has been evident in the rainfall record across the contiguous US and the Midwest, with the past 20 years showing a positive trend in daily amounts >50 mm. Further, some individual stations often report daily totals >250 mm. These extreme events are infrequent and spatially isolated, but they lead to large run-off amounts and flooding of the local area.

Updated climate outlook

These climate trends have been documented for many years and have been summarised by the IPCC (Collins *et al.* 2013) and the 2014 National Climate Assessment (Walsh *et al.* 2014). However, the trends continue, such that the following six key findings have been included in the 2018 National Climate Assessment (Wuebbles *et al.* 2017a):

1. *The global climate continues to change rapidly compared to the pace of the natural variations in climate that have occurred throughout Earth's history. Trends in*

globally averaged temperature, sea level rise, upper-ocean heat content, land-based ice melt, Arctic sea ice, depth of seasonal permafrost thaw, and other climate variables provide consistent evidence of a warming planet. These observed trends are robust and have been confirmed by multiple independent research groups around the world.

2. *The frequency and intensity of extreme heat and heavy precipitation events are increasing in most continental regions of the world. These trends are consistent with expected physical responses to a warming climate. Climate model studies are also consistent with these trends, although models tend to underestimate the observed trends, especially for the increase in extreme precipitation events. The frequency and intensity of extreme high temperature events are virtually certain to increase in the future as global temperature increases. Extreme precipitation events will very likely continue to increase in frequency and intensity throughout most of the world. Observed and projected trends for some other types of extreme events, such as floods, droughts, and severe storms, have more variable regional characteristics.*
3. *Many lines of evidence demonstrate that it is extremely likely that human influence has been the dominant cause of the observed warming since the mid-20th century. Formal detection and attribution studies for the period 1951 to 2010 find that the observed global mean surface temperature warming lies in the middle*

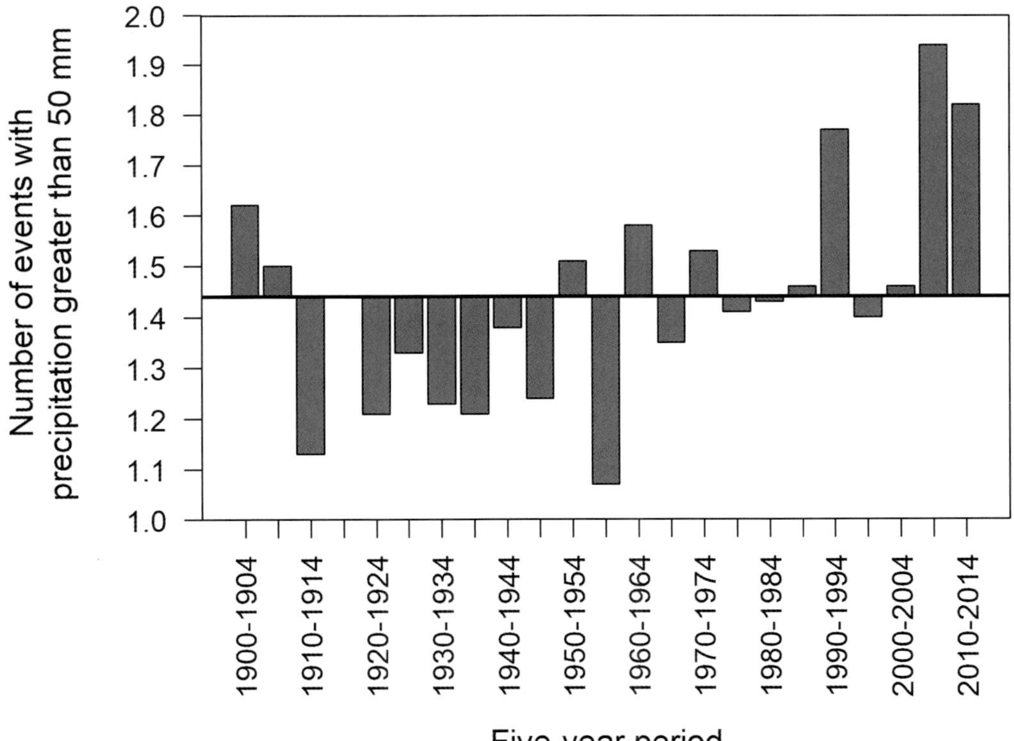

Fig. 2.2. Changes in precipitation intensity for Iowa by 5-year periods from 1900 to 2014 (https://statesummaries.ncics.org/ia).

of the range of likely human contributions to warming over that same period. We find no convincing evidence that natural variability can account for the amount of global warming observed over the industrial era. For the period extending over the last century, there are no convincing alternative explanations supported by the extent of the observational evidence. Solar output changes and internal variability can only contribute marginally to the observed changes in climate over the last century, and we find no convincing evidence for natural cycles in the observational record that could explain the observed changes in climate.

4. *Global climate is projected to continue to change over this century and beyond. The magnitude of climate change beyond the next few decades will depend primarily on the amount of greenhouse (heat-trapping) gases emitted globally and on the remaining uncertainty in the sensitivity of Earth's climate to those emissions. With significant reductions in the emissions of greenhouse gases, the global annually averaged temperature rise could be limited to 3.6°F (2°C) or less. Without major reductions in these emissions, the increase in annual average global temperatures relative to preindustrial times could reach 9°F (5°C) or more by the end of this century.*
5. *Natural variability, including El Niño events and other recurring patterns of ocean–atmosphere interactions, impact temperature and precipitation, especially regionally, over months to years. The global influence of natural variability, however, is limited to a small fraction of observed climate trends over decades.*
6. *Longer-term climate records over past centuries and millennia indicate that average temperatures in recent decades over much of the world have been much higher, and have risen faster during this time period, than at any time in the past 1700 years or more, the time period for which the global distribution of surface temperatures can be reconstructed.*

Potential impact of climate change on soil hydrology

Climate change impacts on soil hydrology encompass all aspects of the water balance in the soil–plant–atmosphere continuum. Broadly, we can take the soil water balance (SWB) as the following:

$$SWB = P + I - E - T - D \qquad [1]$$

where P is precipitation, I is irrigation, E is soil water evaporation, T is transpiration and D is drainage. The trends in climate change have the potential to affect these parameters over both long- and short-term monitoring, and these effects will likely dramatically alter the capacity of the soil to deliver agroecosystem services, especially crop biomass production and regulation of water quality and quantity. Hereafter, we describe how climate change will affect infiltration, evaporation, transpiration, evapotranspiration and soil water storage.

Infiltration

Infiltration is the process by which water enters the soil, and so it is one critical component of soil hydrology that must be understood in the context of climate change. Before water can be used, stored or transported, it must enter the soil, and changes in the seasonality, intensity and frequency of precipitation events will be the dominant factors affecting the potential resupply of water to the soil, especially in rain-fed agricultural systems. The infiltration process is governed by the texture of the soil and the associated pore networks, the size distribution and stability of surface soil aggregates and the soil water content. In general, we can consider that sandy soils would have an infiltration rate in excess of 2.0 cm h^{-1}, loam soils would have an infiltration rate between 0.5 and 1.0 cm h^{-1} and clay soils would have an infiltration rate between 0.1 and 0.5 cm h^{-1} (Johnson 1963). These are general categories; however, they provide a guide as to how different soils will react to precipitation events. For example, many of the soils in the upper Midwest have high loam and/or clay contents and so tend to have low infiltration rates. Conversely, many soils in the Great Plains or along the South Atlantic coast have high sand contents with high infiltration rates.

In addition to soil texture, soil aggregate size and strength play important roles in infiltration (Wacha *et al.* 2018). During a storm event, surface aggregates are directly exposed to the kinetic energy of falling raindrops, which can instigate the breakdown and dislodgement of soil particles from the soil matrix. As rainfall continues, water begins to enter aggregate structures and build up internal pressure. Weak aggregates may collapse, releasing encapsulated material, which can clog pore networks and dampen infiltration, causing overland flow conditions to develop (Barthès and Roose 2002). Size fractions of soil and nutrients can be easily entrained and swept away in run-off and redistributed along the downslope flowpaths (Wacha *et al.* 2020).

During high-intensity events that are predicted to occur more frequently, lower infiltration rates indicate that more water will be lost to run-off and there will likely be more erosion and nutrient loss. Similarly, as the seasonality of precipitation has shifted to more spring rainfall, infiltration will be impeded by the higher probability of falling on saturated soils in the spring, leading to more run-off. However, regardless of region or soil type, the trend of increasing precipitation intensity makes it more likely that rainfall events will occur that exceed infiltration rates of even coarse-textured soil, reducing the amount of water entering the soil profile.

Evaporation

Evaporation is the process of converting liquid water to water vapour. The process of evaporation is driven by atmospheric demand for water and the amount of water available. Atmospheric demand is governed by the temperature, humidity and wind speed, such that higher air temperatures have the capacity to hold more water. This capacity is quantified as the saturation vapour pressure (e_s), which increases exponentially as temperature increases (Eqn 2; Buck 1981):

$$e_s(\text{kPa}) = 0.61121 \exp\left[\frac{17.502T_c}{240.97 + T_c}\right] \quad [2]$$

where T_c is ambient temperature. For example, evaporative demand increases when air temperature increases from 25°C to 28°C because e_s increases from 3.167 to 3.78 kPa. This relationship describes why annual global surface (i.e. both water and land) evaporation will increase with increasing temperatures; however, evaporation is also driven by water availability, so evaporation over land will be linked to precipitation (Trenberth 2011; Collins *et al.* 2013).

Evaporation will be affected by climate change differently across the different regions of the world. In temperate regions, there is a tendency towards increasing annual precipitation and more intense events, so the expectation would be for greater evaporation because of more water and the potential for surface ponding due to slow infiltration. In contrast, in arid areas, there is no trend in increasing precipitation and a large amount of variation among years, and we would expect evaporation to be variable as well. However, the one change in the arid areas is the increase in temperature, which will directly increase evaporation rates, and we could expect that precipitation on the surface would rapidly evaporate from bare soil surfaces. In the more humid areas, evaporation rates will be increased by the increasing temperature trends that increase evaporative demand; however, the trend in precipitation exhibits considerable variability.

Evaporation as part of the soil hydrological process is more complex than a simple balance between precipitation amounts and atmospheric demand. Although changing temperatures and precipitation patterns will certainly affect evaporation, the actual evaporation rate from the soil surface also depends on specific spatiotemporal characteristics, such as the amount of ground cover, the time of day and the time of year. Thus, any meaningful discussion of soil water evaporation in agroecosystems must be understood within the context of plant water use and transpiration.

Transpiration

Transpiration is the process of evaporation of water from plant stomata that has been extracted from the soil via the root system. The process of transpiration involves the complete energy balance of a crop canopy, as well as soil water dynamics, because it relates to soil water storage and the ability of plant roots to explore the soil profile to access available water. Transpiration is necessary to ensure there is adequate water in the plant leaf to sustain physiological functions. For example, photosynthesis declines when water availability decreases because the stomata close, restricting the entry of CO_2 into the leaf. Hatfield and Prueger (2011a) outlined the effect of climate change on plant response and showed leaf- and canopy-level responses to changes in CO_2 and temperature. Increases in CO_2 will decrease water use and increase water use efficiency, whereas increases in temperature will offset the benefit of increased CO_2 because of the increased atmospheric demand. Generally, the effect of climate change on plants will be a balance between the effects of temperature and CO_2 on plant physiological functions, although these relationships will be further complicated by soil water availability.

To understand the effect of climate change on transpiration, there is a need to distinguish between potential and actual transpiration. The most used formula for estimating potential transpiration has been the Penman–Monteith equation, expressed as follows:

$$\lambda E = \frac{(\Delta R_n - G) + \rho C_p \{e_s(T_a) - e\} r_a^{-1}}{\Delta + \gamma\left(1 + {r_c}/{r_a}\right)} \qquad [3]$$

where λE is the latent heat of evaporation (W m^{-2}), Δ the rate of change of saturation vapour pressure with temperature (Pa C^{-1}), R_n is net radiation (W m^{-2}), G is soil heat flux (W m^{-2}), ρ is the density of air (kg m^{-3}), C_p is the specific heat of air at constant pressure (J kg^{-1} C^{-1}), e_s the saturation vapor pressure at a given air temperature (kP$_a$), T_a is the partial pressure of water vapour in air (kP$_a$), r_a is the aerodynamic resistance (s m^{-1}), r_c is the canopy resistance to water vapour transport (s m^{-1}) and γ the psychrometric constant (kPa C^{-1}). This equation has been used by the Food and Agriculture Organization of the United Nations (FAO) in their standard methodology to estimate water use by crops as described by Allen *et al.* (1998). In contrast with the method used by Allen *et al.* (1998), the approach we use estimates water lost from a surface, namely a short grass surface with adjustments made using crop coefficients. The approach using the resistance terms allows for the incorporation of factors linked with climate change. We can use Eqn 3 to estimate the potential transpiration by setting r_c to zero. This approach would provide a theoretical estimate of water evaporation from a surface; however, it does not represent the reality of the coupling of agricultural systems to a changing climate. A changing climate will not significantly affect potential transpiration or evaporation, except through the increasing e_s due to increasing temperatures; however, the major effect of climate will be on actual evapotranspiration, as detailed in the next section.

Evapotranspiration

Evapotranspiration is the combination of transpiration and soil water evaporation. It is part of the surface energy balance that describes the partitioning of energy into latent heat (evapotranspiration), sensible heat (warming the air) and soil heat (warming the soil) or, in the case of plant canopies, warming the foliage; however, if there is water available, latent heat is the largest user of energy. A well-watered crop that completely covers the soil will show >90% of the available energy is used in evapotranspiration. In regions with bare soils for portions of the year, the surface energy balance may be useful in informing how climate change will affect soil temperature and water dynamics through evapotranspiration. For example, the trend of more spring rainfall in some regions like the upper Midwest will maintain higher partitioning to evapotranspiration and delay soil warming later into the season. Alternatively, arid areas will continue to have limited evapotranspiration so that soil warming may increase rapidly. The altered precipitation regimes that result in prolonged droughts may have similar effects on even temperate regions, an effect recently shown by O'Brien and Hatfield (2020), who found that bare soil surfaces under dry conditions showed surface temperatures near 50°C. This effect could be evident in orchards, vineyards or widely spaced vegetable crops in arid environments where the soil water evaporation would be limited, and the increased air temperature and decreased humidity of the air would increase the atmospheric demand, leading to increased transpiration by the crop.

To further understand the influence of climate change on the components of soil hydrology, it is useful to understand the contributions of soil water evaporation and transpiration to total evapotranspiration. To do so, we can expand Eqn 3 to include energy balance and resistance terms for soil water evaporation and the plant canopy components, as shown in Eqn 4:

$$\lambda E = \frac{(\Delta R_n - G) + \rho C_p \{e_s(T_a) - e\} r_a^{-1}}{\Delta + \gamma\left(1 + {}^{r_s}\!/_{r_a}\right)} + \frac{(\Delta R_n) + \rho C_p \{e_s(T_a) - e\} r_a^{-1}}{\Delta + \gamma\left(1 + {}^{r_c}\!/_{r_a}\right)} \qquad [4]$$

where the first part of the equation is for soil water evaporation and the second for transpiration, and where the r_s term represents the resistance to soil water evaporation. The balance between soil water evaporation and transpiration is proportional to the growth of the crop and the amount of ground cover. Diagrammatically, we can evaluate these components by envisioning the energy exchanges over the soil surface and plant canopy. In row crops, as the crop canopy develops and completely covers the soil surface, the soil water evaporation component decreases, and evapotranspiration is dominated by

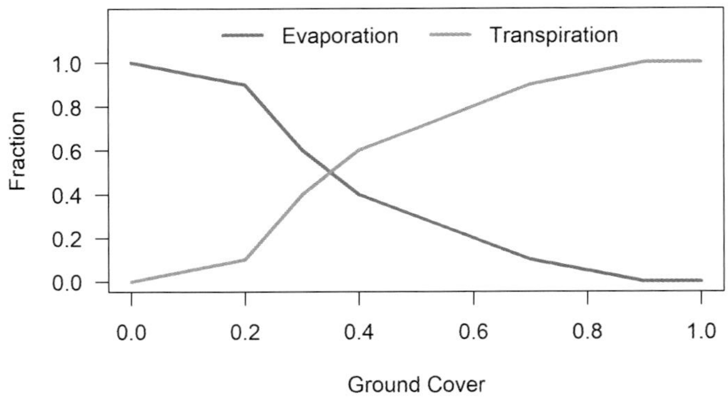

Fig. 2.3. Soil water evaporation and transpiration fractions as a function of ground cover.

transpiration. There are cropped surfaces, such as orchards, vineyards and vegetable crops, in which soil water evaporation will be a major component of the energy balance. We can depict these relationships through the diagram in Fig. 2.3, which shows the balance between soil water evaporation and transpiration as a function of ground cover. As ground cover approaches full cover, soil water evaporation becomes minimal and transpiration is dominant and becomes dependent upon the supply of soil water within the root zone.

To understand the impact of climate change on soil water and evapotranspiration, we have to examine the relationships between: (1) r_s and surface soil water content for evaporation; and (2) r_c and soil profile water content for transpiration. Soil water evaporation under a bare soil with no crop cover will depend heavily upon the texture of the soil because of its role in determining how much soil water is available to evaporate. For example, a fine sand with a water-holding capacity of 0.6 mm cm^{-1} may evaporate all of that water within 1 day under high net radiation, whereas a silt loam soil with a water-holding capacity of 0.17 mm cm^{-1} may sustain an evaporation rate for over 1 day. Lascano and Hatfield (1992) showed how drying a small depth interval in the upper soil surface suppressed evaporation, and this effect will depend upon the ability of the soil to have paths for the movement of water from depth to the soil surface. Soil water evaporation can be divided into three stages: Stage 1, in which the evaporation rate is determined by the energy impinging on the soil surface; Stage 2, in which the rate is controlled by the soil moisture conditions and the soil hydraulic properties; and Stage 3, which is governed by the rate of transfer of water vapour from the deeper depths of the soil profile, and potentially through cracks in the soil (Ritchie 1972; van de Griend and Owe 1994; Porté-Agel *et al.* 2000; Suleiman and Ritchie 2003; Han and Zhou 2013). With increased rainfall during the early season, we could expect Stage 1 evaporation to become more dominant with increased amounts of water vapour evaporated into the atmosphere.

Soil water storage

The dynamics of soil water evaporation and transpiration are directly related to soil water availability, which is a function of texture, organic matter, porosity and pore structure. Differences among textures are presented in Table 2.1. The dynamics of the SWB (Eqn 1) become evident when we couple soil water availability with the water use and precipitation pattern for a given time. The true impact of climate on soil hydrology has to encompass the SWB, which reflects the interaction among all the components that supply water to the plant.

Table 2.1. Soil texture differences among soils for –33 and –1500 kPa soil water pressures and available soil water

Soil texture	Water content (mm cm^{-1})		Available water (mm cm^{-1})
	–1500 kPa	–33 kPa	
Sand	0.4	1.0	0.6
Loamy sand	0.7	1.6	0.9
Sandy loam	0.9	2.1	1.2
Loam	1.2	2.7	1.5
Silt loam	1.5	3.0	1.5
Sandy clay loam	1.8	2.9	1.1
Sandy clay	1.5	2.8	1.3
Clay loam	1.8	3.2	1.4
Silty clay loam	2.0	3.6	1.6
Silty clay	2.0	4.0	2.0
Clay	2.2	4.0	1.8

The ability of the soil to supply water to meet atmospheric demand is dependent upon soil water-holding capacity. For example, a loamy sand has an available water-holding capacity of 0.9 mm cm^{-1}, which is half that of a silty clay (2.0 mm cm^{-1}); so, if we expand the rooting depth to 1 m, we then have 90 mm of water available in a loamy sand, compared with 200 mm in a silty clay. This is the potential size of the pool of water in the soil, and what is available at one point in time represents the replenishment of water via precipitation or irrigation, the infiltration rate of the soil, drainage and the rate of soil water evaporation and transpiration. Differences in the soil will directly affect the ability to supply water for transpiration, and there have been a large number of research studies that have shown water is one of the primary determinants of crop yields across the world (Lobell and Field 2007; Hatfield *et al.* 2011, 2018; Mishra and Cherkauer 2010; Walthall *et al.* 2012; Wang *et al.* 2016). Climate directly affects precipitation, infiltration effectiveness and the rate of evaporation and transpiration, and understanding these changes will help us begin to determine how to mitigate the impacts of climate change.

Potential impacts of climate change on plants

To fully understand the impact of climate change on soil hydrology, we also must understand its effect on plants as part of the hydrological cycle. Plants are a critical part of the hydrological cycle because of their role in evaporation and transpiration, which, in turn, affect the capacity for water storage, infiltration, run-off and drainage. What we do not often consider is the linkage between the effects of climate on plants and the dynamics of the hydrological cycle. In this section, we provide an overview of how climate change affects plants and potential impacts on soil hydrology.

Hatfield *et al.* (2011) and Izaurralde *et al.* (2011) evaluated the impact of climate change on agronomic, range and pasture crops. In summary, increasing temperatures will reduce plant growth to offset the positive effect of increasing CO_2, and the increasing variability in precipitation will increase the variation in plant growth and productivity. The longer growing season will be positive because plants will begin to grow earlier and the season will extend into the autumn to provide more potential plant growth. Warmer temperatures will increase the rate of plant development, leading to a condition in which plants could develop faster, which would affect the rate of water use and the rate at which the ground is covered, which will then affect the balance between soil water evaporation and transpiration (Fig. 2.3).

Increasing temperatures influence the rate of phenological development, and the change in minimum temperatures that has been detected in the climate signal hastens the rate of maturity in all crops. Hatfield (2016) showed that exposure to high temperatures during pollination was detrimental to corn grain yields, but decreased the time of grain filling by several days. This would directly affect the length of time that a crop is extracting soil water from the profile. What is not available is information on any change in the rooting depth or rate of root expansion into the soil as a result of climate change. However, there is an effect of the increasing spring rainfall in the upper Midwest on soil water dynamics. We have found that at the county scale, corn and soybean yields are negatively correlated with above-normal precipitation in the April–June period (J. L. Hatfield, unpubl. data, 2019). These correlations exist because excessive precipitation delays field operations and time of planting, which limits the length of the growing season, in addition to the effects of excess soil water on crop vigour. Another effect was observed by Hatfield and Prueger (2011b), whereby water use patterns during the season showed no effect of soil water availability in the early season; however, during the grain-filling period, soil water availability determined grain yield. This effect can be explained by water use by the crop being small until the beginning of the grain-filling period and, when the soil is extremely wet from the spring rainfalls, the root system required to supply the crop water demands is minimal compared with periods of high water use during the growth cycle.

Need for urgency in climate change adaptation

Both the 2014 and 2018 National Climate Assessments highlight the increasing chance of more extremes in precipitation and high temperatures (Walsh *et al.* 2014; Wuebbles *et al.* 2017a). We have discussed how these changes will interact with soil hydrology and plant production. Clearly, the climate projections indicate that our ability to protect our natural resources will be tested and raise the question about whether our current conservation practices are sufficient to protect against the changes expected in our climate. The findings from the IPCC Land Degradation summary do not portray a very optimistic picture unless aggressive management practices are adopted to protect the soil resource (Olsson *et al.* 2019). Not all these projections have a direct connection to agriculture and natural resource systems; however, all these changes serve as a reminder that we will live in a world with a changing climate.

The understanding that management frameworks need to adapt to changing climate conditions is not new. An assessment undertaken by the Soil and Water Conservation Society (SWCS) in 2003 to document the potential impacts of climate change on soil erosion and run-off from croplands found we need to be attentive to the impact climate change will have on our soil resource (SWCS 2003). This was a forward-looking analysis; however, the degree of climate change is larger than anticipated 15 years ago. The SWCS (2003) report generated three major conclusions to address the problem of climate change and erosion:

1. Immediately update climatic parameters in critical conservation planning tools
2. Undertake targeted investigations to firm up estimates of the damage that would likely occur under simulated or observed climate regimes

3. Evaluate the benefits of building the risk of damage from severe rainstorms into the conservation planning process through risk-based assessments targeted to key conservation systems and environmental outcomes.

We have made progress on many of these efforts; however, the degree of climate change will potentially affect our ability to advance management practices to the level required to address these changes in climate. In the 2018 National Climate Assessment, Gowda *et al.* (2018) summarised two key messages critical to agriculture and the natural resources:

1. Food and forage production will decline in regions experiencing increased frequency and duration of drought. Shifting precipitation patterns, when associated with high temperatures, will intensify wildfires that reduce forage on rangelands, accelerate the depletion of water supplies for irrigation, and expand the distribution and incidence of pests and diseases for crops and livestock. Modern breeding approaches and the use of novel genes from crop wild relatives are being used to develop higher-yielding, stress-tolerant crops.
2. The degradation of critical soil and water resources will expand as extreme precipitation events increase across our agricultural landscape. Sustainable crop production is threatened by excessive run-off, leaching and flooding, which result in soil erosion, degraded water quality in lakes and streams and damage to rural community infrastructure. Management practices to restore soil structure and the hydrological function of landscapes are essential to improve resilience to these challenges.

Moving forward: adaptation strategies to offset climate changes

The effort required to transform our agroecosystems to become more resilient to climate change is underway, and we have made some significant strides. Since the First National Climate Assessment was released in 2000, US agriculture has seen important technological advances in conjunction with a renewed emphasis on conservation management and precision agriculture, especially as those related to climate. Some US growers are becoming increasingly aware of the risks that climate change poses to their soil and water resources, and multiple climate-smart agricultural initiatives (e.g. cover crops, specialised irrigation and nutrient management) are being implemented to help cope with ongoing climatic variability and extreme weather events (Lipper *et al.* 2014). However, no single management practice is a panacea because the efficacy of each in enhancing the resiliency of critical soil and water resources to climate stress varies greatly across the US's diverse agroecosystems (Chatrchyan *et al.* 2017; Roesch-McNally *et al.* 2017). The uncertainly about the impacts of climate change (Morton *et al.* 2017), as well as the effectiveness of management strategies to offset those impacts, has inhibited the widespread adoption of these proposed climate-resilient land management practices. Further, some growers are even taking actions to address climate change that actually stress the systems more, such as installing subsurface tile drainage to remove excess water from the system (Roesch-McNally *et al.* 2017); these drainage systems have the potential to greatly increase nutrient concentrations in surface waters and can contribute to widespread environmental quality issues (Dale *et al.* 2008). Despite these obstacles, the resiliency of soil hydrological function under changing climate conditions may generally be improved by practices that: (1) encourage infiltration while reducing run-off; (2) reduce water lost to evaporation or drainage; and/or (3) increase soil water storage capacity. Although numerous strategies are available to producers to achieve these goals, the best choice must be dictated by local climate trends and site-specific characteristic. Thus, rather than espouse any specific management practices, three endpoints are identified that have the potential to improve soil hydrological function: (1) increasing soil surface cover; (2) diversifying spatiotemporal water uptake regimes; and (3) maximising soil water storage.

Soil surface cover

Conditions at the soil surface are critical in determining how climate change will affect soil hydrology because it is the interface where water infiltrates, evaporates or is transferred laterally. Increasing the proportion of soil covered and the length of time the soil is covered may be beneficial in successfully adapting to both general trends in precipitation and temperature changes, as well as an increased frequency and intensity of extreme events. Soil surface cover, typically from residue or standing vegetation, promotes infiltration by protecting the soil from raindrop impact, which has the potential to break down soil aggregates (Abban *et al.* 2017) and clog soil pores (Wacha *et al.* 2018). Conversely, in prolonged dry periods, surface cover can reduce evaporation (Klocke *et al.* 2009; Swella *et al.* 2015) by serving as a boundary layer that reduces vapour diffusivity. The cover also intercepts incoming radiation, which reduces the amount of energy partitioned to evaporating soil water, as well as moderating extreme soil temperatures (Horton *et al.* 1996; O'Brien and Daigh 2019).

Producers manipulate surface cover in agroecosystems in a variety of ways. For example, tillage is often used as a tool to increase residue decomposition, decrease weed pressure and prepare the seedbed for germination, but incorporating the residue leaves the soil surface exposed (de Almeida *et al.* 2018). Thus, some drawbacks of tillage, including erosion, changes to pore networks and accelerated evaporation (Morris *et al.* 2010; Reicosky 2015), are likely to be exacerbated by the changing climate conditions. Similarly, some practices that remove annual crop biomass for animal feed, bedding or cellulosic bioenergy feedstock also reduce the amount of residue available to cover the surface (Blanco-Canqui *et al.* 2014; Ibrahim *et al.* 2018). The orientation of tillage with respect to flow (e.g. ridge till or contour) has been shown to break up flow hydraulic connectivity and decrease shear stress and the magnitude of erosion (Wacha *et al.* 2020).

Finally, not all crops produce the same amount or quality of residue, so removal and/or incorporation may have varying effects depending on crop sequences. Although tillage and residue removal have the potential to reduce surface cover and increase susceptibility to deterioration under current climate trends, the practices do not necessarily need to be discontinued to maintain resilient systems. Rather, these practices should be paired with other strategies to compensate for lack of soil surface cover.

One of the most common approaches to increase soil surface cover in row–crop systems is to implement cover crops (or catch crops, service crops) to grow after the harvest in autumn, with some overwintering to provide active growth and surface cover the following spring. Cover crops increase both the spatial and temporal coverage of the soil surface and have been associated with climate-resilient targets such as increased infiltration, reduced evaporation and increased water storage capacity (Dabney *et al.* 2001; Kaye and Quemada 2017; Ogilvie *et al.* 2019). Cover crops can also increase biological activity in the soil, which can improve nutrient cycling and promote aggregation processes. Diversifying crop rotations that include sequences with perennial vegetation provides similar benefits by extending the seasonal coverage of the soil surface and promoting favourable growing conditions (Gaudin *et al.* 2015). The addition of ecological infrastructure, such as wind breaks, prairie strips or saturated buffers, protects the soil surface from extreme climate events (Stavi and Lal 2013; Schulte *et al.* 2017), although they encompass less overall area. Similarly, the application of mulches/manures or synthetic mulches has the potential to reduce evaporation and erosion and to moderate temperature fluctuations in targeted at-risk areas, although widespread applications are likely not feasible. Notably, these strategies are not successful in every situation at promoting infiltration, reducing evaporation and minimising run-off; rather, they are tools that producers should use to create diversified systems with improved resiliency based on local trends and conditions.

Diversified water uptake regimes

Agroecosystems that promote active growth and root water uptake by vegetation throughout the year will be better suited to exploit the increased variability in precipitation that is predicted. For example, early spring growth in cover crops or perennial species has the potential to create biomass by using water that otherwise would have been lost to evaporation. Similarly, creating agroecosystems with diversified root structures may allow plants to access water stored at different depths in the profile (Chen and Weil 2011). Overall, diversifying the space and timing of water uptake will create resilience in agroecosystems by minimising the risk of crop failure (Gaudin *et al.* 2015; Davis *et al.* 2017). If all crops demand water at the same time of the year and all are accessing the same depth in the soil profile, then all crops will suffer if water is not present at the time of water demand. However, if multiple crops on the landscape require water at different times and can access different parts of the soil profile, then the odds that water is available increase markedly, likely resulting in at least some successful production.

Strategies to target water uptake for longer in the season typically include adding cover crops, perennial species and diversified crop rotations to agronomic management plans. Both cover crops and perennial species not only have the potential to use soil water before the planting of primary crops, but they also improve water quality by scavenging nutrients that may otherwise be lost via subsurface drainage (Kaspar *et al.* 2007). However, they may also create a risk for yield losses in dry years, so land managers need to monitor soil water conditions and terminate cover crops before extracting too much water (Kaye and Quemada 2017). In addition to extending the time of water uptake, agroecosystems can also use a variety of root structures to expand the area of the soil profile with accessible water. For example, many brassica species have a deep taproot that can access water deeper than many row crops (Chen and Weil 2011). In addition to diversifying row crop rooting structures, agroforestry practices may provide some benefits under climate change (Stavi and Lal 2013). For example, in alley-cropping systems, not only do trees access water far deeper than the crop understorey, but the tree canopy may also provide shade protection to avoid water losses in the understorey during dry years.

A different approach to adapt to the temporal trends in precipitation involves capturing and storing drainage water from heavy spring rains and using that water to irrigate crops later in the season when crop water demand exceeds soil water availability (Baker *et al.* 2012). An added benefit to capturing this water is the interception of nutrients that would otherwise be introduced to surface waters. This practice needs much more investigation before widespread application, but it represents an example of management practices adapting to how changing climate conditions affect soil hydrology and agricultural water management.

Maximising soil water storage

Increasing the amount of water that can be stored in the soil may be the most effective way to adapt to changing precipitation and temperature regimes, but changing this capability can be very slow. Conceptually, increasing soil water in the profile is valuable during wet cycles. Infiltration and storage is maximised; similarly, soils that can store more water can provide plant-available water longer throughout dry periods. Management practices must primarily aim to improve soil porosity and reactive surface area in the soil profile, typically by improving soil aggregation, root growth and soil organic matter. Soil porosity and plant-available water may be improved using cover crops (Basche *et al.* 2016) or perennial species (Daigh *et al.* 2014). Similarly, amendments like biochar or manure are often touted for their potential to improve water-holding capacity, but research findings have been inconsistent (Haynes and Naidu 1998; Atkinson 2018; Irmak *et al.* 2018). Continued research will investigate the best amendments, rates and application conditions to improve soil water storage. Nonetheless, soil water storage is largely governed by soil texture, so the capacity for improvement is low and changes in storage may be slow.

Conclusions

Changes in our climate will affect plant, animal and natural resources, and we must become prepared to understand these dynamics as part of the agroecological system and the influences on soil hydrological processes. Understanding these linkages will help us develop an understanding of the role climate variation plays in the soil hydrological system and how we can more effectively manage the water resources needed for agriculture. The changing seasonality in precipitation will alter the timing of when precipitation events will occur; however, these are location dependent, with a greater increase in spring precipitation across the upper portion of the US. Generally, annual precipitation in the US is increasing, and the storms are becoming more intense, which creates a situation in which soils may not be able to infiltrate the water, leading to more run-off and less storage. Changing temperature patterns will alter the rate of water use through soil water evaporation and transpiration because of changing plant growth patterns. The continued degradation of the soil through our current management practices, such as intense tillage and simplified crop rotations, will increase the vulnerability to climate change. Thus, implementing management practices that maintain soil surface cover throughout the year, increase spatiotemporal diversity of plant water use and maximise soil water storage will help these systems be more resilient in the face of climate change.

References

Abban BK, Papanicolaou AT, Giannopoulos CP, Dermisis DC, Wacha KM, Wilson CG, Elhakeem M (2017) Quantifying the changes of soil surface microroughness due to rainfall impact on a smooth surface. *Nonlinear Processes in Geophysics* **24**, 569–579. doi:10.5194/npg-24-569-2017

Allen RG, Pereira LS, Raes D, Smith M (1998) *Crop Evapotranspiration: Guidelines for Computing Crop Water Requirements*. Food and Agriculture Organization of the United Nations, Rome.

Allstadt AJ, Vavrus SJ, Heglund PJ, Pidgeon AM, Thogmartin WE, Radeloff VC (2015) Spring plant phenology and false springs in the conterminous US during the 21st century. *Environmental Research Letters* **10**, 104008. doi:10.1088/1748-9326/10/10/104008

Atkinson CJ (2018) How good is the evidence that soil applied biochar improves water holding capacity? *Soil Use and Management* **34**, 177–186. doi:10.1111/sum.12413

Baker JM, Griffis TJ, Ochsner TE (2012) Coupling landscape water storage and supplemental irrigation to increase productivity and improve environmental stewardship in the US Midwest. *Water Resources Research* **48**, W05301. doi:10.1029/2011WR011780

Barthès B, Roose E (2002) Aggregate stability as an indicator of soil susceptibility to runoff and erosion; validation at several levels. *Catena* **47**, 133–149. doi:10.1016/S0341-8162(01)00180-1

Basche AD, Kaspar TC, Archontoulis SV, Jaynes DB, Sauer TJ, Parkin TB, Miguez FE (2016) Soil water improvements with the long-term use of a winter rye cover crop. *Agricultural Water Management* **172**, 40–50. doi:10.1016/j.agwat.2016.04.006

Blanco-Canqui H, Ferguson RB, Jin VL, Schmer MR, Wienhold BJ, Tatarko J (2014) Can cover crop and manure maintain soil properties after stover removal from irrigated no-till corn? *Soil Science Society of America Journal* **78**, 1368–1377. doi:10.2136/sssaj2013.12.0550

Buck AL (1981) New equations for computing vapor pressure and enhancement factor. *Journal of Applied Meteorology* **20**, 1527–1532. doi:10.1175/1520-0450(1981)020<1527:NEFCVP>2.0.CO;2

Chatrchyan AM, Erlebacher RC, Chaoprich NT, Chan J, Tobin D, Allred SB (2017) United States agricultural stakeholder views and decisions on climate change. *Wiley Interdisciplinary Reviews: Climate Change* **8**, e469. doi:10.1002/wcc.469

Chen G, Weil RR (2011) Root growth and yield of maize as affected by soil compaction and cover crops. *Soil & Tillage Research* **117**, 17–27. doi:10.1016/j.still.2011.08.001

Collins M, Knutti R, Arblaster J, Dufresne J-L, Fichefet T, Friedlingstein P, Gao X, Gutowski WJ Jr, Johns T, Krinner G *et al.* (2013) Chapter 12 – Long-term climate change: projections, commitments and irreversibility. In *Climate Change 2013: The Physical Science Basis. Contribution of Working Group I to the Fifth Assessment Report of the Intergovernmental Panel on Climate Change.* (Eds TF Stocker, D Qin, G-K Plattner, M Tignor, SK Allen, J Boschung, A Nauels, Y Xia, V Bex and PM Midgley) pp. 1029–1136. Cambridge University Press, Cambridge.

Dabney SM, Delgado JA, Reeves DW (2001) Using winter cover crops to improve soil and water quality. *Communications in Soil Science and Plant Analysis* **32**, 1221–1250. doi:10.1081/CSS-100104110

Daigh, AL, Zhou X, Helmers MJ, Pederson CH, Ewing R Horton R (2014) Subsurface drainage flow and soil water dynamics of reconstructed prairies and corn rotations for biofuel production. *Vadose Zone Journal* **13**, 1–11. doi:10.2136/vzj2013.10.0177

Dale V, Bianchi T, Blumberg A, Boynton W, Conley DJ, Crumpton W, David M, Gilbert D, Howarth RW, Kling C *et al.* (2008) Hypoxia in the northern Gulf of Mexico: An update by the EPA Science Advisory Board. USEPA, Washington, DC, <https://www.epa.gov/ms-htf/hypoxia-northern-gulf-mexico-update-epa-science-advisory-board>.

Davis KF, Seveso A, Rulli MC, D'Odorico P (2017) Water savings of crop redistribution in the United States. *Water (Basel)* **9**, 83. doi:10.3390/w9020083

de Almeida WS, Panachuki E, de Oliveira PTS, da Silva Menezes R, Sobrinho TA, de Carvalho DF (2018) Effect of soil tillage and vegetal cover on soil water infiltration. *Soil & Tillage Research* **175**, 130–138. doi:10.1016/j.still.2017.07.009

Gaudin ACM, Tolhurst TN, Ker AP, Janovicek K, Tortora C, Martin RC, Deen W (2015) Increasing crop diversity mitigates weather variations and improves yield stability. *PLoS One* **10**, e0113261. doi:10.1371/journal.pone.0113261

Gowda P, Steiner JL, Olson C, Boggess M, Farrigan T, Grusak MA (2018) Agriculture and Rural Communities. In *Impacts, Risks, and Adaptation in the United States: Fourth National Climate Assessment, Volume II.* (Eds DR Reidmiller, CW Avery, DR Easterling, KE Kunkel, KLM Lewis, TK Maycock and BC Stewart) pp. 391–437. U.S. Global Change Research Program, Washington, DC. doi:10.7930/NCA4.2018.CH10

Han J, Zhou Z (2013) Dynamics of soil water evaporation during soil drying: laboratory experiment and numerical analysis. *TheScientificWorldJournal* **2013**, 240280. doi:10.1155/2013/240280

Hatfield JL (2016) Increased temperatures have dramatic effects on growth and grain yield of three maize hybrids. *Agricultural & Environmental Letters* **1**, 150006. doi:10.2134/ael2015.10.0006

Hatfield JL, Prueger JH (2011a) Agroecology: implications for plant response to climate change. In *Crop Adaptation to Climate Change*. (Eds SS Yadav, RJ Redden, JL Hatfield, H Lotze-Campen and AE Hall) pp. 27–43. Wiley-Blackwell, Chichester.

Hatfield JL, Prueger JH (2011b) Spatial and temporal variation in evapotranspiration. In *Evapotranspiration: From Measurements to Agricultural and Environmental Applications*. (Ed. GA Gerosa) pp. 3–16. IntechOpen, London.

Hatfield JL, Walthall CL (2015) Meeting global food needs: realizing the potential via genetics × environment × management interactions. *Agronomy Journal* **107**, 1215–1226. doi:10.2134/agronj15.0076

Hatfield JL, Boote KJ, Kimball BA, Ziska LH, Izaurralde RC, Ort D, Thomson AM, Wolfe DW (2011) Climate impacts on agriculture: implications for crop production. *Agronomy Journal* **103**, 351–370. doi:10.2134/agronj2010.0303

Hatfield JL, Wright-Morton L, Hall B (2018) Vulnerability of grain crops and croplands in the Midwest to climatic variability and adaptation strategies. *Climatic Change* **146**, 263–275. doi:10.1007/s10584-017-1997-x

Hayhoe K, Wuebbles DJ, Easterling DR, Fahey DW, Doherty S, Kossin J, Sweet W, Vose R, Wehner W (2018) Our changing climate. In *Impacts, Risks, and Adaptation in the United States: Fourth National Climate Assessment, Volume II*. (Eds DR Reidmiller, CW Avery, DR Easterling, KE Kunkel, KLM Lewis, TK Maycock and BC Stewart) pp. 72–144. U.S. Global Change Research Program, Washington, DC. doi:10.7930/NCA4.2018.CH2

Haynes RJ, Naidu R (1998) Influence of lime, fertilizer and manure applications on soil organic matter content and soil physical conditions – a review. *Nutrient Cycling in Agroecosystems* **51**, 123–137. doi:10.1023/A:1009738307837

Horton R, Bristow KL, Kluitenberg GJ, Sauer TJ (1996) Crop residue effects on surface radiation and energy balance. *Theoretical and Applied Climatology* **54**, 27–37. doi:10.1007/BF00863556

Ibrahim E, Osborne SL, Schumacher TE, Riedell WF (2018) Corn residue removal effects on hydraulically effective macropores. *Communications in Soil Science and Plant Analysis* **49**, 1491–1501. doi:10.1080/00103624.2018.1464187

Irmak S, Sharma V, Mohammed AT, Djaman K (2018) Impacts of cover crops on soil physical properties: field capacity, permanent wilting point, soil-water holding capacity, bulk density, hydraulic conductivity, and infiltration. *Transactions of the ASABE* **61**, 1307–1321. doi:10.13031/trans.12700

Izaurralde RC, Thomson AM, Morgan JA, Fay PA, Polley HW, Hatfield JL (2011) Climate impacts on agriculture: implications for forage and rangeland production. *Agronomy Journal* **103**, 371–381. doi:10.2134/agronj2010.0304

Johnson AI (1963) 'A field method for measurement of infiltration. General ground-water techniques.' Geological Survey Water-Supply Paper 1544-F, United States Government Printing Office, Washington, DC.

Kaspar TC, Jaynes DB, Parkin TB, Moorman TB (2007) Rye cover crop and gamagrass strip effects on NO_3 concentration and load in tile drainage. *Journal of Environmental Quality* **36**, 1503–1511. doi:10.2134/jeq2006.0468

Kaye JP, Quemada M (2017) Using cover crops to mitigate and adapt to climate change. A review. *Agronomy for Sustainable Development* **37**, 4. doi:10.1007/s13593-016-0410-x

Klocke NL, Currie RS, Aiken RM (2009) Soil water evaporation and crop residues. *Transactions of the ASABE* **52**, 103–110. doi:10.13031/2013.25951

Kral-O'Brien KC, O'Brien PL, Harmon JP (2019) Need for false spring research in the Northern Great Plains, USA. *Agricultural & Environmental Letters* **4**, 190025. doi:10.2134/ael2019.07.0025

Kunkel KE, Andsager K, Easterling DR (1999) Long-term trends in extreme precipitation events over the conterminous United States and Canada. *Journal of Climate* **12**, 2515–2527. doi:10.1175/1520-0442(1999)012<2515:LTTIEP>2.0.CO;2

Lascano RJ, Hatfield JL (1992) Spatial variability of evaporation along two transects of a bare soil. *SoilScienceSocietyofAmericaJournal* **56**,341–346.doi:10.2136/sssaj1992.03615995005600020002x

Lipper L, Thornton P, Campbell BM, Baedeker T, Braimoh A, Bwalya M, Caron P, Cattaneo A, Garrity D, Henry K, *et al.* (2014) Climate-smart agriculture for food security. *Nature Climate Change* **4**, 1068–1072. doi:10.1038/nclimate2437

Lobell DB, Field CB (2007) Global scale climate–crop yield relationships and the impacts of recent warming. *Environmental Research Letters* **2**, 014002. doi:10.1088/1748-9326/2/1/014002

Mishra V, Cherkauer KA (2010) Retrospective droughts in the crop growing season: implications to corn and soybean yield in the Midwestern United States. *Agricultural and Forest Meteorology* **150**, 1030–1045. doi:10.1016/j.agrformet.2010.04.002

Morris NL, Miller PCH, Orson JH, Froud-Williams RJ (2010) The adoption of noninversion tillage systems in the United Kingdom and the agronomic impact on soil, crops, and the environment – a review. *Soil & Tillage Research* **108**, 1–15. doi:10.1016/j.still.2010.03.004

Morton LW, Roesch-McNally G, Wilke AK (2017) Upper Midwest farmer perceptions: too much uncertainty about impacts of climate change to justify changing current agricultural practices. *Journal of Soil and Water Conservation* **72**, 215–225. doi:10.2489/jswc.72.3.215

National Oceanic and Atmospheric Administration (NOAA) (2013) 'Regional climate trends and scenarios for the U.S. National Climate Assessment – Part 4. Climate of the U.S. Great Plains.' NOAA Technical Report NESDIS 142-4, NOAA, Washington, DC.

O'Brien PL, Daigh ALM (2019) Tillage practices alter the surface energy balance: a review. *Soil & Tillage Research* **195**, 104354. doi:10.1016/j.still.2019.104354

O'Brien PL, Hatfield JL (2020) Extreme soil temperatures reflect the need to rethink agronomic management. *Agricultural & Environmental Letters* **5**, e20002. doi:10.1002/ael2.20002

O'Brien PL, Hatfield JL, Dold C, Kistner-Thomas EJ, Wacha KM (2020) Cropping pattern changes diminish agroecosystem services in North and South Dakota, USA. *Agronomy Journal* **112**, 1–24. doi:10.1002/agj2.20001

Ogilvie CM, Deen W, Martin RC (2019) Service crop management to maximize crop water supply and improve agroecosystem resilience: a review. *Journal of Soil and Water Conservation* **74**, 389–404. doi:10.2489/jswc.74.4.389

Olsson L, Barbosa H, Bhadwal S, Cowie A, Delusca K, Flores-Renteria D, Hermans K, Jobbagy E, Kurz W, Li D, *et al.* (2019) Land degradation. In *IPCC Special Report on Climate Change, Desertification, Land Degradation, Sustainable Land Management, Food Security, and Greenhouse gas fluxes in Terrestrial Ecosystems.* (Eds PR Shukla, J Skea, EC Buendia, V Masson-Delmotte, H Pörtner, DC Roberts, P Zhai, R Slade, S Connors, R van Dieman, *et al.*) pp. 345–436. Cambridge University Press, Cambridge.

Porté-Agel F, Parlange M, Cahill AT, Gruber A (2000) Mixture of time scales in evaporation: desorption and self-similarity of energy fluxes. *Agronomy Journal* **92**, 832–836. doi:10.2134/agronj2000.925832x

Reicosky DC (2015) Conservation tillage is not conservation agriculture. *Journal of Soil and Water Conservation* **70**, 103A–108A. doi:10.2489/jswc.70.5.103A

Ritchie JT (1972) Model for predicting evaporation from a row crop with incomplete cover. *Water Resources Research* **8**, 1204–1213. doi:10.1029/WR008i005p01204

Roesch-McNally GE, Arbuckle J, Tyndall JC (2017) What would farmers do? Adaptation intentions under a Corn Belt climate change scenario. *Agriculture and Human Values* **34**, 333–346. doi:10.1007/s10460-016-9719-y

Schulte LA, Niemi J, Helmers MJ, Liebman M, Arbuckle JG, James DE, Kolka RK, O'Neal ME, Tomer MD, Tyndall JC, *et al.* (2017) Prairie strips improve biodiversity and delivery of multiple ecosystems service from corn-soybean croplands. *Proceedings of the National Academy of Sciences of the United States of America* **114**, 11247–11252. doi:10.1073/pnas.1620229114

Soil and Water Conservation Society (SWCS) (2003) *Conservation Implications of Climate Change: Soil Erosion and Runoff from Cropland*. SWCS, Ankeny.

Stavi I, Lal R (2013) Agroforestry and biochar to offset climate change: a review. *Agronomy for Sustainable Development* **33**, 81–96. doi:10.1007/s13593-012-0081-1

Suleiman AA, Ritchie JT (2003) Modeling soil water evaporation during second-stage evaporation. *Soil Science Society of America Journal* **67**, 377–386. doi:10.2136/sssaj2003.3770

Swella GB, Ward PR, Siddique KHM, Flower KC (2015) Combinations of tall standing and horizontal residue affect soil water dynamics in rainfed conservation agriculture systems. *Soil & Tillage Research* **147**, 30–38. doi:10.1016/j.still.2014.11.004

Trenberth KE (2011) Changes in precipitation with climate change. *Climate Research* **47**, 123–138. doi:10.3354/cr00953

van de Griend AA, Owe, M (1994) Bare soil surface resistance to evaporation by vapor diffusion under semiarid conditions. *Water Resources Research* **30**, 181–188. doi:10.1029/93WR02747

Wacha K, Papanicolaou A, Giannopoulos C, Abban B, Wilson C, Zhou S, Hatfield JL, Filley TR, Hou T (2018) The role of hydraulic connectivity and management on soil aggregate size and stability in the Clear Creek watershed, Iowa. *Geosciences* **8**, 470. doi:10.3390/geosciences8120470

Wacha KM, Papanicolaou AT, Abban BK, Wilson CG, Giannopoulos CP, Hou T, Filley TR, Hatfield JL (2020) The impact of tillage row orientation on physical and chemical sediment enrichment. *Agrosystems, Geosciences & Environment* **3**, e20007. doi:10.1002/agg2.20007

Walsh J, Wuebbles D, Hayhoe K, Kossin J, Kunkel K, Stephens G, Thorne P, Vose R, Wehner M, Willis J, *et al.* (2014) Our changing climate. In *Climate Change Impacts in the United States: The Third National Climate Assessment*. (Eds JM Melillo, TC Richmond and GW Yohe) pp. 19–67. U.S. Global Change Research Program, Washington, DC. doi:10.7930/J0KW5CXT.

Walthall CL, Hatfield J, Backlund P, Lengnick L, Marshall E, Walsh M, Adkins S, Aillery M, Ainsworth EA, Ammann C, *et al.* (2012) 'Climate change and agriculture in the united states: effects and adaptation.' USDA Technical Bulletin 1935, Washington, DC.

Wang R, Bowling LC, Cherkauer KA (2016) Estimation of the effects of climate variability on crop yield in the Midwest USA. *Agricultural and Forest Meteorology* **216**, 141–156. doi:10.1016/j.agrformet.2015.10.001

Wuebbles DJ, Easterling DR, Hayhoe K, Knutson T, Kopp RE, Kossin JP, Kunkel KE, LeGrande AN, Mears C, Sweet WV, *et al.* (2017a) Our globally changing climate. In *Climate Science Special Report: Fourth National Climate Assessment, Volume I*. (Eds DJ Wuebbles, DW Fahey, KA Hibbard, Dokken DJ, BC Stewart, TK Maycock) pp. 35–72. U.S. Global Change Research Program, Washington, DC. doi:10.7930/J08S4N35.

3

Soil health and soil water

Udayakumar Sekaran, Ali Mehmandoost Kotlar, Sandeep Kumar

Introduction

Soil health is defined as the '*continued capacity of a soil to function as a vital living system within ecosystem boundaries that sustains plants, animals, and humans*' (Natural Resources Conservation Service (NRCS) 2020). Soil water plays a critical role in soil functionality and health due to its deep connection to biogeochemical and hydrological processes. For example, it is essential in crop growth and development, and its deficiency in soils limits crop production (Manici *et al.* 2019). Due to erratic rainfall patterns and associated drought episodes, water availability will be even more challenging in the future due to climate change (Thornton *et al.* 2014). Living organisms require soil water to decompose soil organic matter (SOM) and respire; in plants, soil water is required for the take up of dissolved nutrients and photosynthesis. A balance between soil air and water is an essential prerequisite for microbial processes, and is responsible for maintaining soil health. Soil water is stored and transported within soil pores, defined by pore diameters corresponding to equivalent soil matric potentials. Because soil water is held within the soil pores, the water-holding capacity depends on capillary movement and pore size.

Soil pore size distribution, which is affected by the texture and structure of a soil, determines the amount of water available to plants and soil organisms (Nielsen *et al.* 2002). Field capacity is a soil water content that results in a negligible water drainage rate at the bottom of the soil profile. This amount of water usually remains in the pores with an equivalent soil matric potential of between −6 and −33 kPa after drainage by gravity has ceased. The permanent wilting point is the amount of water that is strongly held on soil mineral particles and is thus unavailable for plants, and corresponds to soil water content at −1500 kPa (Richards 1931). Plant-available water (PAW) is the water in the soil between field capacity and the permanent wilting point. Micropores supply most PAW in the soil. Sandy-textured soils with larger pores (macropores) are not able to hold as much water, and drain excessively with low values of PAW, whereas fine-textured soils with a higher clay content and micropores provide greater PAW. Loamy-textured soils (loams, sandy loams and silt loams) have an ideal combination of mesopores and micropores, and have the greatest PAW (Richards 1931; Vogel 2000).

Soil water retention is a key soil hydrology property that governs soil ecosystem function and affects soil management (Leão and Tuller 2014). Compared with larger sand particles, fine particles like clay and silt have a much larger surface area and cation exchange capacity (CEC), and so a higher water-holding capacity (Six *et al.* 2000; Bruun *et al.* 2010). Soils with a higher CEC have more clay and organic matter, and greater water-holding capacity, than soils with a low CEC. In soil, CEC and specific surface area are the two fundamental properties that control water sorption, especially in dry soils, when cation hydration is the major mechanism of retaining soil water (Khorshidi and Lu 2016). Soil water retention and holding capacity are also highly influenced by the organic matter concentration. In general, as SOM increases, the water-holding capacity and water retention also increase because of the high affinity of water to SOM. According to Minasny and McBratney (2018), the increase in soil water content after the addition of SOM is greatest at 0 kPa (saturation), followed by –10 or –33 kPa (field capacity) and –1500 kPa (permanent wilting point). An increase in SOM improves soil aggregation and structure, and could create larger volumes of macro-, meso- and micropores and increase water infiltration (Minasny and McBratney 2018). Soil aggregates are critical for holding soil water, and SOM is essential for forming aggregates. Soil microbes, especially fungi, play a major role in the stabilisation of aggregates. Arbuscular mycorrhizal fungi (AMF) produce glomalin, which binds soil particles together, protecting them from erosion, and coats hyphae to keep water and nutrients from being lost on the way to and from the plant (Goss *et al.* 2017).

Healthy soil is key to meeting growing global food demand, but few studies have clarified the links between indicators of soil health (physical, chemical and biological) and soil water retention and storage. Theoretical calculations by Libohova *et al.* (2018) showed that volumetric water-holding capacity could increase by 1.5- to 1.7-fold of organic matter weight for a unit increase in organic matter (% weight), which is in agreement with earlier results demonstrating that soil available water increased by 1–4.9 g with every 1-g increase in soil organic carbon (SOC) for every 100 g of soil (Hudson 1994). Determining relationships between soil health parameters and soil water remains a major challenge and is rarely studied; thus, the aim of this chapter is to provide in-depth knowledge about soil health properties and their relationship with soil water based on existing literature.

Soil health assessment includes measurement of different physical, chemical and biological indicators that can potentially describe the health status of soil (NRCS 2020; Fig. 3.1). Assessing changes in the physical, chemical and biological properties of soils, and monitoring the effects of changes in these properties on the soil's capacity to support plant growth and environmental functions, is necessary to ensure the sustainability of a production system (NRCS 2020). Evaluation of soil health is also a target-based problem that cannot be generalised for all kinds of soils; thus, indicators of soil health must be selected according to soil and land use management, not necessarily the indicator itself (Table 3.1).

Soil hydrophysical indicators and their relationship to soil water

Soil physical properties

Soil physical quality has a strong effect on soil water capture and storage, and hence is an important aspect of soil health. McKenzie *et al.* (2011) defined soil physical quality as '*the ability of a given soil to meet plant and ecosystem requirements for water, aeration, and strength over time and to resist and recover from processes that might diminish that ability*'. Soil physical quality can be measured by studying the combination of soil

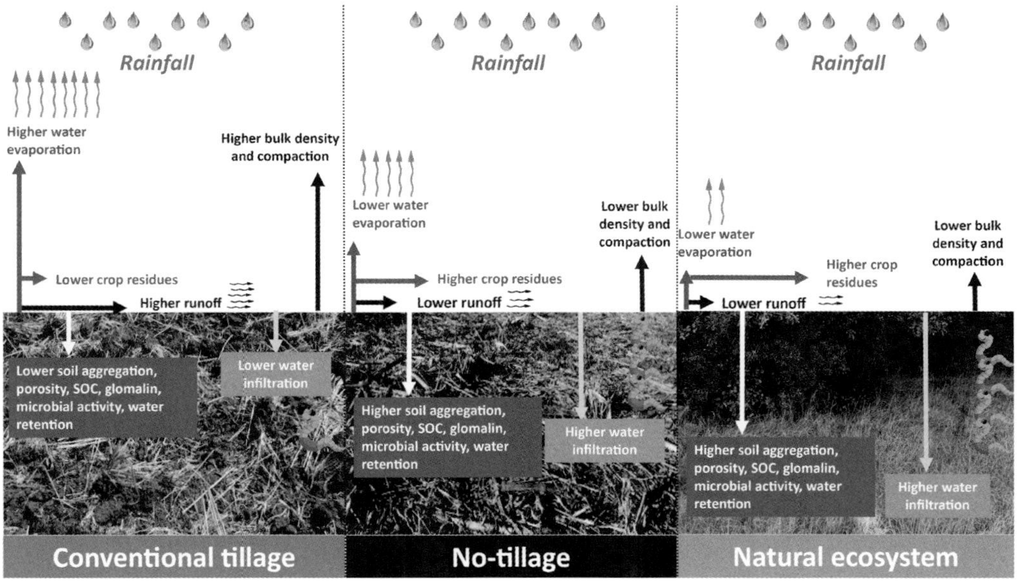

Fig. 3.1. Effects of three selected systems (conventional tillage, no tillage and natural ecosystem) on soil health indicators and soil water. The length of the arrows represents the magnitude of the increase (pointing upwards) or decrease (pointing downwards) in soil health indicators and soil water. SOC, soil organic carbon.

Table 3.1. Soil functions and ecosystem services of soil health indicators

PAW, plant-available water; CEC, cation exchange capacity; SOM, soil organic matter

Soil health indicators	Soil ecosystem services, functions or roles	References
Physical indicators		
Bulk density	Structural support, water and solute movement and aeration	Basso *et al.* (2013)
Aggregate stability	Pore space, root development, water-holding capacity and resistance against erosion	Wang *et al.* (2013)
Available water capacity	Retain soil water and make it available for plant use, biological activity and nutrient transport	Yu *et al.* (2020)
Water infiltration	Soils temporarily store water, making it available for plant and soil organisms	Omondi *et al.* (2016)
Porosity	Respiration, water movement and root development	Abel *et al.* (2013)
Soil structure	Aggregation, porosity and aeration	Abiven *et al.* (2009)
Chemical indicators		
pH	Affects microbial activity, nutrient leaching and availability, toxic elements and soil structure	
Electrical conductivity	Affects plant nutrient availability, activity of soil microorganisms, soil–water balance and nutrient availability	Sheldon *et al.* (2017)
CEC	Affects soil structure stability, nutrient availability and nutrient mobility	Aprile and Lorandi (2012)

continued

Table 3.1. continued

Soil health indicators	Soil ecosystem services, functions or roles	References
Soil nutrients	Nutrient uptake for plant growth and microbial activity	
Biological indicators		
SOM	Soil structural formation, carbon sequestration, microbial activity, PAW, infiltration, nutrient release, formation of organomineral complexes, enzyme activity and mineralisation	Franzluebbers (2002)
Microbial activity or biomass	Carbon cycling, SOM decomposition, nutrient release, pathogen control, soil structure maintenance and nitrogen fixation	Sekaran *et al.* (2020)
Glomalin or soil protein	Soil aggregation and stability, carbon sequestration, microbial activity, water-holding capacity, water retention and SOM content	Liu *et al.* (2020)
Enzyme activity	Nutrient cycling (i.e. carbon, nitrogen, phosphorus and sulfur cycling) and microbial activity	Sekaran *et al.* (2019)
Respiration	Nutrient cycling, microbial activity and SOM content and decomposition	Franzluebbers *et al.* (2002)
Macrofauna	Soil aggregation and stability, porosity, nutrient transport, water infiltration, gaseous exchange, respiration and SOM decomposition	Cheik *et al.* (2019)

indicators related to water flux and storage (e.g. water storage, hydraulic conductivity and infiltration), soil aeration (e.g. porosity), compaction (e.g. bulk density and soil resistance to penetration) and soil structure (e.g. aggregation; Reynolds *et al.* 2002; Bünemann *et al.* 2018). These parameters provide direct and indirect estimates of the ability of a soil to store root-zone water and air necessary for crop growth. Therefore, evaluation of soil physical properties is important for determining the factors that influence soil health. Soil physical indicators and their effects on soil water capture and storage are discussed in detail below.

Water retention, storage and movement

Soil water retention relates to the actual amount of water retained in the soil for crop use. Soil water retention and PAW content are primarily dependent on soil structure (or pore size distribution), texture (or particle size distribution), bulk density and SOC content (Pollacco 2008). Soil water retention characteristics directly determine the amount of air and water that can be retained by the soil at a given matric potential (Dharumarajan *et al.* 2013). The water retained at specific matric potentials is related to soil properties (e.g. organic matter, particle or pore size distribution) and management (Malgwi and Abu 2011).

The hydraulic conductivity of soils governs water infiltration and movement within the soil profile, thus influencing water retention and storage, and reduces run-off after a heavy rainfall or irrigation event. It also influences water and nutrient transport for plant uptake (Omondi *et al.* 2016). Hydraulic conductivity can be enhanced with improved soil macroporosity (Abel *et al.* 2013) and aggregation, thereby improving soil drainage (Omondi *et al.* 2016).

Soil texture and structure

Soil texture controls some ecological and hydrological processes including water intake rates, water retention and storage in the soil, soil aeration and nutrient availability (Arias *et al.* 2005; Cosh *et al.* 2008). Soil texture can also influence water uptake by affecting root growth, and can enhance water and nutrient uptake (Dou *et al.* 2016). Soil texture controls some ecological and hydrological processes, such as water retention and storage, aeration and nutrient availability (Cosh *et al.* 2008). The mineralogy of clay and silt affects soil water retention and storage because these particles chemically interact with water and organic matter. In addition, clay content favours the capillary and adsorption forces. These two forces are responsible for water retention. Thus, soils with attributes that favour these phenomena will retain a greater amount of water (Reichert *et al.* 2009). The highest values for water adsorption capacity, specific surface area and other hydrophysical properties are observed in clay soils, whereas the lowest values are found in sandy and sandy loam soils (coarse texture). Soil with a high clay content has higher specific surface area, soil water content and retention and water adsorption capacity (Amer 2015). Water content in the surface soil changes with the type and content of clay present in the soil. Soil water retention is positively correlated with clay content because the clay fraction favours the generation of capillary forces through the occurrence of micropores and menisci. Capillarity and adsorption are the two major phenomena that determine matric potential and are responsible for soil water retention. Consequently, soils with structures or constituents that favour these phenomena will retain a greater amount of water (Reichert *et al.* 2009).

Soil structure is an important indicator for characterising soil health (Abiven *et al.* 2009). Soil structure can be defined in terms of form and stability, in which the structural form of the soil refers to the heterogeneous arrangement of solid and void spaces that exist at a given time, whereas structural stability is the soil's ability to retain this arrangement when exposed to different stresses (Angers and Caron 1998). Soil structure helps determine water and air movement in the soil, influencing soil water balance and soil workability, and has a significant effect on plant growth and development (Amoakwah *et al.* 2017). The stability of the aggregates and the pores between them affects the movement and storage of water, aeration, erosion, biological activity and plant growth (Amézketa 1999). Soil with a good structure will drain quickly after irrigation or heavy rain, minimising waterlogging. In contrast, poorly structured soil can be waterlogged after irrigation or rain events, which may inhibit plant growth and development and restrict the plant root system (Rai *et al.* 2017). The question is, how do the different structural arrangements we observe in soils affect water influx and movement? Platy structured topsoils can block the entrance of water, cause waterlogging and restrict seed germination due to poor aeration. Conversely, topsoil with a granular structure promotes the easy entry of water and good seed germination. A prismatic structure favours the vertical movement of water, and therefore restricts the supply of water to plant roots. Soil structure is not permanent and can be changed or improved by adopting appropriate management practices, unlike soil texture which cannot be easily changed (Angers and Caron 1998).

Soil structure and SOC are interrelated. Soil aggregate stability is a building block of the soil structure, and SOC acts as a binding and stabilising agent in the formation of soil aggregates (Bronick and Lal 2005). SOC improves aggregate formation, which is followed by an increase in soil water content at field capacity due to the formation of larger pores (Eden *et al.* 2017).

Stable soil aggregates play a key role in the maintenance of soil structure, reduce the impact of raindrops on particle detachment, prevent the formation of surface crusts and seals that reduce wind and water erosion and facilitate water filtration and root development (Wang *et al.* 2013). A well-aggregated soil has better air and water infiltration rates favourable for plant and microbial growth. Furthermore, a well-aggregated soil with improved pore connectivity can enhance oxygen transport in the soils, which is very important for root development. Soils with low oxygen diffusivity (below a threshold value of 0.02) restrict root development (Deepagoda *et al.* 2011), resulting in low root water uptake. Higher root biomass and root activity increase soil aggregate stability, and hence can enhance PAW.

Soil structure is also influenced by the presence of AMF. AMF can improve soil structure by secreting a binding agent called glomalin, which can improve soil structure (Goss *et al.* 2017). Several studies have reported that soils with AMF had better soil aggregation, more porosity and higher saturated hydraulic conductivity than soils without AMF (Augé *et al.* 2001; Hallett *et al.* 2009). These results suggest that AMF enhanced soil aggregate stability, soil water sorptivity (at both –200 kPa and air-dry conditions), unsaturated hydraulic conductivity, water infiltration capacity and the water retention properties of rhizosphere soil. Milleret *et al.* (2009) evaluated the effect of plant inoculation with AMF on soil structure development and soil hydraulic properties, finding that AMF and plant roots had a synergistic effect on stabilising soil structure. Milleret *et al.* (2009) proposed that exudates from both roots and hyphae probably served as carbon sources for other soil microbes, which, in turn, enhanced soil aggregation and porosity and thus altered the soil water retention curve.

Soil bulk density

Soil bulk density is a physical soil health indicator of soil compaction that affects water dynamics, soil erosion, the C and N cycles and plant growth (De Neve and Hofman 2000). Soils with higher bulk density generally have poorer soil structure that limits water and air infiltration and decreases water-holding capacity (Peters *et al.* 2020; Feng *et al.* 2019). These problematic higher bulk densities are created by various management practices, such as animal treading and agricultural machinery. However, the extent of changes in soil bulk density depend on soil texture and soil water content. For example, fine-textured and lower organic matter soils are more vulnerable to compaction than coarse-textured and higher organic matter soils (Wortmann and Jasa 2003; Lipiec *et al.* 2002). Similarly, dry soils are less vulnerable to soil compaction due to higher aggregate stability than moist soils (Mosaddeghi *et al.* 2000).

Compacted soils can increase run-off and erosion by reducing water infiltration into the soil. Bulk density can also affect soil nutrient cycling, and critically, available water content because high bulk density soils can retain water more strongly and impede root penetration, reducing PAW (Fig. 3.2). Not only can the increase in bulk density limit root growth, roots can experience anaerobic conditions due to poor aeration and lower percolation rate.

Soil porosity

Soil porosity is an indicator of soil structural quality (Pires *et al.* 2005). Soil pore characterisation provides an understanding of the retention and movement of water and gas exchange in the soil, as well as relationships between soil structure and soil biological and chemical properties (Pagliai and Vignozzi 2006). Soil pores occurring within (intra) and between (inter) soil aggregates serve as pathways for soil water and air movement (Amoakwah *et al.* 2017). The number of pores and their size, as well as pore

Fig. 3.2. Three-dimensional tomographic images showing soil porosity (dark grey portion of the images) under a natural ecosystem (with higher soil organic carbon) and in croplands (lower soil organic carbon and higher bulk density) located next to each other.

connectivity and tortuosity, influence the movement of water and gas in the soil profile (Osozawa 1998).

Quantification of soil pore space size, shape, continuity, orientation and arrangement allows for defining the complexity of soil structure and how human activity has modified soils (Pagliai and Vignozzi 2006). Different-sized pores in the soil profile have distinct functions in the soil. According to Pires *et al.* (2017), pores with a size equivalent cylindrical diameter of >50 μm are classified as transmission pores, whereas those with a diameter <0.50 μm are classified as residual and bonding pores. The transmission pores are responsible for air movement and the drainage of excess water, whereas the residual pores are responsible for the retention and diffusion of ions in soil solutions (Pires *et al.* 2017).

Soil chemical indicators and their relationship to soil water

Soil electrical conductivity

Soil electrical conductivity (EC) is a measure of the salinity in the soil and is an additional important indicator of soil health. EC affects plant nutrient availability, crop growth, crop yield, microorganism activity, greenhouse gas emissions and soil water balance (Adviento-Borbe *et al.* 2006). Levels of soil EC can act as an indirect indicator of the amount of soil water and water-soluble nutrients (e.g. nitrate-nitrogen) available for plant uptake. The presence of excessive salts like sodium can cause additional problems such as poor soil structure, infiltration, drainage and toxicity for many plants and hinder their growth.

Salinity affects both final soil water content, which plants can extract, and the rate at which plants use that water (Sheldon *et al.* 2017). Dry soils are poor conductors of electricity, and the EC values could represent the volumetric water content, the concentration of electrolytes dissolved in the soil water and the type and amount of clay present (Brevik *et al.* 2006). Ekwue and Bartholomew (2011) observed a strong positive relationship between soil moisture content and EC values for three soil types evaluated in the laboratory and under field conditions. In the Midwest US, Brevik *et al.* (2006) concluded that EC is strongly influenced by soil moisture content.

Soil EC differs depending on the amount of water held by soil particles. Sand, silt and clay have low, medium and high conductivity respectively, and soil EC is consequently strongly correlated with texture, porosity and soil water content. Soil with higher porosity conducts electricity easily, and soils with clay have more total pore space than sandy soils when all other parameters remain constant. Areas under drought have distinct textural differences from soils with excess water, and these can be identified using soil EC measurements. Soils with EC values in the medium range (8–80 mS m^{-1}) are medium-textured, with medium water-holding capacity, and may be the most productive. Because water-holding capacity has the greatest effect on crop yield, the measurement and mapping of soil EC have valuable roles in soil, water and crop management.

Cation exchange capacity

CEC, the soil's ability to supply essential nutrients, such as calcium, magnesium and potassium, for plant growth, is a robust soil health indicator (Khorshidi and Lu 2017). Negatively charged clay and humus particles (colloids) hold soil cations and act as a storehouse of nutrients for plant roots. During nutrient uptake, cations in the soil water replace the cations that are released from the soil colloids. If one cation is present in a higher concentration in soil water, that cation will force out the other cations from the soil colloids and take their place (Yukselen and Kaya 2008). If a soil has a higher concentration of a monovalent cation like sodium, its CEC will be reduced, and up to half the other cations in the soil may be moved to soil water or solution and not held by the soil particles. The cations present in the soil solution are highly susceptible to leaching and being drained away in soil water (Yukselen and Kaya 2008).

Cations present in the soil play a significant role in hydration. The CEC of a soil has been considered as a fundamental property of soil function. Several researchers have shown that water tends to form clusters at high soil water suction values before complete surface coverage of the soil particle (Ferrage *et al.* 2005; Khorshidi *et al.* 2017). Revil and Lu (2013) reported a strong link between soil CEC and soil water retention curves. This indicates that cation hydration is the dominant mechanism for water retention of soil with CEC in the high suction range (Khorshidi and Lu 2017). Several studies have also shown that the swelling potential of soil may be related to CEC and used for soil expansive classification (Yilmaz 2004; Khorshidi *et al.* 2017). In addition, soil CEC greatly influences water sorption, electric and chemical flow, fluid flow and coupled fluid in clayey soils (Revil and Lu 2013).

Soil organic matter

SOM and water repellency

SOM content and composition affect both soil structure and adsorption properties; therefore, changes in SOM that occur due to changes and modifications of management practices may affect water retention (Rawls *et al.* 2003). SOM is a key soil health indicator that provides energy and substrates to support microbial activity, which affects soil water retention, aggregate stability and water infiltration (Franzluebbers 2002; Singh *et al.* 2019). There are two chemistries of organic matter that affect soil water: hydrophobic and hydrophilic. Hydrophobic organic matter can cause water repellency in soils which retards the spontaneous wetting soils, affecting water infiltration. Hydrophilic organic matter increases the soil-water contact angle or magnitude of water repellency (Leelamanie 2011). SOM content and composition are associated with water repellency. Plant and soil microorganisms are the major contributors of SOM, and the type of SOM determines the water-repellent conditions. For example, soil microorganisms exude organic compounds, such as fatty acids, alcohols, esters, alkanes, long-chain aliphatic compounds and amphiphilic substances, that are reported to be responsible for water repellency in soil (Leelamanie 2011). These organic compounds may be present as coatings on mineral particles during natural saprophytic processes and cause soil–water repellency. High soil–water repellency reduces the infiltration rate, which, in turn, enhances water run-off rates and increases erosion risk. However, soil–water repellency can also have positive effects on soil ecosystems, being associated with soil aggregate stability and, thus, carbon sequestration (Olorunfemi *et al.* 2014).

SOM and water content

SOM acts directly as a binding agent or indirectly by promoting soil microbial activity, to enhance soil aggregation and stability, as well as water-holding capacity. Various organic substances involved in aggregate formation can be released during SOM decomposition or produced by microbial activity (Sekaran *et al.* 2019). It is well documented that soil water content can be improved by increasing SOM content (Minasny and McBratney 2018; Desrochers *et al.* 2019). Minasny and McBratney (2018) performed a detailed meta-analysis to determine the relationship between organic matter content and soil water content at saturation, field capacity and permanent wilting point. They found that every 1% increase in organic carbon (10 g C kg^{-1} soil) coincides with an increase in soil water content of 2.9, 1.6, 0.17 and 1.2 mm water per 100 mm soil at saturation, field capacity, permanent wilting point and PAW respectively. This increase is highest in sandy soils, moderate in loams and least in clays. Franzluebbers (2002) concluded that the stratification ratio of SOC could be used as a simple diagnostic tool to identify the effect of soil management strategies on improvements in soil water infiltration, water-holding capacity and PAW.

Soil biological indicators and their relationship to soil water

Macrofauna

A large proportion of macrofauna biomass in soil comprises earthworms and termites; in most ecosystems, earthworms are the major component of soil fauna communities. By accumulating their biogenic structures, such as casts, pellets and galleries, the earthworms have the capacity to modify the soil and crop residue environment (Cheik *et al.* 2019). Ideal soil conditions for earthworm activity are silty soils with high organic matter and water-holding capacity compared with sandy soils. Many studies provide supporting evidence that soil macrofauna play a vital role in soil structure formation, nutrient cycling, water movement and SOM dynamics (incorporation and decomposition) at different spatial and temporal scales through their perturbation and production of various biogenic structures (Locke *et al.* 2013; Table 3.2). Scientists consider these organisms 'ecosystem engineers' due to their direct and indirect effects on the availability of resources for other species as a result of the physical changes they bring about in biotic and abiotic materials (Jones *et al.* 1994). Burrowing and extensive channelling action by earthworms and termites aerates

Table 3.2. Relationships between soil physical indicators and soil hydrological properties

CC, cover crops; CT, conventional tillage; K_{sat}, saturated hydraulic conductivity; MT, minimum tillage; NCC, no cover crops; NT, no tillage

Soil	Soil physical indicators	Soil hydrological properties	System	References
Tifton loamy sand	Lower bulk density (1.71 vs 1.73 g cm^{-3}), higher carbon (7.25 vs 4.69 mg g^{-1})	Higher K_{sat} (33.1 vs 27.0 cm h^{-1}) and volumetric moisture content (0.126 vs 0.113 cm^3 cm^{-3})	CC vs NCC	Hubbard *et al.* (2013)
Nutley–Sinai soil series	Lower bulk density (1.23 vs 1.36 g cm^{-3}) and penetration resistance (1.06 vs 1.78 MPa)	Higher infiltration (322.9 vs 158.2 mm h^{-1}), K_{sat} (344.8 vs 227.9 mm h^{-1}) and soil water retention (0.52 vs 0.46 m3 m^{-3})	High vs low nitrogen rate	Singh *et al.* (2019)
Marietta fine sandy loam	Lower bulk density (1.33 vs 1.48 g cm^{-3}) and penetration resistance (1.09 vs 1.45 MPa)	Higher field capacity (23.03% vs 19.55%), available water content (18.96% vs 14.78%), K_{sat} (1.23 vs 0.63 × 10^{-5} m s^{-1}), infiltration rate (170.7 vs 58.37 mm h^{-1}) and water-stable aggregates (0.78 vs 0.67)	Organic manure vs control	Feng *et al.* (2019)
Egan soil series	Lower bulk density (1.19 vs 1.24 g cm^{-3}), lower penetration resistance (2.02 vs 2.54 MPa)	Higher soil water content (0.44 vs 0.37 m^3 m^{-3})	NT vs CT	Alhameid *et al.* (2019)
Cecil sandy loam	Lower bulk density	Higher soil porosity and K_{sat}	CC vs NCC	Peters *et al.* (2020)
Calloway silt loam	Higher microaggregate carbon content (64% vs 27%)	Higher K_{sat} and volumetric water content (8.6% greater)	NT vs CT	Desrochers *et al.* (2019)
Dundee silt loam	Lower bulk density (1.23 vs 1.26 g cm^{-3}) and higher abundance of earthworms (88 vs 75 m^{-2})	Higher infiltration rate (24.4 mm h^{-1})	MT vs NT	Locke *et al.* (2013)
Dundee silt loam	Higher aggregate stability (73.7% vs 64.9%)	Higher infiltration rate (86.6 vs 17.6 mm h^{-1})	CC vs NCC	Locke *et al.* (2013)

and loosens the soil and improves soil drainage capacity, and creates solid organomineral structures that may persist in the soil for a longer time than the organisms that created them. Earthworms modify the physical structure of soils by producing soil pores and water-stable aggregates, which improves soil water infiltration, drainage and aeration (Cheik *et al.* 2019). In general, soils with a high earthworm population drain 10-fold faster than soils without earthworms (Galanakis 2018). Because earthworms feed on plant residues, they are involved in the decomposition process, nutrient cycling and transportation of nutrients in the soil profile, stimulating root growth and satisfying the water and nutrient requirements of crops.

Cheik *et al.* (2019) investigated the effects of dung beetles, termites and earthworms on soil porosity and saturated hydraulic conductivity (K_{sat}) using X-ray computed tomography. In that study, there was a positive impact of these macrofauna on soil macroporosity and K_{sat}. However, compared with both earthworms and termites, the presence of dung beetles led to the highest K_{sat} values in soils (Cheik *et al.* 2019). This could be due to the harder and larger body diameter of beetles than that of earthworms and termites. However, Cheik *et al.* (2019) concluded that earthworm species play an important role in improving soil water infiltration and other hydraulic properties because of the low and sporadic activity of beetles and termites. Pulleman *et al.* (2005) also described the importance of earthworms in forming microaggregates, enhancing microbial respiration and biomass and protecting SOM within those aggregates, thus increasing water retention and movement (Cheik *et al.* 2019).

Soil health and soil hydrology

There is notable evidence that global warming increases the intensity and frequency of extreme rainfall events or droughts, and has a large effect on the occurrence of hydrological hazards, reduced crop productivity and soil carbon loss (Alfieri *et al.* 2015). Furthermore, severe anthropogenic disturbances, such as poorly managed agroecosystems, disrupt soil hydrology functioning and subsequent life-supporting services. The dominant effect of inherent soil hydraulic properties on various hydrological components has been well documented, beginning from the surface processes including infiltration, evaporation and run-off to storage and root water uptake in the vadose zone to drainage out of the soil profile (Liu *et al.* 2020). Therefore, soil is the foundation for any flexible response to upcoming food, water and energy security challenges. Soil health management aims to promote soil resiliency by improving soil hydrological properties and increasing soil water infiltration and storage by enhancing organic matter and soil aggregates. Soil hydrological health supports soil functions such as biodiversity, filtering and nutrient cycling. Despite the significant effort for proper land use management to optimise ecosystem service delivery, there is still a demand to address agrohydrological responses such as the quantity and quality of evaporation, surface run-off and drainage, as well as root water uptake and crop water use efficiency, with respect to the status of soil health. The following sections address the importance of soil health status as an explicit but understudied issue in principal soil hydrological components occurring at the surface, such as run-off, erosion and infiltration, then crop water and finally the issue of water quality within field and catchment scales.

Soil health and surface hydrological components

Healthy soils with desirable water infiltration and retention protect the farm and watershed from surface run-off and soil erosion. Soil erosion is the major challenge for sustainable soil management, costing the world economy approximately US$8 billion per year and reducing global crop productivity by 22.5 million tonnes (Sartori *et al.* 2019). Accelerated topsoil erosion due to weather extremes such as drought and flood, along with ever-increasing poor soil management in agriculture, has on- and off-site effects on soil functionality and health (Blanco-Canqui and Lal 2009). On-site effects include soil productivity loss due to the removal of organic matter and clay and less PAW, whereas off-site effects are associated with the eutrophication of waterbodies and gaseous emissions (Lal 2014). In 2001 and 2012, cropland, with its high erosion rate that is ~75-fold greater than that of natural forest areas, contributed to 49.7% and 50.5% of the predicted global soil erosion respectively, whereas implementation of conservation agriculture practices reduced global soil erosion by 0.49 and 1.05 Pg year^{-1} respectively (Borrelli *et al.* 2017).

Soil health indicators are potentially cost-effective and less labour intensive alternatives to assess soil erodibility. Soil aggregate stability is a key factor regulating susceptibility to surface erosion. The breakdown of soil aggregates during initial rainfall events after tillage shapes surface crusts and accelerates soil erosion. Barthès and Roose (2002) found that soil loss and run-off depth are strongly linked to stable topsoil aggregates resistant to slaking in experiments performed in France, Cameroon, Benin and Mexico. Reductions in run-off rate and sediment concentrations were also associated with improvements in aggregate stability due to an increased mulch application rate. The reduction in erosion caused by plant roots depends on root traits, such as length and density. Fibrous roots effectively control erosion because they increase soil cohesion to a greater extent than the tap root system (Vannoppen *et al.* 2017). Biodiversity and the population of soil organisms are interconnected with soil loss. Earthworm removal increased soil loss and transport downslope from fine litter by factors of 5.3 and 3.4 respectively in Luquillo experimental forest, Puerto Rico (Larsen *et al.* 2012). Soil erosion models mainly consider only abiotic factors. However, although erosion and earthworm activities are interlinked and a correction factor based on the soil's richness (number of species) and abundance (number of individuals) of earthworms was included in a European soil erosion model, the model showed both reduced and increased effects of earthworms on erosion (Orgiazzi and Panagos 2018).

Soil health and crop water

The complex fate of infiltrated water or soil moisture in the profile depends on soil and rhizosphere hydraulic properties, root distribution and soil lower profile boundary conditions. Under drying conditions, the most recent studies by Liu *et al.* (2020) and Carminati and Javaux (2020) showed the dominant role of soil hydraulic properties, in particular soil moisture, compared with vapour pressure deficit and plant hydraulics in plant transpiration. Healthy soils combat plant dryness stresses by providing sufficient amounts of available water for root uptake and better root distribution through soil structural improvement and reduced surface evaporation. Soils with good tilth and structure provide pore spaces for water, air and roots to move freely, distribute and contact the surrounding microenvironment to absorb moisture. Regardless of soil mineralogy and texture, SOM and physical disturbances are key factors in soil structure (Bronick and Lal 2005). It is widely documented that improving soil health management can increase SOM content and aggregate stability, leading to more water retention and crop water use (Blanco-Canqui and Lal 2009).

Hudson (1994) was among the first to elaborate the feedback process between SOM and PAW (i.e. the difference between water content at field capacity and the wilting point) by showing that an increase in SOM content of between 0.5% and 3% can double the PAW of soils. Later, Olness and Archer (2005) confirmed a 2.5–5% increase in PAW per 1% change in organic carbon. From 17 long-term field experiments, Eden *et al.* (2017) concluded that the application of organic matter improved PAW up to 30%. The effect of SOM on PAW is dependent on soil texture, as demonstrated by Minasny and McBratney (2018) in their meta-analysis, where a 1% increase in SOM increased PAW by 1.94% and 1.41% for sandy and clay soils respectively. Minasny and McBratney (2018) found that the main positive effect of SOM is on soil porosity, improving air-filled capacity and K_{sat}. However, there is an ongoing debate as to the relative role of SOM on PAW and its components. In fact, field capacity and wilting points are more static parameters and PAW does not perfectly state moisture supply to plants in a growing season (Bonfante *et al.* 2020). More importantly, tipping bucket models using the PAW concept consider unavailability of water content above field capacity for plant uptake, whereas van Lier (2017) showed that between 10% and 50% of water transpired by plants is supplied from the moisture between saturation and field capacity. Therefore, studies assessing the benefit of SOM on PAW for plant uptake should now take a step towards considering the dynamic nature of plant water uptake using the process-based simulation models (Bouma 2020).

A holistic approach towards understanding the impact of soil health management on water use efficiency (WUE) is capable of removing the static nature of plant-available related parameters because the efficacy of soil health management practices in WUE depends on climate, soil type and cropping system (Hatfield and Dold 2019). Data from a 15-year field experiment conducted on the Loess Plateau of northern China showed that no-till and residual cover increased the WUE of winter wheat by 15%, on average, whereas no-till was significantly more efficient in the conversion of available water into crop yield than conventional tillage (Li *et al.* 2007). No-till and straw cover did not significantly change the WUE of maize cultivated in the west of China in the first 4 years of a 12-year experiment, whereas in the last 4 years there was a ~26% increase in WUE observed (Wang *et al.* 2014), similar to the general conclusion of a meta-analysis by Wang *et al.* (2018), who showed that significant improvements in WUE for maize under no-till occur after 6 years. The adoption of no-till systems in the semi-arid Pacific north-west US resulted in a 30% increase in WUE of rain-fed winter wheat (Williams *et al.* 2015). The WUE of sunflower, corn, soybean and wheat cultivated in the semi-arid Pampa, Argentina, under no-till conditions improved by 15%, on average, compared with conventional tillage (Noellemeyer *et al.* 2013).

Soil health and water quality

Water quality is tied to the land condition and soil health. The pollution of waterbodies with the excessive use of synthetic fertilisers and pesticides is a critical problem in many cultivated areas; however, conservation agriculture mitigates this issue by improving soil health with reduced erosion and purifying infiltrated water. To ensure increasing food production and economy, approximately 2.72 million tonnes of pesticides are applied globally in agroecosystems each year (Silva *et al.* 2019); this, in turn, can result in serious health and environmental problems. Ideally, pesticides should be degraded to harmless compounds, but slow-rate degradable pesticides tend to leach or be transport by surface run-off. The ability of soils to buffer, immobilise or degrade pesticides depends on the state of soil health. In areas such as Orestimba Creek, California, 76% of pesticide loading is due to high surface run-off, and thus the implementation of buffer zones and filter strips is an effective soil health practice (Zhang *et al.* 2018). Agricultural practices influence non-point source pollution and pesticide–soil interactions. The return of crop residues to cover more than 30% of the soil surface can intercept, on average, more than half of the applied pre-emergence herbicides (Alletto *et al.* 2010). Sugarcane straw and ash, biochar or organic residues have successfully increased pesticide adsorption; however, there are still controversial results regarding the cooler zone created under residues that can diminish pesticide degradation or rapid leaching of pesticide through generated preferential pathways (Alletto *et al.* 2010). The accumulation of phosphorous (P) in many waterbodies from eroded agricultural fields poses threats of algal growth, oxygen depletion and eutrophication. Agriculture contributes to 38% of loaded P in fresh water, of which one-third comes from cereal production (Mekonnen and Hoekstra 2018). Shifting to conservation agriculture to diminish eutrophication comes with trade-offs, and thus requires trials on management, climate and physical variables of the region of interest. For example, Daryanto *et al.* (2017) reviewed 27 studies and concluded that no-till practices reduced particulate P by 55%, whereas they increased dissolved P compared with conventional tilling. Verbree *et al.* (2010) also confirmed increased dissolved P loss on poorly drained soils under no-till, although compensation is possible by manure injection. An increase in dissolved P due to no-till was also the reason for algal blooms in Lake Erie (Ehmke 2013).

Inorganic fertiliser, sewage sludge, waste water and manure are major contributors of heavy metals in agricultural areas. Despite the high sorbing characteristics of metals, they can be mobilised with dispersed colloidal particles such as SOM or bypassing the matrix through preferential flow. An increased frequency of high-intensity rainfall events or irrigation can potentially increase heavy metal and P mobility. Therefore, the systematic implementation of soil health management practices that are effective in reduction in the delivery of sorbed particles is required in addition to edge-of-field monitoring.

Summary and future perspectives

Soil water plays a critical role in soil health and is essential for living organisms to decompose organic matter, respire and survive, and for plants to take up required water and dissolved nutrients. Healthy soil is key to meeting the growing global food demand. The link between soil health and soil water is not widely studied. Thus, this chapter discusses the relationship between various soil health indicators and soil water storage, movement, and WUE, and provides more in-depth information using currently available literature. A proper balance between soil water and air is critical to release nutrients into the soil and for microbial activity. Soil health indicators such as soil texture, structure (aggregation), bulk density, compaction, penetration resistance, pH, CEC, SOM, microbial activity and macrofauna provide direct estimates of the ability of a soil to store the water and air required for crop growth. An increase in SOM improves aggregate stability, soil structure and pore spaces and controls soil erosion, hence enhancing PAW, water infiltration and soil water retention in less healthy soils. For deeper soils, SOM is critical for forming aggregates by enhancing soil microbial activity, especially fungi, which plays a major role in the stabilisation of aggregates. Therefore, the evaluation of soil health indicators in different soil drying and wetting conditions is important for determining the factors that influence soil water availability.

Assessing the changes in soil physical, chemical and biological indicators and monitoring the effects of these changes on a soil's capacity to enhance soil water, support plant growth and environmental functions are necessary to ensure the sustainability of a production system. Apart from lack of operational measurement methodology for soil health, the missing key in soil health and soil water is considering the inherent dynamic of this concept, which can be potentially captured by the available information, ranging from satellite images and remote sensing data to soil global databases. The inclusion of soil health dynamics in distributed physically based agrohydrological models is vital to simulate the real-field condition. The relationship between soil health and soil water, by improving soil climate resiliency against drying conditions and promoting sustainable agriculture and food systems, makes an important contribution to advancing and achieving sustainable goal developments.

References

Abel S, Peters A, Trinks S, Schonsky H, Facklam M, Wessolek G (2013) Impact of biochar and hydrochar addition on water retention and water repellency of sandy soil. *Geoderma* **202–203**, 183–191. doi:10.1016/j.geoderma.2013.03.003

Abiven S, Menasseri S, Chenu C (2009) The effects of organic inputs over time on soil aggregate stability – a literature analysis. *Soil Biology & Biochemistry* **41**, 1–12. doi:10.1016/j.soilbio.2008.09.015

Adviento-Borbe M, Doran J, Drijber R, Dobermann A (2006) Soil electrical conductivity and water content affect nitrous oxide and carbon dioxide emissions in intensively managed soils. *Journal of Environmental Quality* **35**, 1999–2010. doi:10.2134/jeq2006.0109

Alfieri L, Burek P, Feyen L, Forzieri G (2015) Global warming increases the frequency of river floods in Europe. *Hydrology and Earth System Sciences* **19**, 2247. doi:10.5194/hess-19-2247-2015

Alhameid A, Singh J, Sekaran U, Ozlu E, Kumar S, Singh S (2019) Crop rotational diversity impacts soil physical and hydrological properties under long-term no- and conventional-till soils. *Soil Research* **58**, 84–94. doi:10.1071/SR18192

Alletto L, Coquet Y, Benoit P, Heddadj D, Barriuso E (2010) Tillage management effects on pesticide fate in soils. A review. *Agronomy for Sustainable Development* **30**, 367–400. doi:10.1051/agro/2009018

Amer AMA (2015) *Water Vapor Adsorption and Soil Wetting.* InTechOpen, London.

Amézketa E (1999) Soil aggregate stability: a review. *Journal of Sustainable Agriculture* **14**, 83–151. doi:10.1300/J064v14n02_08

Amoakwah E, Frimpong KA, Okae-Anti D, Arthur E (2017) Soil water retention, air flow and pore structure characteristics after corn cob biochar application to a tropical sandy loam. *Geoderma* **307**, 189–197. doi:10.1016/j.geoderma.2017.08.025

Angers DA, Caron J (1998) Plant-induced changes in soil structure: processes and feedbacks. *Biogeochemistry* **42**, 55–72. doi:10.1023/A:1005944025343

Aprile F, Lorandi R (2012) Evaluation of cation exchange capacity (CEC) in tropical soils using four different analytical methods. *Journal of Agricultural Science* **4**, 278–289. doi:10.5539/jas.v4n6p278

Arias ME, González-Pérez JA, González-Vila FJ, Ball AS (2005) Soil health: a new challenge for microbiologists and chemists. *International Microbiology* **8**, 13–21.

Augé RM, Stodola AJ, Tims JE, Saxton AM (2001) Moisture retention properties of a mycorrhizal soil. *Plant and Soil* **230**, 87–97. doi:10.1023/A:1004891210871

Barthès B, Roose E (2002) Aggregate stability as an indicator of soil susceptibility to runoff and erosion; validation at several levels. *Catena* **47**, 133–149. doi:10.1016/S0341-8162(01)00180-1

Basso AS, Miguez FE, Laird DA, Horton R, Westgate M (2013) Assessing potential of biochar for increasing water-holding capacity of sandy soils. *GCB Bioenergy* **5**, 132–143. doi:10.1111/gcbb.12026

Blanco-Canqui H, Lal R (2009) Crop residue removal impacts on soil productivity and environmental quality. *Critical Reviews in Plant Sciences* **28**, 139–163. doi:10.1080/07352680902776507

Bonfante A, Basile A, Bouma J (2020) Exploring the effect of varying soil organic matter contents on current and future moisture supply capacities of six Italian soils. *Geoderma* **361**, 114079. doi:10.1016/j.geoderma.2019.114079

Borrelli P, Robinson DA, Fleischer LR, Lugato E, Ballabio C, Alewell C, Meusburger K, Modugno S, Schütt B, Ferro V (2017) An assessment of the global impact of 21st century land use change on soil erosion. *Nature Communications* **8**, 2013. doi:10.1038/s41467-017-02142-7

Bouma J (2020) Soil security as a roadmap focusing soil contributions on sustainable development agendas. *Soil Security* **1**, 100001. doi:10.1016/j.soisec.2020.100001

Brevik EC, Fenton TE, Lazari A (2006) Soil electrical conductivity as a function of soil water content and implications for soil mapping. *Precision Agriculture* **7**, 393–404. doi:10.1007/s11119-006-9021-x

Bronick CJ, Lal R (2005) Soil structure and management: a review. *Geoderma* **124**, 3–22. doi:10.1016/j.geoderma.2004.03.005

Bruun TB, Elberling B, Chistensen BT (2010) Lability of soil organic carbon in tropical soils with different clay minerals. *Soil Biology & Biochemistry* **42**, 888–895. doi:10.1016/j.soilbio.2010.01.009

Bünemann EK, Bongiorno G, Bai Z, Creamer RE, De Deyn G, de Goede R, Fleskens L, Geissen V, Kuyper TW, Mäder P (2018) Soil quality – a critical review. *Soil Biology & Biochemistry* **120**, 105–125. doi:10.1016/j.soilbio.2018.01.030

Carminati A, Javaux M (2020) Soil rather than xylem vulnerability controls stomatal response to drought. *Trends in Plant Science* **25**, 868–880. doi:10.1016/j.tplants.2020.04.003

Cheik S, Bottinelli N, Minh TT, Doan TT, Jouquet P (2019) Quantification of thee dimensional characteristics of macrofauna macropores and their effects on soil hydraulic conductivity in northern Vietnam. *Frontiers in Environmental Science* **7**, 31. doi:10.3389/fenvs.2019.00031

Cosh MH, Jackson TJ, Moran S, Bindlish R (2008) Temporal persistence and stability of surface soil moisture in a semi-arid watershed. *Remote Sensing of Environment* **112**, 304–313. doi:10.1016/j.rse.2007.07.001

Daryanto S, Wang L, Jacinthe P-A (2017) Impacts of no-tillage management on nitrate loss from corn, soybean and wheat cultivation: a meta-analysis. *Scientific Reports* **7**, 12117. doi:10.1038/s41598-017-12383-7

De Neve S, Hofman G (2000) Influence of soil compaction on carbon and nitrogen mineralization of soil organic matter and crop residues. *Biology and Fertility of Soils* **30**, 544–549. doi:10.1007/s003740050034

Deepagoda TC, Moldrup P, Schjønning P, Kawamoto K, Komatsu T, de Jonge LW (2011) Generalized density-corrected model for gas diffusivity in variably saturated soils. *Soil Science Society of America Journal* **75**, 1315–1329. doi:10.2136/sssaj2010.0405

Desrochers J, Brye KR, Gbur E, Pollock ED, Savin MC (2019) Long-term residue and water management practice effects on particulate organic matter in a loessial soil in eastern Arkansas, USA. *Geoderma* **337**, 792–804. doi:10.1016/j.geoderma.2018.10.027

Dharumarajan S, Singh S, Bannerjee T, Sarkar D (2013) Water-retention characteristics and available water capacity in three cropping systems of Lower Indo-Gangetic Alluvial Plain. *Communications in Soil Science and Plant Analysis* **44**, 2734–2745. doi:10.1080/00103624.2013.803561

Dou F, Soriano J, Tabien RE, Chen K (2016) Soil texture and cultivar effects on rice (*Oryza sativa*, L.) grain yield, yield components and water productivity in thee water regimes. *PLoS One* **11**, e0150549. doi:10.1371/journal.pone.0150549

Eden M, Gerke HH, Houot S (2017) Organic waste recycling in agriculture and related effects on soil water retention and plant available water: a review. *Agronomy for Sustainable Development* **37**, 11. doi:10.1007/s13593-017-0419-9

Ehmke T (2013) A vexing problem: keeping farm-based phosphorus out of Lake Erie. *Crops & Soils magazine* **46**, 4–9. doi:10.2134/cs2013-46-3-1

Ekwue E, Bartholomew J (2011) Electrical conductivity of some soils in Trinidad as affected by density, water and peat content. *Biosystems Engineering* **108**, 95–103. doi:10.1016/j.biosystemseng.2010.11.002

Feng G, Adeli A, Read J, McCarty J, Jenkins J (2019) Consequences of pelletized poultry litter applications on soil physical and hydraulic properties in reduced tillage, continuous cotton system. *Soil & Tillage Research* **194**, 104309. doi:10.1016/j.still.2019.104309

Ferrage E, Lanson B, Sakharov BA, Drits VA (2005) Investigation of smectite hydration properties by modeling experimental X-ray diffraction patterns: Part I. Montmorillonite hydration properties. *The American Mineralogist* **90**, 1358–1374. doi:10.2138/am.2005.1776

Franzluebbers AJ (2002) Water infiltration and soil structure related to organic matter and its stratification with depth. *Soil & Tillage Research* **66**, 197–205. doi:10.1016/S0167-1987(02)00027-2

Franzluebbers K, Franzluebbers AJ, Jawson MD (2002) Environmental controls on soil and whole-ecosystem respiration from a tallgrass prairie. *Soil Society of America Journal* **66**, 254–262. doi:10.2136/sssaj2002.2540

Galanakis CM (2018) *Sustainable Food Systems from Agriculture to Industry: Improving Production and Processing.* Academic Press, Cambridge.

Goss MJ, Carvalho M, Brito I (2017) *Functional Diversity of Mycorrhiza and Sustainable Agriculture: Management to Overcome Biotic and Abiotic Stresses.* Academic Press, London.

Hallett PD, Feeney DS, Bengough AG, Rillig MC, Scrimgeour CM, Young IM (2009) Disentangling the impact of AM fungi versus roots on soil structure and water transport. *Plant and Soil* **314**, 183–196. doi:10.1007/s11104-008-9717-y

Hatfield JL, Dold C (2019) Water-use efficiency: advances and challenges in a changing climate. *Frontiers in Plant Science* **10**, 103. doi:10.3389/fpls.2019.00103

Hubbard RK, Strickland TC, Phatak S (2013) Effects of cover crop systems on soil physical properties and carbon/nitrogen relationships in the coastal plain of southeastern USA. *Soil and Tillage Research* **126**, 276–283. doi:10.1016/j.still.2012.07.009

Hudson BD (1994) Soil organic matter and available water capacity. *Journal of Soil and Water Conservation* **49**, 189–194.

Jones CG, Lawton JH, Shachak M (1994) Organisms as ecosystem engineers. *Oikos* **69**, 373–386.

Khorshidi M, Lu N (2016) Soil water retention-based methodology for the classification of expansive soils. In *Geo-Chicago 2016: Sustainable Geoenvironmental Systems*, 14–18 August 2016, Chicago, IL. American Society of Civil Engineers, <https://ascelibrary.org/doi/pdf/10.1061/9780784480144.086>.

Khorshidi M, Lu N (2017) Intrinsic relation between soil water retention and cation exchange capacity. *Journal of Geotechnical and Geoenvironmental Engineering* **143**, 04016119. doi:10.1061/(ASCE)GT.1943-5606.0001633

Khorshidi M, Lu N, Akin ID, Likos WJ (2017) Intrinsic relationship between specific surface area and soil water retention. *Journal of Geotechnical and Geoenvironmental Engineering* **143**, 04016078. doi:10.1061/(ASCE)GT.1943-5606.0001572

Lal R (2014) Soil conservation and ecosystem services. *International Soil and Water Conservation Research* **2**, 36–47. doi:10.1016/S2095-6339(15)30021-6

Larsen MC, Liu Z, Zou X (2012) Effects of earthworms on slopewash, surface runoff, and fine-litter transport on a humid-tropical forested hillslope in eastern Puerto Rico. In *Water Quality and Landscape Processes of Four Watersheds in Eastern Puerto Rico.* (Eds SF Murphy and RF Stallard) pp. 179–198. U.S. Geological Survey Professional Paper, Reston.

Leão TP, Tuller M (2014) Relating soil specific surface area, water film thickness, and water vapor adsorption. *Water Resources Research* **50**, 7873–7885. doi:10.1002/2013WR014941

Leelamanie D (2011) Changes in soil water content with ambient relative humidity in relation to the organic matter and clay. *Tropical Agricultural Research & Extension* **13**, 6–10. doi:10.4038/tare.v13i1.3130

Li H, Gao H, Wu H, Li W, Wang X, He J (2007) Effects of 15 years of conservation tillage on soil structure and productivity of wheat cultivation in northern China. *Soil Research* **45**, 344–350. doi:10.1071/SR07003

Libohova Z, Seybold C, Wysocki D, Wills S, Schoeneberger P, Williams C, Lindbo D, Stott D, Owens P (2018) Reevaluating the effects of soil organic matter and other properties on available water-holding capacity using the National Cooperative Soil Survey Characterization Database. *Journal of Soil and Water Conservation* **73**, 411–421. doi:10.2489/jswc.73.4.411

Lipiec J, Ferrero A, Giovanetti V, Nosalewicz A, Turski M (2002) Response of structure to simulated trampling of woodland soil. *Advances in GeoEcology* **35**, 133–140.

Liu L, Gudmundsson L, Hauser M, Qin D, Li S, Seneviratne SI (2020) Soil moisture dominates dryness stress on ecosystem production globally. *Nature Communications* **11**, 4892. doi:10.1038/s41467-020-18631-1

Locke MA, Zablotowicz RM, Steinriede RW, Jr, Testa S, Reddy KN (2013) Conservation management in cotton production: Long-term soil biological, chemical, and physical changes. *Soil Science Society of America Journal* **77**, 974–984. doi:10.2136/sssaj2012.0325

Malgwi W, Abu S (2011) Variation in some physical properties of soils formed on a hilly terrain under different land use types in Nigerian Savanna. *International Journal of Soil Science* **6**, 150–163. doi:10.3923/ijss.2011.150.163

Manici L, Castellini M, Caputo F (2019) Soil-inhabiting fungi can integrate soil physical indicators in multivariate analysis of Mediterranean agroecosystem dominated by old olive groves. *Ecological Indicators* **106**, 105490. doi:10.1016/j.ecolind.2019.105490

McKenzie BM, Tisdall JM, Vance WH (2011) Soil physical quality. In *Encyclopedia of Agrophysics.* (Eds J Gliński, J Horabik and J Lipiec) pp. 770–777. Springer Netherlands, Dordrecht.

Mekonnen MM, Hoekstra AY (2018) Global anthropogenic phosphorus loads to freshwater and associated grey water footprints and water pollution levels: a high-resolution global study. *Water Resources Research* **54**, 345–358. doi:10.1002/2017WR020448

Milleret R, Le Bayon R-C, Lamy F, Gobat J-M, Boivin P (2009) Impact of roots, mycorrhizas and earthworms on soil physical properties as assessed by shrinkage analysis. *Journal of Hydrology* **373**, 499–507. doi:10.1016/j.jhydrol.2009.05.013

Minasny B, McBratney A (2018) Limited effect of organic matter on soil available water capacity. *European Journal of Soil Science* **69**, 39–47. doi:10.1111/ejss.12475

Mosaddeghi M, Hajabbasi M, Hemmat A, Afyuni M (2000) Soil compactibility as affected by soil moisture content and farmyard manure in central Iran. *Soil & Tillage Research* **55**, 87–97. doi:10.1016/S0167-1987(00)00102-1

Nielsen MN, Winding A, Binnerup S (2002) 'Microorganisms as indicators of soil health'. NERI Technical Report No. 388, Ministry of the Environment, National Environmental Research Institute, Roskilde.

Noellemeyer E, Fernández R, Quiroga A (2013) Crop and tillage effects on water productivity of dryland agriculture in Argentina. *Agriculture* **3**, 1–11. doi:10.3390/agriculture3010001

Natural Resources Conservation Service (NRCS) (2020) *Soil Health Glossary.* United States Department of Agriculture, <https://www.nrcs.usda.gov/wps/portal/nrcs/detailfull/soils/health/?cid=nrcs142p2_053848#quality>.

Olness A, Archer D (2005) Effect of organic carbon on available water in soil. *Soil Science* **170**, 90–101. doi:10.1097/00010694-200502000-00002

Olorunfemi IE, Ogunrinde TA, Fasinmirin JT (2014) Soil hydrophobicity: an overview. *Journal of Scientific Research and Reports* **3**, 1003–1037. doi:10.9734/JSRR/2014/7325

Omondi MO, Xia X, Nahayo A, Liu X, Korai PK, Pan G (2016) Quantification of biochar effects on soil hydrological properties using meta-analysis of literature data. *Geoderma* **274**, 28–34. doi:10.1016/j.geoderma.2016.03.029

Orgiazzi A, Panagos P (2018) Soil biodiversity and soil erosion: it is time to get married: adding an earthworm factor to soil erosion modelling. *Global Ecology and Biogeography* **27**, 1155–1167. doi:10.1111/geb.12782

Osozawa S (1998) A simple method for determining the gas diffusion coefficient in soils and its application to soil diagnosis and analysis of gas movement in soil. *Bulletin of the National Institute of Agro-Environmental Sciences (Japan)* **15**, 1–66. [In Japanese with an English abstract]

Pagliai M, Vignozzi N (2006) Soil porosity as an indicator of soil health. *Annals of Arid Zone* **45**, 259.

Peters SJ, Saikawa E, Markewitz D, Sutter L, Avramov A, Sanders ZP, Yosen B, Wakabayashi K, Martin G, Andrews JS (2020) Soil trace gas fluxes in living mulch and conventional agricultural systems. *Journal of Environmental Quality* **49**, 268–280. doi:10.1002/jeq2.20041

Pires LF, Bacchi OO, Reichardt K, Timm LC (2005) Application of γ-ray computed tomography to analysis of soil structure before density evaluations. *Applied Radiation and Isotopes* **63**, 505–511. doi:10.1016/j.apradiso.2005.03.019

Pires LF, Borges JA, Rosa JA, Cooper M, Heck RJ, Passoni S, Roque WL (2017) Soil structure changes induced by tillage systems. *Soil & Tillage Research* **165**, 66–79. doi:10.1016/j.still.2016.07.010

Pollacco JAP (2008) A generally applicable pedotransfer function that estimates field capacity and permanent wilting point from soil texture and bulk density. *Canadian Journal of Soil Science* **88**, 761–774. doi:10.4141/CJSS07120

Pulleman M, Six J, Van Breemen N, Jongmans A (2005) Soil organic matter distribution and micro-aggregate characteristics as affected by agricultural management and earthworm activity. *European Journal of Soil Science* **56**, 453–467. doi:10.1111/j.1365-2389.2004.00696.x

Rai RK, Singh VP, Upadhyay A (2017) *Planning and Evaluation of Irrigation Projects: Methods and Implementation.* Academic Press, New York.

Rawls W, Pachepsky YA, Ritchie J, Sobecki T, Bloodworth H (2003) Effect of soil organic carbon on soil water retention. *Geoderma* **116**, 61–76. doi:10.1016/S0016-7061(03)00094-6

Reichert JM, Albuquerque JA, Kaiser DR, Reinert DJ, Urach FL, Carlesso R (2009) Estimation of water retention and availability in soils of Rio Grande do Sul. *Revista Brasileira de Ciência do Solo* **33**, 1547–1560. doi:10.1590/S0100-06832009000600004

Revil A, Lu N (2013) Unified water isotherms for clayey porous materials. *Water Resources Research* **49**, 5685–5699. doi:10.1002/wrcr.20426

Reynolds W, Bowman B, Drury C, Tan C, Lu X (2002) Indicators of good soil physical quality: density and storage parameters. *Geoderma* **110**, 131–146. doi:10.1016/S0016-7061(02)00228-8

Richards LA (1931) Capillary conduction of liquids through porous mediums. *Physics* **1**, 318–333. doi:10.1063/1.1745010

Sartori M, Philippidis G, Ferrari E, Borrelli P, Lugato E, Montanarella L, Panagos P (2019) A linkage between the biophysical and the economic: assessing the global market impacts of soil erosion. *Land Use Policy* **86**, 299–312. doi:10.1016/j.landusepol.2019.05.014

Sekaran U, McCoy C, Kumar S, Subramanian S (2019) Soil microbial community structure and enzymatic activity responses to nitrogen management and landscape positions in switchgrass (*Panicum virgatum* L.). *Global Change Biology. Bioenergy* **11**, 836–851. doi:10.1111/gcbb.12591

Sekaran U, Loya JR, Abagandura GO, Subramanian S, Owens V, Kumar S (2020) Intercropping of kura clover (*Trifolium ambiguum* M. Bieb) with prairie cordgrass (*Spartina pectinata* link.) enhanced soil biochemical activities and microbial community structure. *Applied Soil Ecology* **147**, 103427. doi:10.1016/j.apsoil.2019.103427

Sheldon AR, Dalal RC, Kirchhof G, Kopittke PM, Menzies NW (2017) The effect of salinity on plant-available water. *Plant and Soil* **418**, 477–491. doi:10.1007/s11104-017-3309-7

Silva V, Mol HG, Zomer P, Tienstra M, Ritsema CJ, Geissen V (2019) Pesticide residues in European agricultural soils – a hidden reality unfolded. *The Science of the Total Environment* **653**, 1532–1545. doi:10.1016/j.scitotenv.2018.10.441

Singh N, Dhaliwal JK, Sekaran U, Kumar S (2019) Soil hydrological properties as influenced by long-term nitrogen application and landscape positions under switchgrass seeded to a marginal cropland. *Global Change Biology. Bioenergy* **11**, 1026–1040. doi:10.1111/gcbb.12611

Six J, Paustian K, Elliott ET, Combrink C (2000) Soil structure and organic matter I. Distribution of aggregate-size classes and aggregate-associated carbon. *Soil Science Society of America Journal* **64**, 681–689. doi:10.2136/sssaj2000.642681x

Thornton PK, Ericksen PJ, Herrero M, Challinor AJ (2014) Climate variability and vulnerability to climate change: a review. *Global Change Biology* **20**, 3313–3328. doi:10.1111/gcb.12581

van Lier QJ (2017) Field capacity, a valid upper limit of crop available water? *Agricultural Water Management* **193**, 214–220. doi:10.1016/j.agwat.2017.08.017

Vannoppen W, De Baets S, Keeble J, Dong Y, Poesen J (2017) How do root and soil characteristics affect the erosion-reducing potential of plant species? *Ecological Engineering* **109**, 186–195. doi:10.1016/j.ecoleng.2017.08.001

Verbree DA, Duiker SW, Kleinman PJ (2010) Runoff losses of sediment and phosphorus from no-till and cultivated soils receiving dairy manure. *Journal of Environmental Quality* **39**, 1762–1770. doi:10.2134/jeq2010.0032

Vogel H (2000) A numerical experiment on pore size, pore connectivity, water retention, permeability, and solute transport using network models. *European Journal of Soil Science* **51**, 99–105. doi:10.1046/j.1365-2389.2000.00275.x

Wang J-G, Li Z-X, Cai C-F, Yang W, Ma R-M, Zhang G-B (2013) Effects of stability, transport distance and two hydraulic parameters on aggregate abrasion of Ultisols in overland flow. *Soil & Tillage Research* **126**, 134–142. doi:10.1016/j.still.2012.09.005

Wang Q, Lu C, Li H, He J, Sarker KK, Rasaily RG, Liang Z, Qiao X, Li H, Mchugh ADJ (2014) The effects of no-tillage with subsoiling on soil properties and maize yield: 12-year experiment on alkaline soils of northeast China. *Soil & Tillage Research* **137**, 43–49. doi:10.1016/j.still.2013.11.006

Wang Y, Zhang Y, Zhou S, Wang Z (2018) Meta-analysis of no-tillage effect on wheat and maize water use efficiency in China. *The Science of the Total Environment* **635**, 1372–1382. doi:10.1016/j.scitotenv.2018.04.202

Williams JD, Wuest SB, Robertson DS (2015) Soil water and water-use efficiency in no-tillage and sweep tillage winter wheat production in northeastern Oregon. *Soil Science Society of America Journal* **79**, 1206–1212. doi:10.2136/sssaj2014.12.0494

Wortmann CS, Jasa PJ (2003) *Management to Minimize and Reduce Soil Compaction*. Institute of Agriculture and Natural Resources, Lincoln, <https://extensionpublications.unl.edu/assets/html/g896/build/g896.htm>.

Yilmaz I (2004) Relationships between liquid limit, cation exchange capacity, and swelling potentials of clayey soils. *Eurasian Soil Science* **37**, 506–512.

Yu X, Zhou W, Chen Y, Wang Y, Cheng P, Hou Y, Wang Y, Xiong X, Yang L (2020) Spatial variation of soil properties and carbon under different land use types on the Chinese Loess Plateau. *Science of the Total Environment* **703**, 134946. doi:10.1016/j.lwt.2020.110651

Yukselen Y, Kaya A (2008) Suitability of the methylene blue test for surface area, cation exchange capacity and swell potential determination of clayey soils. *Engineering Geology* **102**, 38–45. doi:10.1016/j.enggeo.2008.07.002

Zhang X, Luo Y, Goh KS (2018) Modeling spray drift and runoff-related inputs of pesticides to receiving water. *Environmental Pollution* **234**, 48–58. doi:10.1016/j.envpol.2017.11.032

4

Conservation tillage and soil water

R. Louis Baumhardt, Robert J. Lascano

Introduction

Tillage practices are a major component of soil management that directly affect field hydrology, whereas fluctuating climates generally reflect climate change. In the context of this chapter, we define climate change as events that drastically alter expected patterns of mainly precipitation and ambient temperature. For example, in terms of rainfall, these anomalies translate to frequent droughts and storms of greater duration and intensity (Allen *et al.* 2018), which may exceed soil infiltrability. The pattern of precipitation is linked to air temperature, and thus drought is usually accompanied by higher air temperatures. These climate changes affect the water and energy balance of a region. Periods of low and of high rainfall affect the infiltration, run-off and storage of water in the soil to different degrees, and thus the roles of tillage and crop residues have to adapt to the prevailing climate. The problem here is the uncertainty of forecasting the weather. However, in the case of a prolonged drought, the use of crop residues may suffer from limited biomass production and tillage would probably be restricted to soil disturbance during planting. Alternatively, rain anomalies appearing as wet periods may be accompanied by a greater reliance on specialised conventional tillage (CT) applications to promote infiltration through detaining and minimising water run-off (Baumhardt *et al.* 2020a).

In the semi-arid US Great Plains, reduced and no tillage (NT) practices have been primarily directed towards soil conservation by controlling either wind or water erosion by retaining or adding residue coverage compared with CT practices, which bury some fraction of the crop residue. The type, depth and degree of soil disturbance differ between CT tools and conservation tillage, which may vary in the geometry or the amount of residue retained. Residue management directly affects soil carbon through tillage disturbance and related carbon loss when carbon is combined with oxygen through microbial activity to form carbon dioxide, which is released back to the atmosphere. Where climate fluctuations may be dominated by semi-arid conditions, the significant differences in soil organic carbon (SOC) attributable to tillage effects that often develop over decade-long periods generally occur near the surface (Schwartz *et al.* 2015). The role of CT in contributing to SOC or the role of conservation tillage in mitigating a shift in SOC may be manifested in the soil water-holding capacity, which increases with increasing SOC (Emerson 1995) depending on the texture of the soil (Rawls *et al.* 2003).

In a drier, changing climate, the important advantage of conservation over CT is greater soil water storage during fallow periods that increases grain yields of wheat (*Triticum aestivum* L.) and sorghum (*Sorghum bicolor* L.) from ~45% to 66% (Greb 1979; Unger and Baumhardt 1999). Generally, the effects of conservation tillage, which retains surface residue, on field hydrology are to increase the infiltration of precipitation by intercepting rain and to reduce evaporative losses of water by intercepting irradiant energy and modifying the energy balance at the soil surface. In this chapter, we initially describe and discuss conservation and CT practices and equipment, as well as the field water balance governing soil water storage, which includes water inputs and losses. Processes and mechanisms of conservation or CT and management practices are sufficiently detailed to provide inferences that may support their optimisation during a gradually changing climate.

Specifying conventional and conservation tillage

The Conservation Technology Information Centre (CTIC 2002) defines conservation tillage as any tillage and planting system that maintains 30% or more soil surface cover with crop residue after planting to reduce soil erosion by water. Where soil erosion by wind is the primary concern, conservation tillage systems are defined as those that maintain at least 1.12 Mg ha^{-1} of flat small grain residue equivalent on the surface throughout the critical wind erosion period (CTIC 2002). However, semi-arid dryland crops occasionally may not produce enough residue to meet the required 'small grain equivalent' by conservation tillage and may be better characterised by the Soil Tillage Intensity Rating (STIR), a numerical index representing the degree of disturbance caused by tillage. Conservation tillage as specified by the US Department of Agriculture Natural Resources Conservation Service (USDA NRCS 2017a) includes: NT (STIR ≤15); strip or ridge tillage (STIR ≤15), in which only one-third of the row is disturbed; and reduced or mulch tillage, with full-width residue-preserving tools like chisel or sweep ploughs (STIR ≤60). Reduced tillage usually with chisels retains between 15% and 30% of the residue cover (USDA NRCS 2017c) or 0.56–1.12 Mg ha^{-1} of flat small grain residue equivalent. CT systems (STIR >100) retain <15% residue cover, requiring a single pass with the mouldboard plough or two passes with an offset disk to bury >90% of the residue cover (USDA NRCS 2017b). Since 2004, conventional 'bare soil' tillage is gradually being replaced by some type of conservation tillage as the preferred soil management system, but that replacement varies widely in the US by region and crop, ranging from 65–70% replacement for corn and soybeans in 2016 to 40% for cotton (Claassen *et al.* 2018). Globally, conservation agriculture, which used no or minimum tillage to preserve soil mulch cover from crop residue or cover crops of diversified rotations or sequences, averaged 12.5% of all crop land (Kassam *et al.* 2019). Kassam *et al.* (2019) also showed gradual adoption of conservation agriculture since 2008 that varied according to region, ranging from 1.1% of crop land in Africa up to 63.2% in South America, with intermediate values of 28.1% in North America and 45.5% in Australia and New Zealand combined. The authors concluded that conservation agriculture is the future of sustainable agriculture (Kassam *et al.* 2019).

Tillage has evolved from an early agrarian interest to improve soil conditions for growing select plants and control of competing weeds using tools that likely began as a stick and evolved into more sophisticated implements. For example, the offset disc (Fig. 4.1) or mouldboard plough are commonly accepted tools used for CT, with the latter inverting the furrow slice to bury 90–100% of the surface residue (Unger *et al.*

Fig. 4.1. Residue retained after a single operation using inversion-like disk tillage is ~30–35% of the undisturbed field that would be expected with no tillage. Photograph courtesy of the US Department of Agriculture Natural Resources Conservation Service.

2012). In the early 1900s, CT practices adopted the 'scientific agriculture' paradigm of Campbell (1907) for semi-arid regions that relied on frequent disk ploughing. As a result, practically all crop residues were incorporated and the soil surface was pulverised with the goal of creating a dust mulch to disrupt or impede capillary water movement in order to reduce water evaporation, and thus conserve soil water. Dust mulches remain a popular CT practice in the Pacific Northwest for use in the wheat–fallow cropping system as a means to limit evaporation and retain soil water for planting, as was also achieved using improved residue management practices (Schillinger 2001). Riar *et al.* (2010) compared fallow tillage methods in the Pacific Northwest and also observed that dust mulches had soil water contents and grain yields similar to or slightly better than other tillage practices. Tillage practices proposed by Campbell (1907) enjoyed success before the 1930s during years of favourable precipitation while producing wheat for European countries engaged in World War I (Baumhardt 2003). However, that dust mulch management paradigm rendered the soil susceptible to wind erosion and contributed to the 'Dust Bowl' on some 16 million ha after climatic conditions entered drought lasting 10 years (Baumhardt and Blanco-Canqui 2014).

Conservation tillage practices developed to ameliorate 'Dust Bowl' conditions successfully produced larger soil aggregates that resisted wind erosion and preserved residue cover on >30% of the soil and achieved a corresponding soil loss ratio averaging <25% of that for bare soil (Unger *et al.* 2012). A particularly successful tillage tool was developed after an unintended discovery by Fred Hoeme, an Oklahoma farmer; when following his protracted field storage of a road scarifier for the county, Hoeme chose to move the scarifier overland to his equipment lot (Benton 1987). The subsequent wheat crop planted on the scarified strip benefited from improved rain infiltration and water conservation that resulted in wheat with the highest grain yield on the Hoeme farm and led to the development and eventual production of the Graham Hoeme plough (Peterson 1948). The Graham-Hoeme plough (Fig. 4.2) avoided the formation of the hard plough sole common to the mouldboard or disk ploughs, and provided narrow channels for water to enter the lower profile of typically hard soils during periods of drought. The Graham-Hoeme plough enjoyed popularity as 'The plough that saved the Plains' (Stewart *et al.* 2010) and converted the family name into a verb for performing chisel tillage to control blowing dust and improve infiltration (i.e. during the 1930s, many producers would 'Hoeme' their fields).

Conservation tillage integrates practices that maximise residue retention and weed control first mechanically and then chemically. Mechanical weed control on the Great Plains initially began with single straight blade cultivators that were subsequently modified using various V-angle sweep ploughs (Fig. 4.3) that could undercut residue and weeds to sever roots but limited residue disturbance (Allen and Fenster 1986). The resulting stubble mulch farming program relied on various combinations of chisel and sweep tools for mechanical weed control while minimising residue disturbance for optimum residue

Fig. 4.2. The Graham-Hoeme plough with chisels on display at the US Department of Agriculture Agricultural Research Service Conservation and Production Research Laboratory (CPRL) at Bushland, Texas, since 2000. Inset, a photograph from 1949 showing a Graham Hoeme plough with chisels and knives being used in fallowed sorghum stubble before autumn planting to winter wheat. Photograph courtesy of the CPRL photograph archives.

Fig. 4.3. Shown is a typical medium format 10.8-m wide sweep plough constructed from three 1.5-m and four 1.7-m wide V-shaped blades that overlap sufficiently to insure weed control while limiting the number of shanks and coulters that can displace residue. This plough advances from early designs like the 1949 Dempster 101 carrier frame (inset) configured with five 0.76-m V-shaped sweep blades mounted with 0.1-m blade overlap and rolling coulters ahead of each plough shank to cut through residue. Photograph courtesy of Conservation and Production Research Laboratory Bushland, Texas, photograph archives.

retention (Fenster 1960). Chemical weed control for NT was made possible by the introduction of chemistries for weed control beginning in the 1940s (Phillips and Phillips 1984), but continues to evolve with developing weed resistance and new control materials. Herbicide development permitting NT began in earnest during the 1960s with weed control strategies that, for example, often combined both selective soil-active and contact-type herbicides (Wiese *et al.* 1967). Water conservation during cropping sequences with successful herbicide combination weed control strategies improved NT grain yield and residue production (Unger and Wiese 1979).

Two main differences separating CT using mouldboard or disc primary tillage and multiple secondary tillage operations from reduced tillage, which eliminates primary and some secondary tillage, or conservation tillage by strip tillage, ridge tillage and NT are the disposition of crop residues and the degree of soil disturbance. Post-harvest residues retained at the soil surface at planting will vary with tillage tool (Table 4.1), but range from 10% with mouldboard ploughing to 80–90% overwinter weathering with NT. The cumulative impact of as many as five sweep tillage operations commonly used for stubble mulch management can reduce natural weathered residue cover to ~50% at best. The corresponding STIR that

Table 4.1. Soil Tillage Intensity Rating (STIR) and percentage of surface residue cover remaining after one operation with various tillage implements and estimated for seasonal weathering (no tillage)

Tillage implement	% Residue cover remaining[A]	STIR[B]
Mouldboard plough	0–10	65
Offset disk	25	39
Tandem disk	30–70	33
One-way disk	55–80	
Chisel straight points (0.30 m)	60–80	62
Sweep plough (stubble mulch)	60–90	26
Paraplough, sub-soiler (depth 0.35 m)	70–90	18–21
Planters and drills	60–95	2–4
Natural no-till overwinter weathering		
Summer harvest small grains	70–90	0–2
Fall harvest summer grains	80–95	0–2

[A]After Unger *et al.* (2012)
[B]From Wisconsin Soil Nutrient Application Planner (https://snapplus.wisc.edu/planning/matching-snapplus-tillages).

indexes the degree of soil disturbance ranges from a low of 0–2 for annual NT disturbance up to 65 for a single mouldboard plough operation to 0.20 m (Table 4.1). The effects of tillage on soil physical properties vary to some extent with the degree of soil disturbance.

Tillage practices associated with reduced residue cover and increased STIR rating due to greater disturbance would intuitively affect those parameters reflecting soil porosity and potential hydraulic conductivity through those pores. That is, tillage disturbance absent or reduced with conservation tillage increases soil porosity, but at the expense of any natural soil structure and potential channels for distributing water (Govaerts *et al.* 2007; Machado *et al.* 2015). In a review article assessing tillage effects on soil hydraulic properties, Strudley *et al.* (2008) considered over 200 studies to document improved flow with NT that achieved macropore connectivity despite reduced total porosity or the resulting higher bulk density depending on tillage duration. In another study, Blanco-Canqui *et al.* (2017) characterised mouldboard, chisel, disk and NT effects on ponded and tension infiltration, saturated hydraulic conductivity and bulk density of a smectitic silty clay loam. Although they reported significantly greater ponded infiltration for mouldboard tillage compared with the other types of tillage due to macropore flow, tension infiltration eliminated macropore flow and, consequently, any differences due to tillage (Blanco-Canqui *et al.* 2017).

Most of the soil-disturbing tillage practices will manifest immediate and pronounced effects on soil properties, but Strudley *et al.* (2008) reported that those hydraulic impacts were somewhat short lived and indistinguishable from unaccounted management or spatial variability. Cooper *et al.* (2020) conducted a long-term comparison of the effects of management with conventional mouldboard tillage, reduced tillage with discs and tines and NT on soil properties in large fields northeast of London, UK, over a period of 5 years. During that time, soil physical and chemical properties did not differ between CT and conservation tillage practices; however, profit increased 13% due to improved operational efficiency and increased crop yields with conservation tillage. Cooper *et al.* (2020) concluded that the short-term impact of conservation tillage alone did not effectively improve environmental sustainability over CT practices. Alternatively, Baumhardt *et al.* (2008)

measured increased ponded infiltration and decreased soil bulk density and penetration resistance 30 years after profile-modifying mouldboard tillage to a depth of 0.7 m. Nevertheless, other overriding factors, such as surface seal formation, may determine the significance of conservation tillage effects on field hydrology.

Reduced or NT conservation tillage may involve occasional tillage operations that are inconsistent with preserving residue and avoiding soil disturbance but are necessary for crop production. For example, a lister or chisel plough may be used to roughen the soil surface and produce resistant >0.84-mm aggregates that are not susceptible to wind erosion or to trap saltating soil to control active soil erosion (Fig. 4.4) as an 'emergency' form of conservation tillage (Baumhardt and Blanco-Canqui 2014). The occasional use of soil-disturbing tillage in otherwise NT management to relieve emerging production constraints was identified by Dang *et al.* (2015b) as a 'strategic tillage' application. Dang *et al.* (2015a) described strategic tillage agronomic benefits as methods for the occasional control of herbicide-resistant weeds or correction of near-soil-surface stratification of nutrients and disruption of soil compaction. Strategic tillage is used to correct soil compaction attributed to cattle trampling, which decreases soil surface permeability and increases run-off and may consequently reduce both irrigated and dryland crop yields (Winter and Unger 2001; Baumhardt *et al.* 2011). In a review article, Blanco-Canqui and Wortmann (2020) noted numerous other benefits from occasional tillage or strategic tillage, including reduced compaction or stratification and help with weed control in an integrated weed management program. However, occasional tillage or strategic tillage did not necessarily reverse benefits gained by using NT or reduced conservation tillage.

Fig. 4.4. Emergency tillage is an example of strategic tillage for the immediate control of wind erosion by roughening the surface with ridges and producing large aggregates not susceptible to movement. Photograph courtesy of the US Department of Agriculture Agricultural Research Service Engineering and Wind Erosion Research Unit, Manhattan (KS, USA).

Tillage and water sources governing field water balance

The effect of tillage on stored soil water (*SW*) for dryland crop production is attributed to the combined water sources (inputs) of precipitation (*P*), surface inflow (*SI*) and upward groundwater capillary flow (*CF*), as modified by run-off (*RO*) and its hydrological abstractions or losses. Those abstractions include water infiltration (*I*), subsurface drainage (*D*) and evapotranspiration (*ET*; i.e. the sum of water evaporation from the soil (*E*) and from the crop (*T*)) processes according to a simplified water balance, with units of millimetres for all terms, as follows:

$$SW = (P + SI + CF) - (RO + I + D + ET) \quad [1]$$

Precipitation is exclusively an atmospheric water source that is not directly affected by soil management, although tillage and cropping systems may be adjusted for expected seasonal precipitation and in relation to a climatic phenomenon like the El Niño Southern Oscillation phase (Jones *et al.* 2000; Richards *et al.* 2014; Baumhardt *et al.* 2016, 2020b). Precipitation is subject to plant canopy interception and surface detention, which are not represented in Eqn 1 because the considered processes of water infiltration (*I*) and water evaporation from plant (*T*) or soil surfaces (*E*) usually account for this water. Detention storage that directly affects run-off, and consequently surface infiltration, has been managed with tillage that forms small ephemeral dams as basins within a furrow or furrow dyking to prevent run-off (Jones and Baumhardt 2003). This detention storage method often increases cumulative water infiltration, stored soil water and crop growth and yield due to greater water evaporation from the plant or *T* (Gerard *et al.* 1984; Nuti *et al.* 2009), but not always (Baumhardt *et al.* 1993b; Bryant *et al.* 2019). Slope-minimising contour-farmed terraces also increase opportunity time for water infiltration (Finnell 1930) and were found to reduce cumulative annual run-off by 30–35% (Baumhardt *et al.* 2020a). Either contour terracing or furrow dyking can be effective tillage practices for increasing stored soil water provided the rate of precipitation is greater than that of water infiltration and generates run-off or surface inflow.

Surface inflow is water made available because landscape and sloping topography are conducive to run-off collection and management provided the soil properties of contributing watersheds are not subject to erosion. Rainwater harvesting as surface inflow has been effective with limited annual precipitation by using water-spreading dykes and micro-catchments, like those in the Negev Desert (Bernstein *et al.* 1974). To prevent occasional flooding of closed-end level terraces, Zingg and Hauser (1959) converted the lower third of the terrace area into level contour benches for spreading water from the rest of the sloping terrace catchment (i.e. watershed). Hauser (1968) documented that annually cropped conservation benches produced competitive dryland yields due to surface inflow that were approximately 50% greater than the mean yield of graded and level terraces cropped in a less intensive rotation. More recently, sorghum grain yields increased progressively across the terrace from that of the upper watershed receiving no surface inflow but contributing run-off to the higher-yielding portion of the watershed that received surface inflow and contributed run-off to the highest-yielding terrace bench receiving surface inflow from the watersheds but not contributing to run-off (Bandaru *et al.* 2006). Comparisons of the conventional sloping terrace with conservation bench terrace systems by Sharda *et al.* (2013) showed an 18-year mean reduction in run-off from 35% to ~7% of the 880 mm mean annual rainfall. Those water savings were in addition to a reduction in soil loss from ~8 to 1.0 Mg ha^{-1} for the conventional and conservation bench terrace systems respectively.

The remaining water source of Eqn 1 is from capillary flow via soil pores, from typically shallow water tables, and contributes to water evaporation from the plant surface (*T*). The groundwater contribution to crop water use and any impact on crop yield depends on groundwater depth, increasing yields for shallow water table depths that are never >2.5 m. For example, under arid conditions, shallow water tables can provide up to 60–70% of the crop water needs depending, in part, on soil texture, water table salinity and depth, ranging from 0.8 to 1.9 m (Grismer and Gates 1988). During two growing seasons with precipitation of 125 and 363 mm, Wu *et al.* (2015) estimated groundwater at depths of 0.7–1.2 m contributed ~50% of the calculated crop evapotranspiration, concluding that semi-arid Inner Mongolia has a high failure risk during growing seasons with limited precipitation or greater groundwater depth. Depending on the crop, desirable shallow groundwater conditions to promote capillary rise for crops (Mueller *et al.* 2005; Satchithanantham *et al.* 2014; Kroes *et al.* 2018) may be counter to drainage requirements if needed, which was noted in a corn study (Zipper *et al.* 2015). Primary tillage principally affects soil to depths of 0.15–0.3 m, with the exception of tillage using the paraplough or V-Ripper sub-soiler, which affect soils at depths ranging from 0.3 to 0.6 m, whereas the target operation depth for many secondary tillage operations may be between 0.1 and 0.15 m (USDA NRCS 2017c). Tillage effects do not directly impact the depths of most shallow water tables compared in the studies above and thus would have a negligible effect on water balance in Eqn 1.

Tillage and hydrological processes governing field water balance

The various water sources contributing to soil water, *SW*, in Eqn 1 include *P*, *SI* and *CF* as adjusted by deducting often interrelated hydrological processes of *RO*, *I*, *D* and *ET*. Tillage has no direct effect on the field water balance contributions from *P* or the *CF* that originates from water tables at depths below the depth of tillage disturbance, but source *SI* is provided from nearby *RO* in amounts depending on tillage practice. Adjustments to the field water balance sources include tillage-dependent *D*, *RO* with related *I* and the *ET* process.

Drainage

Subsurface *D* in Eqn 1, or the downward movement of water through the soil profile beyond the root zone that does not return by capillary water movement usually from a shallow water table, will not contribute to *SW* or crop *ET*. As noted when describing tillage practices, chisel ploughing to break up flow-limiting soil layers and increased residue cover with strip tillage or NT promote greater infiltration than bare soil mouldboard ploughing. For example, the amount of drainage increased in the order of mouldboard ploughing < chisel tillage < ridge tillage < NT (Kanwar *et al.* 1997). It is not surprising that regions having mean annual precipitation (MAP) exceeding crop potential ET (PET) often require drainage of excess water for sustainable crop production, such as the managed tile systems or network of canals used in the Midwest US (Strock *et al.* 2010). Follett *et al.* (2012) listed the ratio of MAP:PET as consistently >1.0 for multiple locations from humid/sub-humid climates of Midwestern US states where tile drainage is not uncommon. In contrast, the MAP:PET ratio for locations on the Great Plains were fractions varying from 0.41 at Bushland, Texas, in the south-west to 0.81 or 0.75 for less arid climates of Lincoln, Nebraska, and Mandan, North Dakota, respectively, suggesting sometimes negligible potential deep drainage in the south-western Great Plains. Using methods quantifying chloride profiles as described by Scanlon *et al.* (2008), Baumhardt *et al.* (2017) determined significantly greater mean annual deep drainage rates from NT of 14 mm, compared with 2 mm for stubble mulch, over a 30-year, long-term experiment. Baumhardt *et al.* (2017) attributed greater drainage through a clay loam soil to the typically wetter soil profile for NT compared with stubble mulch at this semi-arid location. Scanlon *et al.* (2008) quantified chloride profiles for a dryland farmed bench terrace with the same clay loam soil, which features a very slowly permeable subsoil layer at a depth of 0.15–0.7 m, and had been either normally tilled with or without a one-time, deep (~0.7 m), mouldboard ploughing 36 years earlier. Scanlon *et al.* (2008) estimated mean annual deep drainage increased 600% due to that mouldboard ploughing disturbance of the soil profile and, possibly, higher ponded infiltration rates (Baumhardt *et al.* 2008).

Run-off

Any precipitation, including rainfall, that does not enter the soil through the infiltration process or contribute to surface inflow for adjoining areas is lost as run-off. Independent of the tillage or cropping system but linked to increased run-off is greater landscape slope, which diminishes the opportunity time for infiltration, and promotes contour farming and terracing (Thompson and Sudduth 2018). Another soil property independent of tillage practice that reflects macroporosity is the coarseness of the soil texture; that is, less run-off would be expected for sand than for clay, which has reduced macroporosity. Factors independent of the soil are rain amount or intensity either separately (Jones *et al.* 1985) or in association with the El Niño Southern Oscillation warm sea surface temperature phase (Baumhardt *et al.* 2020b). Additional precipitation factors affecting run-off were identified in Canada when Tiessen *et al.* (2010) concluded snowmelt run-off from conventional primary tillage plus multiple secondary operations did not differ in amount from conservation tillage with limited secondary tillage. Their data showed that the run-off of rain increased significantly with CT, illustrating infiltration and run-off differences due to tillage are diminished where precipitation was snow because of slow snowmelt and no crust formation (Tiessen *et al.* 2010).

The recognised benefit of the residues retained with CT to improve rain infiltration and reduce run-off is broadly realised depending on contrasting tillage practices. The humid subtropical climate of the southern US often produces run-off from NT that is significantly less than that from conventional mouldboard plough–disc harrow tillage on loessial fragipan soils in north Mississippi cornfields harvested for silage or grain (McDowell and McGregor 1984). In the semi-arid inland Pacific Northwest, run-off from conventional chisel or mouldboard tillage was an order of magnitude larger from CT than from NT under wheat-fallow (Williams *et al.* 2014). Williams *et al.* (2014) attributed reduced runoff with NT to residue cover. Nevertheless, exceptions to reduced run-off with residue retaining NT were reported by Jones *et al.* (1994) and Baumhardt *et al.* (2017), possibly as a result of often overlooked, but not unique, soil properties of increased surface density or the profile soil water content that are counter to greater infiltration.

The benefits of NT residue to intercept raindrop impact that forms soil crusts and reduces run-off can be enhanced with disrupting tillage that preserves some residue cover. A study evaluating crop performance and run-off from chisel- and no-tilled watersheds annually cropped to soybeans and corn in a humid climate of the US Midwest found less run-off was produced than measured for a rotation that added wheat to make a 3-year rotation and substituted disk tillage to avoid adverse consequences of mouldboard ploughing (Shipitalo and Edwards 1998). That study was revisited by Shipitalo *et al.* (2013)

to find chisel tillage had 10% less run-off than NT and 25% less than the disked rotation. A clearer exception to the expected benefit of NT residue management to decrease run-off was shown for large (>2 ha) paired watersheds in a wheat–sorghum–fallow rotation and managed with conventional stubble mulch tillage or NT (Jones *et al.* 1994; Baumhardt *et al.* 2017). Not only was measured run-off greater for NT than stubble mulch management, but the findings also corroborated simulated rain infiltration measurements by Jones *et al.* (1994), Baumhardt and Jones (2002) and Baumhardt *et al.* (2012) showing the effects of surface compaction and a higher soil profile water content. Likewise, Lindstrom *et al.* (1984) reported that soil surface compaction with NT was responsible for reduced infiltration of simulated rain despite residue cover in the northwest US Corn Belt.

Infiltration

Deforestation and the loss of soil-covering litter during Euro-American settlement in the south-eastern US resulted in catastrophic water-driven soil erosion that, for example, led to the 'Badlands of Mississippi' and eventual return of land to forested conditions (Dabney *et al.* 2012). Lowdermilk (1930) reported forest litter reduced both run-off and soil eroded, concluding, in part, that suspended particles in run-off water sufficiently sealed the surface soil pores to cause marked differences in the rate of infiltration between the litter-covered and bare soils. Similarly, Duley and Russel (1939) improved water conservation using straw soil cover regardless of strategic tillage compared with tilled land and no straw, whereas that same amount of straw ploughed into the soil produced only a limited increase in soil water conserved. Duley and Russel (1939) concluded that using crop residue as a soil cover was a practical method to reduce run-off and soil erosion that also increased crop yields. Building on these findings, Duley (1940) investigated the infiltration of sprinkled water into five soils that were bare or mulch covered and reported water intake rates that were consistently lower for bare soils. When calculated as a portion of the infiltration into mulched soil, bare soil infiltration ranged from 18% to 76% because of a thin compact layer that formed at the surface that was not covered by residue. Studies by McIntyre (1958), Morin and Benyamini (1977) and Wakindiki and Ben-Hur (2002), among others, have also described this infiltration-limiting thin soil surface layer with lower porosity and saturated hydraulic conductivity than the underlying bulk soil matrix, known as a seal or crust.

Seal formation begins as drops of rain or irrigation water impact a bare soil surface to disrupt aggregates and detach soil particles that are carried with colloids into the bulk soil by infiltrating water (Unger *et al.* 2012). Subsequent soil particle deposition occludes surface conducting pores and gradually reduces the seal hydraulic conductivity and infiltration, which leads to surface ponding and accelerated aggregate slaking/dispersion. The ratio of the saturated hydraulic conductivity to seal layer thickness or seal conductance can decrease rapidly during raindrop impact, converting the initial bulk soil condition to the terminal seal as the cumulative kinetic energy increases (Fig. 4.5). The plotted surface conductance was estimated at incipient ponding, when the infiltration rate equals the rainstorm intensity, from the rainstorm intensity divided by the subsurface soil suction calculated using a simplified solution of Richards' equation developed by Parlange *et al.* (1984). Baumhardt *et al.* (1991) demonstrated that seal conductance decreased with increasing cumulative rainstorm kinetic energy and decreasing intensity. The estimated seal conductive properties have been used when modelling seal effects on rain infiltration to better understand the formation process in relation to rainfall and soil characteristics (Baumhardt *et al.* 1990; Assouline and Mualem 1997).

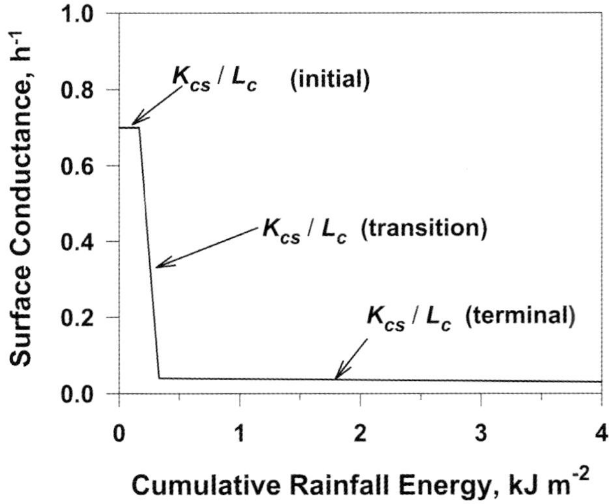

Fig. 4.5. Change in the ratio of the surface hydraulic conductivity (K_{cs}) and the thickness of a rain-formed soil seal (L_c) shown as a function of the rainstorm cumulative kinetic energy, after Baumhardt *et al.* (1991).

Conservation tillage may retain variable amounts of residue for reduced tillage that increase to a maximum with NT, resulting in progressively more cover for intercepting the raindrop impact that reduces cumulative kinetic energy during a rainstorm. Cumulative simulated rain infiltration increased with increasing residue cover through 80%, although cumulative infiltration with 40% cover appeared ~15% lower (Li *et al.* 2001) and is similar to the 40–60% residue coverage needed to maintain infiltration (Baumhardt and Schwartz 2004). Strip and ridge conservation tillage similarly retains residue that is unevenly distributed over the soil surface but, in the case of a strip ridge-tilled sandy loam soil, infiltration with full or partial cover differed only from the bare soil (Baumhardt and Lascano 1996). Any residue cover asymmetry associated with conservation tillage did not eliminate the beneficial increased rain infiltration that contributes to soil water.

Evaporation and surface energy–water balance

The effects of crop residues and conservation tillage on water conservation have been reviewed by many authors (Steiner 1989, 1994; Carter 1994; Díaz-Zorita *et al.* 2002; Erenstein 2002; Alvarez and Steinbach 2009; Nyakudya and Stroosnijder 2015; Kader *et al.* 2017; and others). For example, in humid climates, the use of mulches and residue incorporation is recommended (Carter 1994; Díaz-Zorita *et al.* 2002), along with minimum tillage (Alvarez and Steinbach 2009), as a means of conserving water and reducing soil erosion. Recent advances in mulching show its effect on reducing the evaporation of water from the soil, supressing weed growth, controlling soil structure and influencing soil microorganisms (Kader *et al.* 2017). In a semi-arid climate, Nyakudya and Stroosnijder (2015) studied CT compared with conservation tillage and concluded that the adoption of either tillage depended on seasonal rainfall. However, under dryland conditions, conservation tillage increased crop yields and productivity (Asmamaw 2017). In general, crop residues, regardless of climate, are conducive to the conservation of water by supressing evaporative losses from the soil. However, the specific amount of water saved is a function of many variables, and thus difficult to evaluate and calculate. The main effect of surface residues is to modify the energy and water balance of the soil surface, as shown in Fig. 4.6 (Lascano *et al.* 1987, 1994; Lascano and Baumhardt 1996).

The left-hand side of Fig. 4.6 shows the terms of the energy balance, whereas the right-hand side shows the water balance components. The effect of a surface residue on infiltration (Inf) and soil water evaporation (E_s) is a function of many variables that are related to each other because they both independently contribute to the soil water balance. Further, to establish a cause and effect of a given residue with specific properties (e.g. architecture

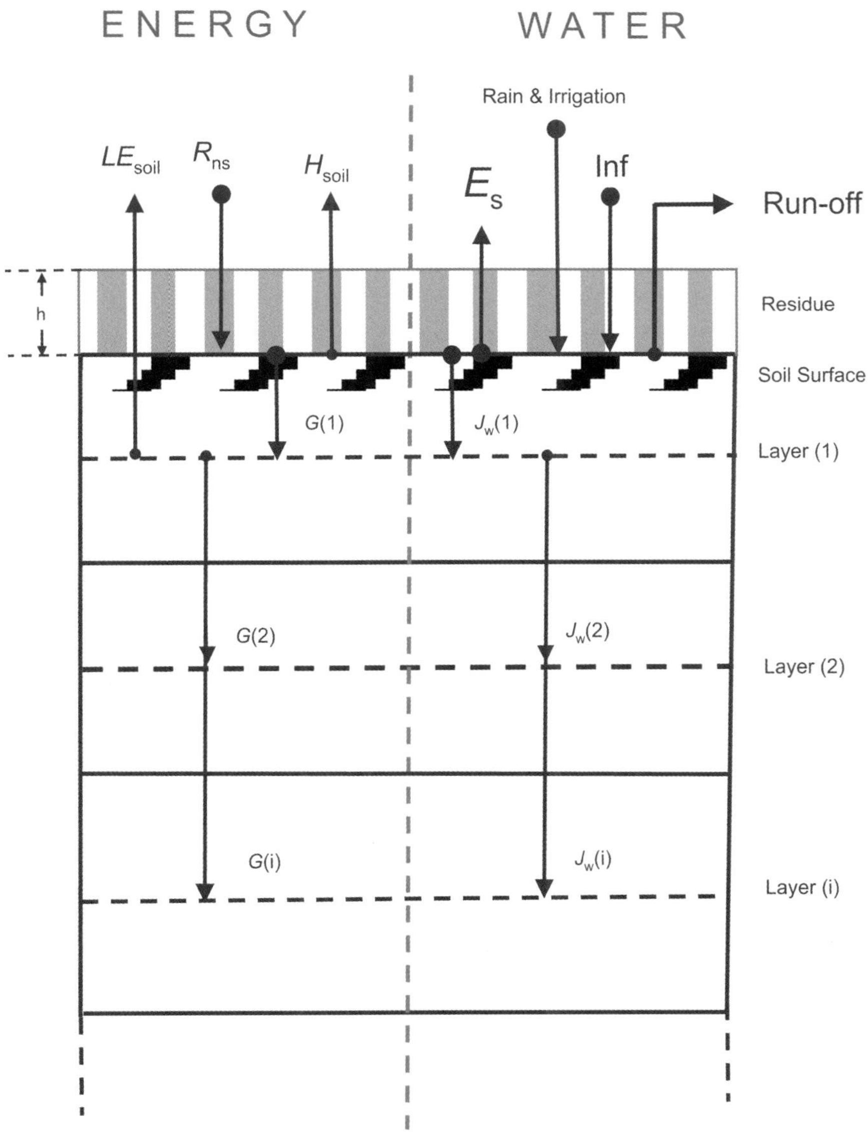

Fig. 4.6. The energy and water balance of a soil surface with a crop residue. The terms of the energy balance are given on the left-hand side, whereas those of the water balance are given on the right-hand side. E_s, soil water evaporation; *G*, the soil heat flux; h, height; H_{soil}, sensible heat flux into the air; Inf, infiltration; J_w, flow of water; LE_{soil}, latent heat flux into the air; R_{ns}, net irradiance at the soil surface.

[height and density] and colour) on increasing infiltration and retarding soil water evaporation requires information on the other terms of the energy and water balance. It is this complexity that has prompted the application of simulation models, which are used to evaluate the effects of surface residues on infiltration and soil water evaporation. Models that mechanistically describe the energy and water balance of surface residues are given by Bristow *et al.* (1986), Lascano and Van Bavel (1986), Lascano *et al.* (1987), Flerchinger and Saxton (1989), Bussière and Cellier (1994), Bristow and Horton (1996), Horton *et al.* (1996), Lascano and Baumhardt (1996), Aiken *et al.* (1997), Gonzalez-Sosa *et al.* (1999), Flerchinger *et al.* (2003), Scopel *et al.* (2004), Kozak *et al.* (2007a, 2007b), Kucharik and Twine (2007), Wilson *et al.* (2012) and Lagos *et al.* (2013). For our purpose, we first describe the energy balance, and thereafter the water balance components. Both balances are given by the sum of input and outputs that are discussed with greater detail in publications by Van Bavel and Hillel (1976), Lascano and Van Bavel (1986), Lascano *et al.* (1987) and Lascano and Baumhardt (1996).

The energy balance of a soil surface with a residue (Fig. 4.6) is given by:

$$R_{ns} + H_{soil} + LE_{soil} + G = 0 \qquad [2]$$

where R_{ns} is the net irradiance at the soil surface, H_{soil} is the sensible heat flux into the air, LE_{soil} is the latent heat flux into the air and G is the soil heat flux (all terms in Eqn 2 are in W/m^2).

The effect of the residue (Fig. 4.6) is to alter the geometry of the soil surface by moving the plane of radiation absorption upward and by slowing the downward movement of absorbed energy to the soil surface due to the low conductivity of the residue. Surface residues reduce the net irradiance at the soil surface by increasing the reflection of incoming short-wave irradiance and decreasing emissivity (Sharratt and Campbell 1994; Sauer *et al.* 1998). Crop residue affects both the latent (LE_{soil}) and sensible (H_{soil}) heat flux in two ways. First, the roughness length, related to the height of the residue, increases, affecting the fluxes through changes in the aerodynamic boundary layer; second, the temperature and water potential of the soil surface are modified, affecting both gradients of humidity and temperature (Tanner and Shen 1990; Lascano and Baumhardt 1996). The soil heat flux (G) is also altered because the thermal conductivity of the residue is lower than that of a mineral soil (Hillel *et al.* 1975). The surface residue represents an additional resistance (r_v) to H_{soil} and LE_{soil}. This resistance has been measured for a chopped corn residue by Tanner and Shen (1990) and for a wheat stubble by Heilman *et al.* (1992).

The resistance term, r_v, measured for a chopped corn residue (Tanner and Shen 1990) is:

$$r_v = \frac{1}{[0.0022(1+0.89u_a)]} \qquad [3]$$

where u_a is the wind speed (m/s) 1.0 m above the top of the residue. The r_v measured in a wheat stubble (Heilman *et al.* 1992) is given by:

$$r_v = \frac{416.67}{(1+0.35u_a)} \qquad [4]$$

A plot of r_v as a function of wind speed is shown for the chopped corn and wheat stubble residues in Fig. 4.7. These values show that at wind speeds <0.1 m s^{-1}, r_v is approximately 400 s m^{-1} for both the corn stubble and wheat residue, and that as wind speed increases to 4.0 m/s the values of r_v decrease to 100 s m^{-1} for the corn stubble residue and to 170 s m^{-1}

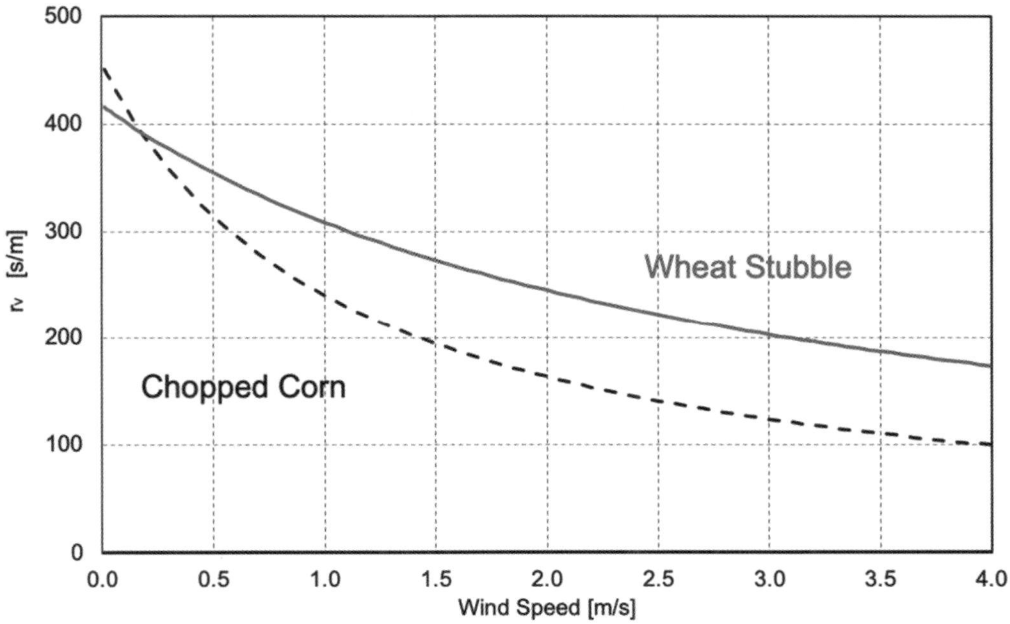

Fig. 4.7. Resistance term (r_v) as a function of wind speed measured over a corn residue (Tanner and Shen 1990) and over wheat stubble (Heilman *et al.* 1992).

for the wheat stubble. For wind speeds between 1.5 and 4.0 m s^{-1}, the wheat stubble resistance is, on average, approximately 80 s m^{-1} larger than for the corn residue.

With the addition of irrigation, the water balance of a soil surface covered by a residue is given by the sum of inputs and outputs (Fig. 4.6) and is similar to Eqn [1]:

$$E_s + (Rain + Irrigation) + I + RO = 0 \qquad [5]$$

where E_s is the evaporation of water from the soil surface, *Rain* and *Irrigation* (sum = *P*) are inputs of water, *I* is the amount of rain and irrigation water that enters the soil as infiltration and *RO* represents the amount of water that cannot be stored in the soil and is lost due to run-off (all terms in Eqn 5 are in mm).

The hydrological characteristics of the soil are given by the saturated hydraulic conductivity of the soil (K_{sat}) and the soil hydraulic properties, namely the pressure potential (ψ) as a function of the soil volumetric water content (θ_v) and the hydraulic conductivity (K), also a function of the soil water content. These two functions, namely $\psi = f(\theta_v)$ and $K = f(\theta_v)$, are measured inputs to the model. The flow of water, J_w, is calculated using Darcy's Law and the vertical flow of water into each soil layer is calculated using the arithmetic average of *K*. The flux of water of the first soil layer equals the sum of the evaporation rate and the infiltration rate. If the rainfall rate is not zero, it is compared to the infiltration capacity and, unless it is greater, it is credited to the topmost soil layer; if it is greater, the excess is accumulated as detention up to a maximum value and the excess is equated to the run-off rate. If there is no rain and or irrigation, the infiltration rate equals the infiltration capacity. The detention and infiltration capacity are inputs to the model. For additional details, readers are referred to Lascano *et al.* (1987, 1994).

The surface residue affects the soil temperature and soil water content of a soil profile and thus is linked to the partitioning of the surface energy balance, as shown by Lascano *et al.* (1994) and Lascano and Baumhardt (1996). The principal benefit to a cotton crop planted in a terminated wheat residue was the effect of the residue to decrease soil water evaporation (*E*) in exchange for increased plant water evaporation (*T*); however, the sum of the losses (*ET*) throughout the season remained the same compared with a conventionally planted cotton crop (Lascano and Baumhardt 1996). That is, rates of water loss from the soil and plant are modified due to the residue, but the integral, seasonal *ET*, remains the same.

In the Texas High Plains, a practice known as strip tillage (Bolton and Booster 1981) plants cotton into a terminated wheat residue killed using herbicides. This cropping system (i.e. terminated wheat–cotton) is designed to take advantage of autumn rains to establish the wheat cover crop and spring rains to grow the cotton. Briefly, winter wheat is planted in late autumn after the cotton is harvested, and in early spring the wheat is killed prior to producing stem nodes with herbicide and cotton is planted into the terminated wheat residue cover. The advantage of this cropping system is to reduce evaporative losses of water from the soil (Unger and Parker 1976) and increase the infiltration of rain (Baumhardt *et al.* 1993a). In addition, the standing wheat residue provides shelter from high wind speeds and blowing sand, which are prevalent at planting (Baker 2007).

Summary and conclusions

Conservation tillage was developed as a response to control soil erosion by both wind and water, but also improved soil water storage. Conventional tillage affects the degree of soil disturbance and the amount of residue cover depending on the implement used or its tool design. Both conservation tillage and CT modify all elements of the field water balance that governs soil water storage, but especially the very active hydrological processes of water infiltration and evaporation. It may be true that the infiltration processes are of paramount importance in sub-humid or humid climates, whereas evaporation is of greater concern in semi-arid climates where the potential evaporation of water considerably exceeds the precipitation amount. Climate change, however, may shift those interests.

The development of CT implements and related cropping management systems is as diverse as the crops produced and soils farmed. By contrast, conservation tillage varies largely in proportion to residue retention, increasing with fewer tillage operations and correlated soil disturbance that decreases to zero with NT. In general, surface residues provide a cover that intercepts the energy driving the evaporative loss of soil water and affects the formation of crusts at the surface by reducing the kinetic energy of raindrops. The net effect is to increase the capacity of the soil to store water from precipitation by affecting the infiltration rate and the redistribution of water in the profile, which leads to a reduction of water lost by run-off or to water that is diverted and infiltrates elsewhere. The result of these two processes on field hydrology is to govern plant-available soil water, which is directly related to crop water productivity as modified by internal profile drainage or any capillary water contributions.

As the anticipated changes in climate are realised, those resulting consequences, namely overall warmer conditions with reduced precipitation from less frequent but likely more severe rain events (Allen *et al.* 2018), will highlight the more critical contribution of conservation tillage to improving hydrological processes. The effect of CT or conservation tillage (NT, reduced tillage) on crop residue and soil disturbance manifested in soil infiltration and evaporation processes may contribute to or offset climate impacts on field hydrology and soil water storage for crop use. We conclude that greater retention of crop residues at the soil surface and a reduction in disturbance through broader application of NT and reduced tillage is a more viable soil management adaptation for producing crops during a changing climate.

References

Aiken RM, Flerchinger GN, Farahani HJ, Johnson KE (1997) Energy balance simulation for surface soil and residue temperatures with incomplete cover. *Agronomy Journal* **89**, 404–415. doi:10.2134/agronj1997.00021962008900030007x

Allen RR, Fenster CR (1986) Stubble-mulch equipment for soil and water conservation in the Great Plains. *Journal of Soil and Water Conservation* **41**, 11–16.

Allen MR, Dube OP, Solecki W, Aragón-Durand F, Cramer W, Humphreys S, Kainuma M, Kala J, Mahowald N, Mulugetta Y, *et al.* (2018) IPCC 2018: summary for policymakers. In *Global Warming of 1.5°C. An IPCC Special Report on the Impacts of Global Warming of 1.5°C Above Pre-Industrial Levels and Related Global Greenhouse Gas Emission Pathways, in the Context of Strengthening the Global Response to the Threat of Climate Change, Sustainable Development, and Efforts to Eradicate Poverty.* (Eds V Masson-Delmotte, P Zhai, H-O Pörtner, D Roberts, J Skea, PR Shukla, A Pirani, W Moufouma-Okia, C Péan, R Pidcock, et al.) Intergovernmental Panel on Climate Change, World Meteorological Organization, Geneva, <https://www.ipcc.ch/sr15/chapter/spm/>.

Alvarez R, Steinbach HS (2009) A review of the effects of tillage systems on some soil physical properties, water content, nitrate availability and crops yield in the Argentine pampas. *Soil & Tillage Research* **104**, 1–15. doi:10.1016/j.still.2009.02.005

Asmamaw DK (2017) A critical review of the water balance and agronomic effects of conservation tillage under rain-fed agriculture in Ethiopia. *Land Degradation & Development* **28**, 843–855. doi:10.1002/ldr.2587

Assouline S, Mualem Y (1997) Modeling the dynamics of seal formation and its effect on infiltration as related to soil and rainfall characteristics. *Water Resources Research* **33**, 1527–1536. doi:10.1029/96WR02674

Baker JT (2007) Cotton seedling abrasion and recovery from wind-blown sand. *Agronomy Journal* **99**, 556–561. doi:10.2134/agronj2006.0256

Bandaru V, Stewart BA, Baumhardt RL, Ambati S, Robinson CA, Schlegel A (2006) Growing dryland grain sorghum in clumps to reduce vegetative growth and increase yield. *Agronomy Journal* **98**, 1109–1120. doi:10.2134/agronj2005.0166

Baumhardt RL (2003) The Dust Bowl era. In *Encyclopedia of Water Science.* (Eds BA Stewart and TA Howell) pp. 187–191. Marcel-Dekker Inc., New York.

Baumhardt RL, Blanco-Canqui H (2014) Soil: conservation practices. In *Encyclopedia of Agriculture and Food Systems*, Vol. 5. (Ed. N van Alfen) pp. 153–165. Elsevier, San Diego.

Baumhardt RL, Jones OR (2002) Residue management and paratillage effects on some soil properties and rain infiltration. *Soil & Tillage Research* **65**, 19–27. doi:10.1016/S0167-1987(01)00273-2

Baumhardt RL, Lascano RJ (1996) Rain infiltration as affected by wheat residue amount and distribution in ridged tillage. *Soil Science Society of America Journal* **60**, 1908–1913. doi:10.2136/sssaj1996.03615995006000060041x

Baumhardt RL, Schwartz RC (2004) CRUSTS/structural. In *Encyclopedia of Soils in the Environment*, Vol. 1. (Ed. D. Hillel) pp. 347–356. Elsevier, Oxford.

Baumhardt RL, Römkens MJM, Whisler FD, Parlange J-Y (1990) Modeling infiltration into a sealing soil. *Water Resources Research* **26**, 2497–2505. doi:10.1029/WR026i010p02497

Baumhardt RL, Römkens MJM, Parlange J-Y, Whisler FD (1991) Predicting soil-surface seal conductance from incipient ponding and infiltration data. *Journal of Hydrology* **128**, 277–291. doi:10.1016/0022-1694(91)90142-5

Baumhardt RL, Keeling JW, Wendt CW (1993a) Tillage and residue effects on infiltration into soils cropped to cotton. *Agronomy Journal* **85**, 379–383. doi:10.2134/agronj1993.00021962008500020038x

Baumhardt RL, Wendt CW, Keeling JW (1993b) Tillage and furrow diking effects on water balance and yield of sorghum and cotton. *Soil Science Society of America Journal* **57**, 1077–1083. doi:10.2136/sssaj1993.03615995005700040033x

Baumhardt RL, Jones OR, Schwartz RC (2008) Long-term effects of profile-modifying deep plowing on soil properties and crop yield. *Soil Science Society of America Journal* **72**, 677–682. doi:10.2136/sssaj2007.0122

Baumhardt RL, Schwartz RC, MacDonald JC, Tolk JA (2011) Tillage and cattle grazing effects on soil properties and grain yields in a dryland wheat–sorghum–fallow rotation. *Agronomy Journal* **103**, 914–922. doi:10.2134/agronj2010.0388

Baumhardt RL, Johnson GL, Schwartz RC (2012) Residue and long-term tillage and crop rotation effects on simulated rain infiltration and sediment transport. *Soil Science Society of America Journal* **76**, 1370–1378. doi:10.2136/sssaj2011.0331

Baumhardt RL, Mauget SA, Schwartz RC, Jones OR (2016) El Niño Southern Oscillation effects on dryland crop production in the Texas High Plains. *Agronomy Journal* **108**, 736–744. doi:10.2134/agronj2015.0403

Baumhardt RL, Schwartz RC, Jones OR, Scanlon BR, Reedy RC, Marek GW (2017) Long-term conventional and no-tillage effects on field hydrology and yields of a dryland crop rotation. *Soil Science Society of America Journal* **81**, 200–209. doi:10.2136/sssaj2016.08.0255

Baumhardt RL, Dockal JR, Johnson GL, Brauer DK, Schwartz RC (2020a) Controlling stormwater runoff that limits water availability and dryland crop productivity. *Frontiers in Sustainable Food Systems* **4**, 533687. doi:10.3389/fsufs.2020.533687

Baumhardt RL, Marek G, Brauer D (2020b) Trends in runoff from dryland, cropped fields on the Texas High Plains, and implications for their management. *Frontiers in Sustainable Food Systems* **4**, 529319. doi:10.3389/fsufs.2020.529319

Benton L (1987) Interview of Kate Graham on the plow that saved the plains in Faces of Amarillo, 1887–1987 [videorecording-VHS]. Amarillo College Television Production, Amarillo.

Bernstein L, Cooley KR, Dobrenz AK, Kassander AR, Jr, Lepley LK, Massengale MA, Riley JJ, Welchert WT, Wilson LG (1974) *More Water for Arid Lands: Promising Technologies and Research Opportunities.* National Academy of Sciences, Washington, DC.

Blanco-Canqui H, Wortmann CS (2020) Does occasional tillage undo the ecosystem services gained with no-till? A review. *Soil & Tillage Research* **198**, 104534. doi:10.1016/j.still.2019.104534

Blanco-Canqui H, Wienhold BJ, Jin VL, Schmer MR, Kibet LC (2017) Long-term tillage impact on soil hydraulic properties. *Soil & Tillage Research* **170**, 38–42. doi:10.1016/j.still.2017.03.001

Bolton FE, Booster DE (1981) Strip-till planting in dryland cereal production. *Transactions of the ASAE. American Society of Agricultural Engineers* **24**, 59–62. doi:10.13031/2013.34199

Bristow KL, Horton R (1996) Modeling the impact of partial surface mulch on soil heat and water flow. *Theoretical and Applied Climatology* **54**, 85–98. doi:10.1007/BF00863561

Bristow KL, Campbell GS, Papendick RI, Elliott LF (1986) Simulation of heat and moisture transfer through a surface residue–soil system. *Agricultural and Forest Meteorology* **36**, 193–214. doi:10.1016/0168-1923(86)90035-3

Bryant CJ, Krutz LJ, Nuti RC, Truman CC, Locke MA, Falconer L, Atwill RL, Wood CW, Spencer GD (2019) Furrow diking as a mid-Southern USA irrigation strategy: soybean grain yield, irrigation water use efficiency, and net returns above furrow diking costs. *Crop, Forage & Turfgrass Management* **5**, 180076. doi:10.2134/cftm2018.09.0076

Bussière F, Cellier P (1994) Modification on of the soil temperature and water content regimes by a crop residue mulch: experiment and modelling. *Agricultural and Forest Meteorology* **68**, 1–28. doi:10.1016/0168-1923(94)90066-3

Campbell HH (1907) *A Complete Guide to Scientific Agriculture as Adapted to the Semi-arid Regions.* The Campbell Soil Culture Co., Lincoln.

Carter MR (1994) A review of conservation tillage strategies for humid temperate regions. *Soil & Tillage Research* **31**, 289–301. doi:10.1016/0167-1987(94)90037-X

Claassen R, Bowman M, McFadden J, Smith D, Wallander S (2018) 'Tillage intensity and conservation cropping in the United States.' EIB-197. U.S. Department of Agriculture, Economic Research Service, <www.ers.usda.gov/webdocs/publications/90201/eib-197.pdf?v=7027.1>.

Cooper RJ, Hama-Aziz ZQ, Hiscock KM, Lovett AA, Vrain E, Dugdale SJ, Sünnenberg G, Dockerty T, Hovesen P, Noble L (2020) Conservation tillage and soil health: lessons from a 5-year UK farm trial (2013–2018). *Soil & Tillage Research* **202**, 104648. doi:10.1016/j.still.2020.104648

CTIC (Conservation Technology Information Centre) (2002) Tillage type definitions. CTIC, West Lafayette, <https://www.ctic.org/resource_display/?id=322&title=Tillage+Type+Definitions>.

Dabney SM, Shields FD, Binger RL, Kuhnle RA, Rigby JR (2012) Watershed management for erosion and sedimentation control case study Goodwin Creek, Panola County, MS. In *Advances in Soil Science, Soil Water and Agronomic Productivity.* (Eds R Lal and BA Stewart) pp. 539–568. Taylor & Francis, Boca Raton.

Dang Y, Moody P, Bell M, Seymour N, Dalal R, Freebairn D, Walker S (2015a) Strategic tillage in no-till farming systems in Australia's northern grain-growing regions: II. Implications for agronomy, soil and environment. *Soil & Tillage Research* **152**, 115–123. doi:10.1016/j.still.2014.12.013

Dang YP, Seymour NP, Walker SR, Bell MJ, Freebairn DM (2015b) Strategic tillage in no-till farming systems in Australia's northern grains growing regions: I. Drivers and implementation. *Soil & Tillage Research* **152**, 104–114. doi:10.1016/j.still.2015.03.009

Díaz-Zorita M, Duarte GA, Grove JH (2002) A review of no-till systems and soil management for sustainable crop production in the subhumid and semiarid Pampas of Argentina. *Soil & Tillage Research* **65**, 1–18. doi:10.1016/S0167-1987(01)00274-4

Duley FL (1940) Surface factors affecting the rate of intake of water by soils. *Soil Science Society America Proceedings* **4**, 60–64. doi:10.2136/sssaj1940.036159950004000C0011x

Duley FL, Russel JC (1939) The use of crop residues for soil and moisture conservation. *Journal – American Society of Agronomy* **31**, 703–709. doi:10.2134/agronj1939.00021962003100080007x

Emerson WW (1995) Water-retention, organic-C and soil texture. *Soil Research* **33**, 241–251. doi:10.1071/SR9950241

Erenstein O (2002) Crop residue mulching in tropical and semi-tropical countries: an evaluation of residue availability and other technological implications. *Soil & Tillage Research* **67**, 115–133. doi:10.1016/S0167-1987(02)00062-4

Fenster CR (1960) Stubble-mulching with various types of machinery. *Soil Science Society America Proceedings* **24**, 518–523. doi:10.2136/sssaj1960.03615995002400060030x

Finnell HH (1930) The moisture saving efficiency of level terraces under semiarid conditions. *Journal – American Society of Agronomy* **22**, 522–529. doi:10.2134/agronj1930.00021962002200060006x

Flerchinger GN, Saxton KE (1989) Simultaneous heat and water model of a freezing snow–residue–soil system. I. Theory and development. *Transactions of the ASAE. American Society of Agricultural Engineers* **32**, 0565–0570. doi:10.13031/2013.31040

Flerchinger GN, Sauer TJ, Aiken RA (2003) Effects of crop residue cover and architecture on heat and water transfer at the soil surface. *Geoderma* **116**, 217–233. doi:10.1016/S0016-7061(03)00102-2

Follett RF, Stewart CE, Pruessner EG, Kimble JM (2012) Effects of climate change on soil carbon and nitrogen storage in the US Great Plains. *Journal of Soil and Water Conservation* **67**, 331–342. doi:10.2489/jswc.67.5.331

Gerard CJ, Sexton PD, Conover DM (1984) Effects of furrow diking, subsoiling, and slope position on crop yields. *Agronomy Journal* **76**, 945–950. doi:10.2134/agronj1984.00021962007600060019x

Gonzalez-Sosa E, Braud I, Thony J-L, Vauclin M, Bessemoulin P, Calvet CJ (1999) Modelling heat and water exchanges of fallow land covered with plant-residue mulch. *Agricultural and Forest Meteorology* **97**, 151–169. doi:10.1016/S0168-1923(99)00081-7

Govaerts B, Fuentes M, Mezzalama M, Nicol JM, Deckers J, Etchevers JD, Figueroa-Sandoval B, Sayre KD (2007) Infiltration, soil moisture, root rot and nematode populations after 12 years of different tillage, residue and crop rotation managements. *Soil & Tillage Research* **94**, 209–219. doi:10.1016/j.still.2006.07.013

Greb BW (1979) Technology and wheat yields in the Central Great Plains: commercial advances. *Journal of Soil and Water Conservation* **34**, 269–273.

Grismer M, Gates T (1988) Estimating saline water table contributions to crop water use. *California Agriculture* **42**, 23–24.

Hauser VL (1968) Conservation bench terraces in Texas. *Transactions of the ASAE. American Society of Agricultural Engineers* **11**, 0385–0386. doi:10.13031/2013.39417

Heilman JL, Mclnnes KJ, Gesch RW, Lascano RJ (1992) Evaporation from ridge-tilled soil covered with herbicide-killed winter wheat. *Soil Science Society of America Journal* **56**, 1278–1286. doi:10.2136/sssaj1992.03615995005600040045x

Hillel DI, Van Bavel CHM, Talpaz H (1975) Dynamic simulation of water storage in fallow soil as affected by mulch of hydrophobic aggregates. *Soil Science Society of American Proceedings* **39**, 826–833. doi:10.2136/sssaj1975.03615995003900050015x

Horton R, Bristow KL, Kluitenberg GJ, Sauer TJ (1996) Crop residue effects on surface radiation and energy balance – review. *Theoretical and Applied Climatology* **54**, 27–37. doi:10.1007/BF00863556

Jones OR, Baumhardt RL (2003) Furrow dikes. In *Encyclopedia of Water Science.* (Eds. BA Stewart and TA Howell) pp. 317–320. Marcel-Dekker Inc., New York.

Jones OR, Eck HV, Smith SJ, Coleman GA, Hauser VL (1985) Runoff, soil, and nutrient losses from rangeland and dry-farmed cropland in the Southern High-Plains. *Journal of Soil and Water Conservation* **40**, 161–164.

Jones OR, Hauser VL, Popham TW (1994) No-tillage effects on infiltration, runoff, and water conservation on dryland. *Transactions of the ASAE. American Society of Agricultural Engineers* **37**, 473–479. doi:10.13031/2013.28099

Jones JW, Hansen JW, Royce FS, Messina CD (2000) Potential benefits of climate forecasting to agriculture. *Agriculture, Ecosystems & Environment* **82**, 169–184. doi:10.1016/S0167-8809(00)00225-5

Kader MA, Senge M, Mojid MA, Ito K (2017) Recent advances in mulching materials and methods for modifying soil environment. *Soil & Tillage Research* **168**, 155–166. doi:10.1016/j.still.2017.01.001

Kanwar RS, Colvin TS, Karlen DL (1997) Ridge, moldboard, chisel, and no-till effects on tile water quality beneath two cropping systems. *Journal of Production Agriculture* **10**, 227–234. doi:10.2134/jpa1997.0227

Kassam A, Friedrich T, Derpsch R (2019) Global spread of conservation agriculture. *The International Journal of Environmental Studies* **76**, 29–51. doi:10.1080/00207233.2018.1494927

Kozak JA, Aiken RM, Flerchinger GN, Nielsen DC, Ma L, Ahuja LR (2007a) Comparison of modeling approaches to quantify residue architecture effects on soil temperature and water. *Soil & Tillage Research* **95**, 84–96. doi:10.1016/j.still.2006.11.006

Kozak JA, Ahuja LR, Green TR, Ma L (2007b) Modelling crop canopy and residue rainfall interception effects on soil hydrological components for semi-arid agriculture. *Hydrological Processes* **21**, 229–241. doi:10.1002/hyp.6235

Kroes J, Supit I, van Dam J, van Walsum P, Mulder M (2018) Impact of capillary rise and recirculation on simulated crop yields. *Hydrology and Earth System Sciences* **22**, 2937–2952. doi:10.5194/hess-22-2937-2018

Kucharik CJ, Twine TE (2007) Residue, respiration, and residuals: evaluation of a dynamic agroecosystem model using eddy flux measurements and biometric data. *Agricultural and Forest Meteorology* **146**, 134–158. doi:10.1016/j.agrformet.2007.05.011

Lagos LO, Martin DL, Verma SB, Irmak S, Irmak A, Eisenhauer D, Suyker A (2013) Surface energy balance model of transpiration from variable canopy cover and evaporation from residue-covered or bare soil systems: model evaluation. *Irrigation Science* **31**, 135–150. doi:10.1007/s00271-011-0298-9

Lascano RJ, Baumhardt RL (1996) Effects of crop residue on soil and plant water evaporation in a dryland cotton system. *Theoretical and Applied Climatology* **54**, 69–84. doi:10.1007/BF00863560

Lascano RJ, Van Bavel CHM (1986) Simulation and measurement of evaporation from a bare soil. *Soil Science Society of America Journal* **50**, 1127–1133. doi:10.2136/sssaj1986.03615995005000050007x

Lascano RJ, Van Bavel CHM, Hatfield JL, Upchurch DR (1987) Energy and water balance of a sparse crop: simulated and measured soil and crop evaporation. *Soil Science Society of America Journal* **51**, 1113–1121. doi:10.2136/sssaj1987.03615995005100050004x

Lascano RJ, Baumhardt RL, Hicks SK, Heilman JL (1994) Soil and plant evaporation from strip tilled cotton: measurement and simulation. *Agronomy Journal* **86**, 987–994. doi:10.2134/agronj1994.00021962008600060011x

Li Y, Tullberg JN, Freebairn DM (2001) Traffic and residue cover effects on infiltration. *Australian Journal of Soil Research* **39**, 239–247. doi:10.1071/SR00017

Lindstrom MJ, Voorhees WB, Onstad CA (1984) Tillage system and residue cover effects on infiltration in North Western corn belt soils. *Journal of Soil and Water Conservation* **39**, 64–68.

Lowdermilk WC (1930) Influence of forest litter on runoff, percolation, and erosion. *Journal of Forestry* **28**, 474–491.

Machado S, Pritchett L, Petrie S (2015) No-tillage cropping systems can replace traditional summer fallow in north-central Oregon. *Agronomy Journal* **107**, 1863–1877. doi:10.2134/agronj14.0511

McDowell LL, McGregor KC (1984) Plant nutrient losses in runoff from conservation tillage corn. *Soil & Tillage Research* **4**, 79–91. doi:10.1016/0167-1987(84)90018-7

McIntyre DS (1958) Permeability measurements of soil crusts formed by raindrop impact. *Soil Science* **85**, 185–189. doi:10.1097/00010694-195804000-00002

Morin J, Benyamini Y (1977) Rainfall infiltration into bare soils. *Water Resources Research* **13**, 813–817. doi:10.1029/WR013i005p00813

Mueller L, Behrendt A, Schalitz G, Schindler U (2005) Above ground biomass and water use efficiency of crops at shallow water tables in a temperate climate. *Agricultural Water Management* **75**, 117–136. doi:10.1016/j.agwat.2004.12.006

Nuti RC, Lamb MC, Sorenson RB, Truman CC (2009) Agronomic and economic response to furrow diking tillage in irrigated and non-irrigated cotton (*Gossypium hirsutum* L.). *Agricultural Water Management* **96**, 1078–1084. doi:10.1016/j.agwat.2009.03.006

Nyakudya IW, Stroosnijder L (2015) Conservation tillage of rainfed maize in semi-arid Zimbabwe: a review. *Soil & Tillage Research* **145**, 184–197. doi:10.1016/j.still.2014.09.003

Parlange J-Y, Hogarth WL, Parlange MB (1984) Optimal analysis of the effect of a surface crust. *Soil Science Society of America Journal* **48**, 494–497. doi:10.2136/sssaj1984.03615995004800030004x

Peterson E (1948) This plow saves rain: the Graham-Hoeme machine looks like it's here to stay even though its grandpappy was a road scarifier. *Farm Journal* **72**, 89–90.

Phillips ER, Phillips SH (Eds) (1984) *No-tillage Agriculture, Principles and Practices.* Van Nostrand Reinhold Co., New York.

Rawls WJ, Pachepsky YA, Ritchie JC, Sobecki TM, Bloodworth H (2003) Effect of soil organic carbon on soil water retention. *Geoderma* **116**, 61–76. doi:10.1016/S0016-7061(03)00094-6

Riar DS, Ball DA, Yenish JP, Wuest SB, Corp MK (2010) Comparison of fallow tillage methods in the intermediate rainfall inland Pacific Northwest. *Agronomy Journal* **102**, 1664–1673. doi:10.2134/agronj2010.0054

Richards RA, Hunt JR, Kirkegaard JA, Passioura JB (2014) Yield improvement and adaptation of wheat to water-limited environments in Australia – a case study. *Crop and Pasture Science* **65**, 676–689. doi:10.1071/CP13426

Satchithanantham S, Krahn V, Sri Ranjan R, Sager S (2014) Shallow groundwater uptake and irrigation water redistribution within the potato root zone. *Agricultural Water Management* **132**, 101–110. doi:10.1016/j.agwat.2013.10.011

Sauer TJ, Hatfield JL, Prueger JH, Norman JM (1998) Surface energy balance of a corn residue-covered field. *Agricultural and Forest Meteorology* **89**, 155–168. doi:10.1016/S0168-1923(97)00090-7

Scanlon BR, Reedy RC, Baumhardt RL, Strassberg G (2008) Impact of deep plowing on groundwater recharge in a semiarid region: case study, High Plains, Texas. *Water Resources Research* **44**, W00A10. doi:10.1029/2008WR006991

Schillinger WF (2001) Minimum and delayed conservation tillage for wheat–fallow farming. *Soil Science Society of America Journal* **65**, 1203–1209. doi:10.2136/sssaj2001.6541203x

Schwartz RC, Baumhardt RL, Scanlon BR, Bell JM, Davis RG, Ibragimov N, Jones OR, Reedy RC (2015) Long-term changes in soil organic carbon and nitrogen under semiarid tillage and cropping practices. *Soil Science Society of America Journal* **79**, 1771–1781. doi:10.2136/sssaj2015.06.0241

Scopel E, da Silva FAM, Corbeels M, Affholder F, Maraux F (2004) Modelling crop residue mulching effects on water use and production of maize under semi-arid and humid tropical conditions. *Agronomie, EDP Sciences* **24**, 383–395. doi:10.1051/agro:2004029

Sharda VN, Sena DR, Shrimali SS, Khola OPS (2013) Effects of an intercrop-based conservation bench terrace system on resource conservation and crop yields in a sub-humid climate in India. *Transactions of the ASABE* **56**, 1411–1425.

Sharratt B, Campbell GC (1994) Radiation balance of a soil–straw surface modified by straw color. *Agronomy Journal* **86**, 200–203. doi:10.2134/agronj1994.00021962008600010035x

Shipitalo MJ, Edwards WM (1998) Runoff and erosion control with conservation tillage and reduced-input practices on cropped watersheds. *Soil & Tillage Research* **46**, 1–12. doi:10.1016/S0167-1987(98)80102-5

Shipitalo MJ, Owens LB, Bonta JV, Edwards WM (2013) Effect of no-till and extended rotation on nutrient losses in surface runoff. *Soil Science Society of America Journal* **77**, 1329–1337. doi:10.2136/sssaj2013.01.0045

Steiner JL (1989) Tillage and surface residue effects on evaporation from soils. *Soil Science Society of America Journal* **53**, 911–916. doi:10.2136/sssaj1989.03615995005300030046x

Steiner JL (1994) Crop residue effects on water conservation. In *Managing Agricultural Residues.* (Ed. PW Unger) pp. 41–76. CRC Press, Lewis Publishers, Chelsea.

Stewart BA, Baumhardt RL, Evett SR (2010) Major advances of soil and water conservation in the U.S. Southern Great Plains. In *Soil and Water Conservation Advances in the United States.* Special Publication 60. (Eds TM Zobeck and WF Schillinger) *pp. 103–129.* Soil Science Society America, Madison.

Strock JS, Kleinman PJA, King KW, Delgado JA (2010) Drainage water management for water quality protection. *Journal of Soil and Water Conservation* **65**, 131A–136A. doi:10.2489/jswc.65.6.131A

Strudley MW, Green TR, Ascough JC, II (2008) Tillage effects on soil hydraulic properties in space and time: state of the science. *Soil & Tillage Research* **99**, 4–48. doi:10.1016/j.still.2008.01.007

Tanner CB, Shen Y (1990) Water vapor transport through a flail-chopped corn residue. *Soil Science Society of America Journal* **54**, 945–951. doi:10.2136/sssaj1990.03615995005400040001x

Thompson A, Sudduth K (2018) Terracing and contour farming. In. *Precision Conservation: Geospatial Techniques for Agricultural and Natural Resources Conservation.* Agronomy Monograph 59. (Eds JA Delgado, GF Sassenrath and T Mueller) pp. 151–163. American Society of Agronomy,

Crop Science Society of America and Soil Science Society of America, Madison, WI, USA. doi:10.2134/agronmonogr59.2016.0010

Tiessen KHD, Elliott JA, Yarotski J, Lobb DA, Flaten DN, Glozier NE (2010) Conventional and conservation tillage: influence on seasonal runoff, sediment, and nutrient losses in the Canadian prairies. *Journal of Environmental Quality* **39**, 964–980. doi:10.2134/jeq2009.0219

Unger PW, Baumhardt R (1999) Factors related to dryland grain sorghum yield increases: 1939 through 1997. *Agronomy Journal* **91**, 870–875. doi:10.2134/agronj1999.915870x

Unger PW, Parker JJ (1976) Evaporation reduction from soil with wheat, sorghum, and cotton residues. *Soil Science Society of America Journal* **40**, 938–942. doi:10.2136/sssaj1976.03615995004000060035x

Unger PW, Wiese AF (1979) Managing irrigated winter wheat residues for water storage and subsequent dryland grain sorghum production. *Soil Science Society of America Journal* **43**, 582–588. doi:10.2136/sssaj1979.03615995004300030030x

Unger PW, Baumhardt RL, Arriaga FJ (2012) Mulch tillage for conserving soil water. In *Advances in Soil Science.* (Eds R Lal R and BA Stewart) pp. 427–453. Taylor & Francis, Boca Raton.

US Department of Agriculture, Natural Resources Conservation Service (USDA NRCS) (2017a) Residue and tillage management no-till/strip-till (329) and reduced-till (345). Iowa job sheet. USDA NRCS, Des Moines, <https://efotg.sc.egov.usda.gov/references/public/IA/Residue_and_Tillage_Management_329_345_JS_2017_06.pdf>.

US Department of Agriculture, Natural Resources Conservation Service (USDA NRCS) (2017b) Soil Tillage Intensity Rating (STIR) for different cropping systems and different tillage systems. USDA NRCS, Des Moines, <https://www.nrcs.usda.gov/wps/portal/nrcs/detail/null/?cid=nrcs144p2_056365>.

US Department of Agriculture, Natural Resources Conservation Service (USDA NRCS) (2017c) Tillage equipment: pocket identification guide. USDA NRCS, Des Moines, <https://www.nrcs.usda.gov/Internet/FSE_DOCUMENTS/nrcseprd1323703.pdf>.

Van Bavel CHM, Hillel DI (1976) Calculating potential and actual evaporation from a bare soil surface by simulation of concurrent flow of water and heat. *Agricultural Meteorology* **17**, 453–476. doi:10.1016/0002-1571(76)90022-4

Wakindiki IIC, Ben-Hur M (2002) Soil mineralogy and texture effects on crust micromorphology, infiltration, and erosion. *Soil Science Society of America Journal* **66**, 897–905. doi:10.2136/sssaj2002.8970

Wiese AF, Burnett E, Box JE, Jr (1967) Chemical fallow in dryland cropping sequences. *Agronomy Journal* **67**, 175–177. doi:10.2134/agronj1967.00021962005900020016x

Williams JD, Wuest SB, Long DS (2014) Soil and water conservation in the Pacific Northwest through no-tillage and intensified crop rotations. *Journal of Soil and Water Conservation* **69**, 495–504. doi:10.2489/jswc.69.6.495

Wilson TB, Meyers TP, Kochendorfer J, Anderson MC, Heuer M (2012) The effect of soil surface litter residue on energy and carbon fluxes in a deciduous forest. *Agricultural and Forest Meteorology* **161**, 134–147. doi:10.1016/j.agrformet.2012.03.013

Winter SR, Unger PW (2001) Irrigated wheat grazing and tillage effects on subsequent dryland grain sorghum production. *Agronomy Journal* **93**, 504–510. doi:10.2134/agronj2001.933504x

Wu Y, Liu T, Paredes P, Duan L, Pereira LS (2015) Water use by a groundwater dependent maize in a semi-arid region of Inner Mongolia: evapotranspiration partitioning and capillary rise. *Agricultural Water Management* **152**, 222–232. doi:10.1016/j.agwat.2015.01.016

Zingg AW, Hauser VL (1959) Terrace benching to save potential runoff for semi-arid land. *Agronomy Journal* **51**, 289–292. doi:10.2134/agronj1959.00021962005100050013x

Zipper SC, Soylu ME, Booth EG, Loheide SP, II (2015) Untangling the effects of shallow groundwater and soil texture as drivers of subfield-scale yield variability. *Water Resources Research* **51**, 6338–6358. doi:10.1002/2015WR017522

5

Crop diversification and soil water

Aaron Lee M. Daigh

Introduction

Sustaining both biodiversity and freshwater resources is one of the most pressing issues humankind currently faces. The management of biodiversity and freshwater resources on the world's 1.4 billion ha of arable cropland (i.e. 11% of the total land area) is paramount in the face of a changing climate and a vastly growing and hungry human population (World Bank Open Data 2020a). However, mechanisation (Isbell *et al.* 2019) and larger farms (Rodríguez and Wiegand 2009; Martin *et al.* 2020) decrease plant species diversity and overall biodiversity. Meanwhile, the current world population of 7.67 billion (World Bank Open Data 2020b) requires a total of 9.20 trillion m^3 of water annually to produce our desired daily diets (Falkenmark and Rockström 2004), not including what currently goes to food waste or the expected future increase in adequate daily calorie intake (Davis *et al.* 2014). As a result, agricultural food production requires >85% of our annual fresh water consumption (Falkenmark and Rockström 2004; D'Odorico *et al.* 2020). Therefore, of great interest, is the question, 'How does diversification of crops interact with the water in agricultural landscapes, namely in the soil where plants acquire their water?'

This chapter discusses the implications of crop species diversification via the use of cover crops, diversified crop rotations, intercropping, relay cropping and alley cropping on soil water, hydraulic properties and the water balance. While doing so, there is an emphasis on water use efficiency (WUE) for cover crops in regions with limited and abundant precipitation. Insights are also provided on the adoption of cover crops and other diversified cropping systems with reduced or no-till systems for managing soil water.

Growing body of literature

From 1990 to 2020, there has been tremendous growth in the scientific literature on crop diversification and soil water. Clarivate Analytics' Web of Science, in its Core Collection (https://clarivate.com/webofsciencegroup/solutions/web-of-science-core-collection/), shows over 13 000 published scientific articles on the topic (Plate 4). Since 2012, researchers have added over 500 new contributions each year, with a rate exceeding 1000 contributions in 2019. The largest percentage of these works (30%) comes from the US, with another 12–3% each coming from, in descending order, China, Australia, Canada, India, Germany, Brazil, France, Spain, Italy and England (https://clarivate.com/webofsciencegroup/solutions/web-of-science-core-collection/). However, Clarivate Analytics' Web of Science is only one database. A more inclusive yet less transparent search in Google Scholar using the same terms listed in Plate 4 shows nearly 18 000 documents.

With over 13 000 articles and many other relevant works, a full, encompassing and exhaustive coverage of the global literature is beyond the scope of this chapter. Instead, it focuses on large-scale production systems where crop species diversity is at its lowest state. In doing so, it provides examples and situations in arid to humid regions and hot to frigid climates around the world. With such a rich body of literature, generalisations are made where the literature is consistent, but innately vast omissions are left.

Crop species diversity in agroecosystems

Crop species diversity tends to be a function of the scale of crop production. In general, species diversity per unit of land decreases as the crop production system increases in scale. Large-scale systems include commercial food production and commodity crops, whereas small-scale systems can include subsistence farms (i.e. the producer's primary food source with some potential for local trade or sell), consumer horticulture (e.g. home gardens, kitchen gardens, home agroforestry) and community gardens (Kaihura and Stocking 2003; Vanlauwe *et al.* 2014; Whitney 2018). There are also systems at intermediate scales, including cooperative produce farms and part-time (or so-called 'hobby') production farms (Katchova 2010; Anigbogu *et al.* 2016; Prylipko *et al.* 2016).

In small-scale systems, a relatively short distance is typical between the crop's production and its consumption or end user. Communities that rely on these systems innately need a variety of foods and fibre to become available throughout the year (Johnson 2009; Abebe *et al.* 2010; Mavengahama *et al.* 2013). Therefore, crop species diversity is relatively high and governed by community needs and preferences (Zaldivar *et al.* 2002). In contrast, large-scale mechanised commercial systems harvest only one or, in some geographic regions, two or three crops annually. The need for mechanisation increases as these farms become larger and the shipping distance for harvested products become greater. The costs and size of specialised farm equipment (e.g. tractors, tillage implements, planters, agrochemical applicators, combines) to prepare and harvest large areas efficiently, so as to prevent harvest and post-harvest losses, tend to decrease the variety of crops grown on a given farm. Although crop diversity decreases, the economies-of-scale associated with large equipment and vast sections of land allow for greater economic returns for producing only a few crops (Key 2019; Vitale *et al.* 2019). Thus, the largest of farms in mechanised regions tend to be those with the lowest diversity of crops.

Large-scale production systems

Large-scale farms are typically located next to other similar farms of scale, in oftentimes near-continuous expanses of land, on every continent with arable land (Kassam *et al.* 2019). Of all the scales and types of cropping systems that currently exist around the world, these are innately the regions of the lowest, or largest drop in, crop diversity in terms of both their spatial and temporal characteristics (e.g. monocropping with only a single crop and a fallow period each year; Abebe *et al.* 2010; Aguilar *et al.* 2015). Over the past 30 years, crop diversity in these areas has continued to decrease, while also clustering together to form large regions of homogeneous crop production (Aguilar *et al.* 2015). Aguilar *et al.* (2015) report that some areas within the continuous US are experiencing an increase in crop species diversity. However, these areas are fewer and smaller in size than regions that are decreasing in crop diversity. In central and southern Brazil, the opposite has been occurring since the early 2000s (DePaula and Fortes 2019). Nevertheless, tremendous amounts of interest and effort have recently gone into innovating new ways to increase crop diversity in these large systems, as will become evident throughout this chapter (Malézieux *et al.* 2009). Given the expansiveness of these operations, the efforts to diversify crops can have significant implications for water on the landscape and subsequent feedbacks on a crop's survival, productivity and yields.

Soil water content and crop yields in diversified agroecosystems

Positive, neutral or negative effects?

All crops take up water from the soil and transpire. After the crop's senescence or termination, the remaining residue serves as a mulch to the soil surface. Therefore, the life cycle of any crop also has a cyclical effect of reducing and conserving soil water (Bodner *et al.* 2015). The success of one crop's life cycle can also alter the network of soil pores and their effective water conductivities (Lu *et al.* 2020). This creates a feedback loop on how rainwater can effectively infiltrate, be retained within soil aggregates and then be available for subsequent crops.

Different crops may exist at various levels of 'closeness', with some growing immediately next to others (e.g. centimetres apart), in patches, alleys and corridors (e.g. metres to decametres apart), whereas others are in neighbouring fields (e.g. hectometres to kilometres apart). All of these may occur in coordinated rotational phases. Crop species diversity has gone through cycles of regression and retroregression and is now emerging into numerous levels of management practice complexities (e.g. cover crops, integrated crop–livestock systems, restored prairies for bioenergy feedstock crops) so as to leverage species diversity for maximum ecosystem services while balancing agronomic profits (Davis *et al.* 2012; Sanderson *et al.* 2013; Aguilar *et al.* 2015; Isbell *et al.* 2017). The above- and below-ground morphology of each crop affects how it responds to a field's antecedent moisture and drainage conditions (Fidantemiz *et al.* 2019; Nichols *et al.* 2019) and subsequent impacts on soil water. If tillage is used in the system, then the mode, depth and aggressiveness will also have contributing effects (Daigh and DeJong-Hughes 2017; O'Brien and Daigh 2019). The spatial and temporal arrangement of species within a cropping system combined with the tillage system has a major influence on soil hydrology and water balance dynamics.

So, what are the mechanisms behind positive, neutral and negative effects of a crop on soil water content? The negative effect is the most intuitive. The crop roots must take up water to supply their metabolic demands through a passive process where potential energy gradients drive a flux of water through the soil–plant–atmosphere continuum. This effectively removes water from the soil.

The most notable positive effect can develop with dead crop residues that are not tilled into the soil, where they are left on the soil surface as a mulch. The thickness of the crop residue layer governs water vapour diffusivity more than the type and morphology of the crop residue (Steiner 1989; Flury *et al.* 2009), with thicker residue layers limiting evaporative losses more efficiently than thin layers with full or partial cover (Chung and Horton 1987; McMaster *et al.* 2000). This results in the conservation of soil water, and thus a positive effect. Moreover, if the management system is able to significantly increase the soil's organic matter over subsequent cropping seasons, then the water retention properties of the soil may also increase (Poeplau and Don 2015; Basche *et al.* 2016). Therefore, both short- and long-term positive effects on storing soil water can be observed in a cropping system if the tillage regime and erosional status also facilitate these processes.

The positive effects from a living crop may not be as obvious as those of the dead crop because the crop is still taking up water. However, when the crop is transpiring, the above-ground biomass also provides leaf cover of the soil surface, thereby providing shade, lowering incident radiant energy, lowering soil surface temperature and lifting the atmospheric boundary conditions in which convective and turbulent air flows are damped to become more like a vapour diffusing dominated zone within the canopy (Shuttleworth and Wallace 1985; Horton 1989; Sauer *et al.* 2007; Kool *et al.* 2014). The aerodynamics among living and dead crop canopies are also distinct. Full stands of dead canopies behave more like a sparsely covered rough soil surface than that of a living canopy, which has substantially higher drag coefficients and momentum roughness lengths (Sauer *et al.* 1996). This cascade of processes reduces the amount of evaporation from the soil. The reduction in soil evaporation can be larger than the crop's transpiration while the crop is still relatively young with low to moderate biomass. When this drop in evaporation is larger than transpiration rates, a positive effect on soil water occurs (Daigh *et al.* 2014a).

When neutral effects are observed, it may be the case that a soil is at the intersection of the processes described above. Therefore, a cancelling or interference effect provides what appears to be a neutral outcome on soil water. In the case of cover crops, there simply may be so little plant growth that any differences are below detection limits and negligible. However, an alternative process can also be occurring. For instance, if soil water is taken up by plants at a rate that should induce a negative effect on soil water content but that rate is similar to the soil's unsaturated hydraulic conductivity, then the water removed by roots may be readily replaced by deeper soil water. For this to occur, an ample supply of water needs to be present within the soil's pore characteristic length (i.e. a shallow water table; Or *et al.* 2015). In this situation, the cropping system is undergoing a dynamic series of water transport processes even though sensors or physical samples for soil water contents would suggest a static system.

Cover crops

Cover crops are well known to have positive, negative and neutral effects on soil water in the surface of agricultural fields. These effects often interact with the frequency of precipitation, vapour deficits and previous crop management (Daigh *et al.* 2014a, 2014b; Bodner *et al.* 2015; Basche *et al.* 2016; Rankoth *et al.* 2019). However, most studies suggest that positive or neutral effects are the norm, as described below. In general, this appears to be due to the fact that cover crops are innately grown during periods of the year when growing conditions are less than ideal. Therefore, the cover crops tend to not live long enough to maximise transpiration rates.

However, negative effects on soil water occur when the cover crop biomass or leaf area exceeds some threshold (i.e. when transpiration rates likely exceed the soil hydraulic conductivity for redistributing deeper water to the surface). Literature on these effects can easily be found from experimental sites across the world (see below), such as in the US (Western, Great Plains, Midwestern and Eastern regions), throughout Europe, in the southern interior of South America, in South and East Asia and in Australia. Many researchers have reported that when a cover crop exceeds 1–3 Mg biomass ha^{-1}, then the cover crop starts taking up enough soil water to have a negative effect and may be detrimental to the subsequent cash crop (Clark *et al.* 1991; Krueger *et al.* 2011; Daigh *et al.* 2014a, 2014b). For example, in the Midwestern US, Qi *et al.* (2011) reported that rye cover crops terminated before 1 Mg biomass ha^{-1} had a slightly positive effect on soil water and no effect on the subsequent maize and soybean yields. In contrast, a cover crop of kura clover produced 3.8 Mg biomass ha^{-1} and had a negative effect on soil water and reduced maize yields by 71% (Qi *et al.* 2011). Similarly, in eastern Australia, Whish *et al.* (2009) reported that French white millet cover crops terminated before 50% soil cover had no detrimental consequences on soil water and the subsequent wheat cash crop, but slightly improved both outcomes.

If the cover crop biomass is limited or controlled (i.e. chemical or physical termination by the farmer) to be below this threshold, then neutral or positive effects on soil water and the subsequent cash crop yields are typically obtained (Qi *et al.* 2011; Daigh *et al.* 2014a; Liebig *et al.* 2015; Alvarez *et al.* 2017; Kaye and Quemada 2017; Marcillo and Miguez 2017; Barker *et al.* 2018; Meyer *et al.* 2020). If the cover crop is terminated close to this threshold and allowed to remain on the soil surface as a mulch, then a subsequent positive effect on soil water emerges (Daigh *et al.* 2014a; Basche *et al.* 2016). If the cover crop is ploughed under as a green manure, then the mulching benefits on soil water are negated (Meyer *et al.* 2020). If crops are irrigated, such as many regions with Mediterranean climates, then a preceding cover crop's negative effect on soil water is mitigated (Shackelford *et al.* 2019). In addition, residue retention in hot dry Mediterranean climates (i.e. western US, western Australia, southern Europe) and hot humid tropical climates (i.e. southern Brazil) can limit the duration that terminated cover crop residues cover the surface before decomposing. In these situations, the conservation of water by a mulching effect will be minimal, if it exists at all, and leave no notable effect on the soil water balance and subsequent crops (Ward *et al.* 2012; Teixeira *et al.* 2014). In addition, grasses are likely preferred to legume cover crops when the goal is water conservation. Grasses accumulate biomass quicker than legumes and provide greater ground cover to reduce evaporation and soil erosion. In contrast, legumes need to be left longer in the field to acquire the desired biomass, which extends the duration that the plant is transpiring. This leads to legumes (e.g. field pea, faba bean, common vetch) inducing lower soil water contents by the time they can provide adequate ground cover (Wunsch *et al.* 2017). Although legumes provide nitrogen to the soil, this means that the residue decomposes quicker than that of grasses and the nitrogen concentrations are oftentimes too little to justify the depletion of water. Therefore, if legumes are desired, then planting them in a strategic mix with grasses is preferred so that they can leverage multiple benefits.

The precise amount of cover crop biomass needed for these disparities is influenced by the weather a region experiences, the amount and thickness of existing residue from the preceding crops, the site's hydraulic properties and drainage class (Baumhardt and Jones 2002; Mulumba and Lal 2008; Daigh *et al.* 2014a; Liebig *et al.* 2015) and the cover crop's WUE and functional traits for water competition or symbiosis (i.e. concentration reduction vs pre-emption strategies, root-shoot signalling, hydraulic lift). When soil water is significantly affected by cover crops, most reports in the scientific literature (i.e. all references in the above paragraphs) list differences within 0.02–0.10 $m^3\ m^{-3}$ of the control plots that did not have cover crops. The greatest change in soil water typically occurs in medium-textured soils, where the bounds for plant-available water (i.e. the difference between field capacity and permanent wilting point) is also the largest, in warm climates and in landscapes with good drainage. The smallest change in soil water occurs in both sand and clay soils due to the innately lower total water retention capacity and unsaturated hydraulic conductivities respectively.

In most situations across the world, cover crops are grown in sequence with one cash crop, where the two mimic a two-phase crop rotation. The caveat lies in the fact that cover crops do not complete their lifecycles (i.e. they freeze, are desiccated or where they are intentionally terminated at an early growth stage) and are innately grown during non-ideal conditions when fields would otherwise have been fallow. Therefore, they never reach their maximum transpiration rates. However, there is a large variety of new intercropping systems for cover crops emerging that require investigation (see below).

Crop rotations

Over successive years, the diversification of full-season crop rotations will affect soil water content and the amount of water available to subsequent crops. The crops in a rotation may successively drawdown and deplete deep or shallow zones of soil water depending on whether they deploy a concentration reduction or pre-emption strategy respectively, as well as their WUE. Plants that deploy a concentration reduction strategy can tolerate lower water potential energies than their competitors, even though their water uptake rates are relatively equal. Plants that deploy a pre-emption strategy have a high root density at shallower-than-optimum depths, so to take up available water at a faster rate than their competitors. These strategies on plant competition traits for water are further described in Craine *et al.* (2005) and Craine and Dybzinski (2013). Because plants take up water close to the root tips, the zone where they tend to have the highest density will be where a systematic drawdown of soil water will be the highest, and this will depend on the root's stage of ontogenesis and how long the roots are allowed to take up water. Because full-season crops will take up more total water than cover crops, as described above, the rainfall to replenish soil water is essential. The frequency, amount and intensity of a region's subsequent rainfall will either completely alleviate, somewhat moderate or leave the soil at risk for low water supplies to the subsequent crop (Manns and Martin 2018). Therefore, the success of each crop phase in a rotation depends on having enough water in the soil depths where the roots (e.g. seedlings and mature plants) are taking up water.

Water use from one phase of a cropping system can impact the water available for the subsequent crop (Anderson 2005), even if there is no rotation at all. In systems that are water limited, the crop yields can serve as a 'sensor' for the growing season's available water. For example, the long-term continuous monocrops of annual spring wheat in semi-arid south-west Saskatchewan, Canada, are dependent on timely rainfall to replenish soil water that was depleted during the previous year (He *et al.* 2013). When these rainfall events are absent, wheat yields can be reduced by more than 60%. Similar to the spring wheat in Canada, continuous monocrops of winter wheat in the semi-arid Central Great Plains in the US are also dependent on timely precipitation. Farmers use a fallow season to allow replenishing of soil water and to reduce the variability among the subsequent crops. However, leveraging the characteristics of a diversified crop rotation can improve water relationships and help stabilise crop yields. By diversifying the crop rotation with maize, proso millet and sunflowers, the annual grain production can increase 60–100%, as well as boost winter wheat yields over time (Anderson *et al.* 1999). Strategic placement of each crop phase is important for water consumption, as well as other variables, including nutrient management and disease vectors. For example, winter wheat yields can be stabilised the most over time by just introducing the low-water-use millet into a two-phase rotation. However, the wheat yields can be maximised by having a four-crop phase rotation. If sunflower is included in these rotations, it can have a detrimental consequence on the water use of subsequent crops, particularly if the sunflower phase experienced infrequent rainfall, because of its high water requirements. Therefore, placing a high WUE crop after sunflower is needed to stabilise crop yields. In the frigid soils of the Northern Great Plains of the US, Lenssen *et al.* (2018) observed increased soil water and crop WUE in diversified pea rotations that included spring wheat and forage barley. Similar trends are well documented in the wheat-based systems of Western Australia, where break crops (grains, legumes, alfalfa and various forms of pasture) between the wheat phases result in the better use of water, nutrients, disease and weed suppression, as well as in improved yields (French *et al.* 2015; Malik *et al.* 2015). In the semi-arid Loess Plateau of China, Huang *et al.* (2003) also observed that diversified wheat rotations with maize, pea and broomcorn millet had no effect on soil water at the time of wheat planting, although the diversified rotation significantly increased annual yields and crop WUE.

Across Brazil's central grain belt, the once monocropping of soybean hectares has been largely converted into a double-cropping system, with the sowing of soybean during the rainy summers and maize during the dry winters (DePaula and Fortes 2019). In some areas, a third forage crop or cover crop, with grasses such as signal grass (*Brachiaria brizantha*), Congo grass (*Brachiaria ruziziensis*) and pearl millet (*Pennisetum americanum* (L.) Leeke) or legumes such as jack bean (*Canavalia ensiformes* (L.) DC) and sunn hemp (*Crotalaria juncea* L.), is also included to conserve soil from erosion (Teixeira *et al.* 2014; Garbelini *et al.* 2020). However, many of the drier and sandier areas of the north-eastern region of Brazil lack the available soil water to make the economics balance for the double cropping systems. In the wet clayey soils (e.g. southern Brazil and extending down into Paraguay and northern Argentina) that retain adequate water, these agronomic advancements allowed producers to forgo tillage and other traditional means for weed control so that the second crop could be sowed in time before the weather became too dry during the winter months.

Anderson *et al.* (2017) cautioned that these double-cropping systems may leave these regions vulnerable to the effects of the El Niño South Oscillation (ENSO) because these crops are highly dependent on the region's climate to replenish soil water. In their analysis, Anderson *et al.* (2017) suggested that the ENSO could jeopardise the success of the second crop phase. If true, then the shorter-season variety of the first crop phase would innately produce lower total yields than the longer-season varieties used in monocropping. Because crop yields are a linear or curvilinear function of cumulative crop transpiration (Purcell *et al.* 2007), the shorter-season varieties innately produce lower yields. In the states of Victoria and South Australia, the climate risks to cropping systems are of paramount importance when deciding on crop rotations and break crops between wheat phases (Whitbread *et al.* 2015). In these areas, annual crop failure due to low precipitation is not uncommon, and the economic outcome of these systems balance on making the most out the good years and making sure the soil conditions are leveraged as much as possible when these years do occur (Whitbread *et al.* 2015). Therefore, short-term (i.e. one to three cycles of the rotation) and long-term (i.e. decadal) planning is required to ensure sustainability of the cropping system in terms of available soil water resources and crop water requirements.

Intercropping and relay cropping

Intercropping is the intentional cultivation of two or more crops simultaneously in relatively close proximity (e.g. at some scale within a single field). Intercropping is commonly used in two ways: (1) all crops provide a harvestable product; or (2) some crops provide a harvestable product, whereas the remaining crops provide some agronomic, environmental or ecological service. Examples of the former include alternating sets of rows (e.g. one set being one to two passes of the farmer's planter) planted to a grain, legume, forage or trees (e.g. nuts, fruit, syrup, timber). Examples of the latter include cover crops planted in the inter-rows of a cash crop or a nurse/companion crop to help establish a semiperennial cash crop (e.g. oats planted as a nurse crop with alfalfa).

Relay cropping is a blending of intercropping and crop rotation (i.e. double cropping). Essentially, the second crop is interseeded into a pre-established, yet unharvested, first crop (Tanveer *et al.* 2017). In frigid soils with a short growing season, it is often recommended that cover crops are interseeded into an established cash crop (Bich *et al.* 2014; Wick 2018; Johnson *et al.* 2021). Although the phrase 'relay cover crops' is not found in the literature, this precisely describes interseeded cover crops. In addition, when a perennial crop or permanent groundcover system is interseeded with rows or alleys of an annual or semiperennial crop, these are practically relay cropping systems too. However, these are more often referred to as a form of permaculture, agroforestry or alley cropping.

The variety of intercropping and relay crop systems is vast. However, the trends among these systems, in terms of their effects on soil water and crop water use, are similar to those of the cover crops and crop rotation systems covered in the previous sections. In general, the diversified crop system has either a neutral or positive effect on soil water and crop water use, as long as the logistics and placement of the crops are strategised (Rahman *et al.* 2017; Tan *et al.* 2020). Below are just a few interesting examples of the wide range of possibilities for intercropping and relay crop systems.

In the Midwestern US, there has been a growing interest in the use of permanent groundcover with relays of annual row crop systems interseeded into bare strips. Little actual field research has been performed on this topic. However, Bartel *et al.* (2020) used a calibrated numerical simulation of permanent groundcover (i.e. Kentucky bluegrass and creeping red tall fescue) with a relay of maize interseeded annually into bare strips of soil. A rough visualisation of these systems can be imagined if you consider a typical maize monocrop with sod turf grown permanently between the rows of maize. In their simulations, Bartel *et al.* (2020) observed that these systems were able to provide many positive ecosystem services while having little expected effects on water competition to the maize crop. This was the case as along as the permanent groundcover was adequately suppressed to avoid water competition late in the maize life cycle. This is representative of many intercropping and relay systems. As along as the competition for water resources is low during one crop's life stage that requires maximum transpiration (i.e. reproductive stages), then negative consequences may be avoided. In the case of relayed annual crops, where two crops are grown in the same year, the relay crop must not be planted too early in order to avoid: (1) large amounts of water uptake that compete with the previously established and more mature crop; and (2) the relay crop becoming water stressed and stunting growth as it progresses along its own growth stages. If the relay crop is planted too late, then adequate soil water may not be available if the annual rainfall becomes infrequent. Therefore, precise timing, plant densities and row spacing are needed to facilitate maximum yields, while also allowing for adequate water, nutrient and light resources to be temporarily shared among the two crops.

However, some systems do not compete for the same water resources, but behave in a symbiotic manner. Select plant species have a trait that allows them to extract water several metres below the soil surface during the day. Then, during the night, when the hydraulic gradient diminishes, a portion of the lifted water that is still contained within the roots is extruded into the dry shallow soil (i.e. 'hydraulic lift'). As described in Bogie *et al.* (2018, 2019), this results in a symbiosis among plants species and is a sort of bioirrigation. In the Sahel, Bogie *et al.* (2018, 2019) observed that millet yields increased 900% solely due to the hydraulic lift and bioirrigation of the shrub *Guiera senegalensis* that was intercropped with the millet. This process is suspected to regularly occur in many crops (e.g. alfalfa, pearl millet, maize, tomatoes), various semi-arid and arid shrubs (e.g. sagebrush, *G. senegalensis*) and many types of trees and palms (e.g. sugar maple, apple trees, date palm; Horton and Hart 1998; Bogie *et al.* 2018). The symbiosis occurs because the two species do not compete for the same age of water. Deeper water is older than shallow water. If the two crops are taking up the different-age waters during the day, then no competition occurs. Because plants growing in dry conditions tend to increase their root densities once they find deep water (Fidantemiz *et al.* 2019; Nichols *et al.* 2019), they can lift ample supplies of water to the surface.

Agroforestry alley cropping is another system that diversifies the growing of agricultural products and ecosystem services, and improves water resources for a piece of land (Ilstedt *et al.* 2007; Kuyah *et al.* 2016; Wolz *et al.* 2018). For example, the pairing of species may include Port Jackson willow with sorghum, black walnut or poplar for timber with maize, soybean or wheat, or polycultures of chestnut, hazelnut, black currant and elderberry with alfalfa, among others (Lehmann *et al.* 1998; Lovell *et al.* 2018; Wolz *et al.* 2018). In these systems, the tree hedgerows extend roots deep into the subsoil and then laterally underneath the alley where the crops are grown (Lehmann *et al.* 1998; Bargués Tobella *et al.* 2017; Bayala and Prieto 2020). The trees acquire most of their water by drawing down from the subsoil, whereas the crop acquires most of its water from the topsoil and upper portion of the subsoil. Although water competition among the trees and crop can occur (Jose *et al.* 2000), and the dynamic differs among wet and dry seasons (Bargués Tobella *et al.* 2017), older groves of trees will progressively take up water from deeper depths until old water becomes their primary source (Zhang *et al.* 2017a; Wolz *et al.* 2018). When this occurs, competition among the two species likely ceases. In addition, in the meta-analysis by Ilstedt *et al.* (2007), the infiltration capacity of agricultural fields increased 300% when agroforestry systems were introduced to the landscape. Therefore, a balance between enhanced infiltration of rainwaters and competition for water resources is likely in a strategically designed system.

Root system effects on soil pore networks and hydraulic properties

Over successive years, the diversification of crop species will impact soil water content. Two processes are responsible for this: (1) the accumulated drawdown of soil water due to plant uptake, as noted above; and (2) the systematic change in soil hydraulic properties by each crop's root growth characteristics. These processes occur in sequence and are dynamic with the crop's life stage. Crop roots physically change soil. Most notably, these changes may be expressed as alterations in the macroporosity, connectivity and length of preferential flow channels, aggregation, organic matter content, rhizosphere wettability and total porosity. These physical and chemical changes have emergent effects on processes such as infiltration, transpiration and drainage.

The literature is very clear on the effects of crops, and plants overall, on soil physical and hydraulic properties. A soil with a history of little to no crop diversity (i.e. bare ground and/or a monocrop system) will gain greater distributions of pore sizes, continuity and connectivity, and range of individual pore topologies (i.e. fractal dimensions), when the farmer increases the diversity of root system architectures (i.e. diversity of crop species with fibrous and tap root morphologies, concentration reduction and pre-emption strategies and root–shoot signalling responses to environmental variables) grown in that soil (Dabney 1998; Unger and Vigil 1998; Udawatta *et al.* 2008; Ilstedt *et al.* 2007; Kumar *et al.* 2008; Skaggs and Shouse 2008; Udawatta and Anderson 2008; Bengough 2012; Hallett *et al.* 2013; Daigh *et al.* 2014b, 2018a, 2020; Basche *et al.* 2016; Hunt 2016; Basche and DeLonge 2017, 2019; Manns and Martin 2018; Acharya *et al.* 2019; Alhameid *et al.* 2019; Lu *et al.* 2020; Meyer *et al.* 2020). A greater number of cycles of diverse plants will lead to more complex and stable soil pore networks, until a new equilibrium is reached (Fig. 5.1).

The only question is the degree of change that will occur. Innately, this is governed by the antecedent state of the soil pore network (i.e. initial conditions), the degree of differences in the new root system architectures compared with the previous crops and, perhaps most importantly, the degree of growth the new crops can achieve (e.g. partial or stunted growth before termination or death vs reaching full rigorous maturity or completing its life cycle). In turn, soils gain greater accessibility to infiltrate and drain water when needed, retain water when needed and bear heavy loads when needed. For example, Basche and DeLonge (2017), Ilstedt *et al.* (2007) and Blanco-Canqui *et al.* (2011) reported that the infiltration capacity of plots with cover crops can increase by >300% compared with plots without cover crops. Similarly, Blanco-Canqui *et al.* (2015) reviewed cover crop studies in temperate climates and noted that cover crops can infiltrate and capture up to 80% more of run-off water. In addition, Daigh *et al.* (2014b) reported that water stored for crop use that otherwise would have gone to drainage increased by 37–46% in plots with cover crops. Moreover, many farmer anecdotes can be heard around the world, along with experimental evidence, that enhanced crop diversity creates soils that are strong and hold up their agricultural equipment better than fields that have been monocropped and aggressively tilled for residue management (Chamen *et al.* 2003; Blanco-Canqui *et al.* 2011). These characteristics and their effects on soil water are not only gained in the vertical direction,

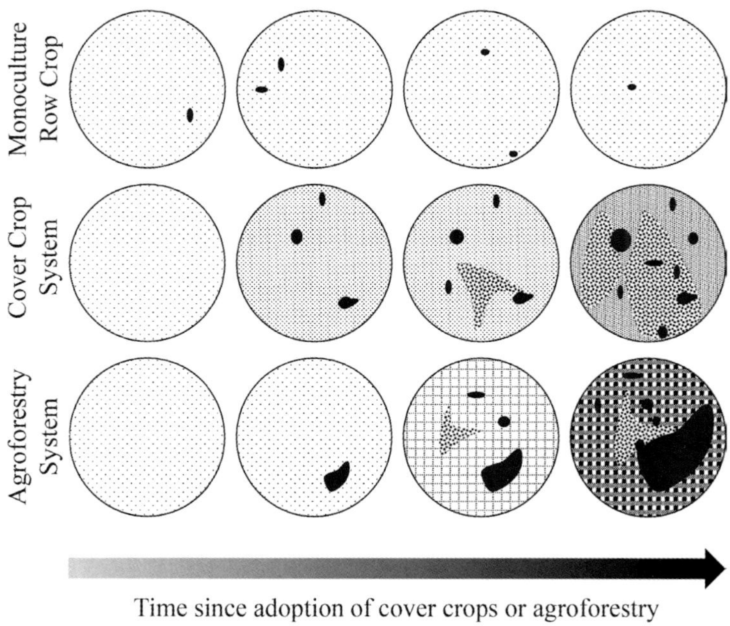

Fig. 5.1. Conceptualisation of soil pore development under monoculture row crops, cover cropped systems and agroforestry systems. White areas within each circle are soil minerals and the black areas are air-filled pores.

but are also distinct horizontally across the fields, where plants may create small, localised mosaic zones in the soil where their root systems modulate soil hydraulic properties and water contents (Daigh *et al.* 2018b). However, small-scale local changes in the subsoils have a magnifying effect that emerge when the processes are scaled up to the watershed level (Dabney 1998).

These desirable pore networks are dependent on the reduction, or elimination, of mechanical disturbances (i.e. tillage and wheel traffic compaction). Mechanical disturbance to these root-derived pore networks disrupts their positive effects on the cropping system (i.e. crop productivity, resource use efficiency and farm inputs and logistics) and ecosystem services (i.e. water and air quality, erosion reduction, flood and drought control, trophic level benefits and environmental externalities). If farmers do not reduce or forgo mechanical disturbances to the soil while diversifying the crop species, then they are reduced to relying solely on accumulating notable levels of soil organic matter (i.e. a much slower process), if possible, which will produce a moderate fraction of benefits than the additional root-derived pore network. In many cases, diversification of crop species can provide an opportunity for farmers to convert their tillage systems to a reduced or no-till system. This is the case in Brazil, where the adoption of a double-crop system simultaneously facilitated the wide-scale, mass adoption of no-till management (Teixeira *et al.* 2014; DePaula and Fortes 2019). In addition, in regions prone to periodic excessively wet harvest conditions, a diversified crop rotation with a shorter-season grain can help shift harvest dates to avoid wheel traffic compaction, and potentially hundreds of millions of dollars (US currency) in economic costs to regional farmers over several years (Chamen *et al.* 2015; Daigh *et al.* 2020).

Diversification effects on soil water balance components

The previous two sections covered how diversification affects soil water content (i.e. the storage component of the soil water balance) and soil physical and hydraulic properties that govern how water flows within the soil. Those attributes are what crop root systems experience, given that precipitation or irrigation waters, and in some cases groundwater (e.g. deep rooted shrubs and trees), are the sources of water input into these systems. However, those attributes do not provide information on the effects of crop diversification on the fate of water (i.e. that which leaves the soil). This section focuses on those components of the water balance, namely drainage and evapotranspiration, with particular emphasis on the role of crop WUE.

Drainage

Deep drainage, or percolation, is the water that leaches from the subsoil and is not expected to return or be accessible by the crop's root zone. Innately, the drawdown of soil water by transpiring crops will result in more space for storing infiltrated water and will tend to decrease deep drainage (Daigh *et al.* 2014b; Basche and DeLonge 2017; Meyer *et al.* 2020). Often, diversified systems that produce large quantities of root biomass at depth will tend to use water in the subsoil while having little to no noticeable effect on soil water in the topsoil (Lehmann *et al.* 1998; Brye *et al.* 2000; Daigh *et al.* 2014b; Bodner *et al.* 2015; Wolz *et al.* 2018). Therefore, studies that only monitor shallow soil water do not notice the zone where some plants are acquiring their water. In terms of relay cropping, the case of older crops preferentially using deeper soil water can help prevent water competition with newly planted seedlings.

However, relatively high rainfall intensity can induce preferential flows in large, well-connected pores. If those pores were created by crops that deploy a concentration-reduction strategy for water competition in water-limited conditions, then the long, deep, vertical and branching root channels are expected to facilitate drainage. In contrast, if those pores were created by crops deploying a pre-emption strategy, either by direct expression of the genetic traits or indirectly due to nutrient-limited conditions (i.e. most crops use a pre-emption strategy for nutrient competition), then the shallow, horizontal roots channels of low fractal dimensions are expected to limit drainage (Craine *et al.* 2005; Craine and Dybzinski 2013; Hunt 2016). Therefore, crop selection and nutrient management can help improve the drainage capacity and aeration of soils that have limited internal pore networks for drainage (i.e. drainage issues are not due to underlying impermeable layers and perched water tables). As noted in the previous section, these changes to the subsoil will also have magnified effects on a region's hydrology when new diversified systems undergo large-scale mass adoption (Dabney 1998; Daigh *et al.* 2014b; Teixeira *et al.* 2014; DePaula and Fortes 2019).

Evapotranspiration and crop WUE

Crop species vary in their maximum WUE during non-stressed conditions, as well as their sensitivity to having lower WUE during water- and nutrient-limited conditions (Zegada-Lizarazu and Iijima 2005; Nielsen *et al.* 2015b). In southern Italy, sorghum (a C_4 species) produced twice as much above-ground biomass than sunflower, wheat and chickpea (all C_3 species) while also consuming less water (Steduto and Albrizio 2005). Overall, a well-watered and fertilised sorghum crop had a WUE of 32.9 g biomass m^{-2}, compared with a WUE of 13.4 g biomass m^{-2} for the other crops, when the WUE was normalised for the reference evapotranspiration rate (Steduto and Albrizio 2005). However, the logistics and designs of diversified cropping systems to the local climate are likely the most important aspects of managing water in diversified cropping systems (Hatfield and Dold 2019). For example, the proportion of soil water that goes to evaporation loss versus productive transpiration is influenced by the amount of crop residues left from previous crops, the spacing of plant rows and the duration until canopy closure, the width of alleys and the age of the shrubs/trees in hedgerows, the frequency of the season's rainfalls and the expected amount of soil water drawdown from the previous crop.

In a meta-analysis by Zhang *et al.* (2017b), mulches on the soil surface were able to increase WUE in maize systems by 61% in the semi-arid Loess Plateau in China due to the positive effect of the mulches on soil water (i.e. conserving water by reducing evaporative losses) and increasing the amount available to the crop. Zhang *et al.* (2017b) reported that WUE for maize in 44 studies ranged from 0.30 to 5.70 kg m^{-3} for yields ranging between 1.0 and 14.6 Mg ha^{-1} under evapotranspiration conditions ranging from 158 to 660 mm H_2O. Similar observations of WUE in maize have been reported in Pakistan (a 35% increase) when wheat crop residues were at 5 Mg ha^{-1} (Ali *et al.* 2018). In Argentina, maize WUE increased up to 17% when narrower row spacing was used to reduce evaporative losses from the soil surface before canopy closure (Barbieri *et al.* 2012). In Spain, compared with a barley monocrop, barley rotated with wheat and rapeseed or wheat and vetch increased in yield and WUE, whereas wheat yields and WUE were the same regardless of rotation diversity (Álvaro-Fuentes *et al.* 2009). The 6-year mean water use in the wheat systems ranged from 300 to 345 mm, whereas the water use in the barley systems ranged from 288 to 297 mm. However, the WUE of barley is higher than that of wheat. In the study of Álvaro-Fuentes *et al.* (2009), the 6-year mean WUE in the wheat systems ranged from 7.2 to 8.7 kg wheat grain mm^{-1} ha^{-1} (21.7–29.5 kg wheat biomass mm^{-1} ha^{-1}), whereas the 6-year mean WUE in barley systems ranged from 9.1 to 10.9 kg barley grain mm^{-1} ha^{-1} (25.8–32.8 kg barley biomass mm^{-1} ha^{-1}). However, Álvaro-Fuentes *et al.* (2009) also observed that the rapeseed and vetch phases failed during 80% and 35% of years, presumably due to low temperatures and soil water at the time of sowing. In semi-arid Mongolia, long-term intercropping of oats and vetch produced lower yields for each species than monocropping (i.e. 0.66 and 0.96 Mg ha^{-1} vs 1.16 and 2.25 Mg ha^{-1} for oats and vetch respectively), but there were no differences when yields were expressed in terms of an equivalent land ratio (i.e. 1.62 and 1.70 Mg ha^{-1} for intercropped and monocropping respectively; Zhang *et al.* 2019). Although the water use was the same among all the systems (235 mm), the intercropping system had slightly dryer soil below a depth of 70 cm and had lower WUE (i.e. 0.22–0.30 vs 0.42–0.82 for oats and vetch respectively; Zhang *et al.* 2019). Thus, there was no benefit to the system compared with monocropping, even though the amount of precipitation was clearly the limiting factor on total yields.

In terms of cover crops, there is a large variety of potential species. The United States Department of Agriculture (USDA) created an open-access interactive chart of 66 cover crop species to help farmers understand their characteristics, including relative water use (USDA Agricultural Research Service Northern Great Plains Research Laboratory (ARS NGPRL) 2018). At a glance, readers can learn that safflower, triticale and sainfoin have a high water use and are categorised as a warm-season broadleaf, cool-season grass and cool-season legume respectively (USDA ARS NGPRL 2018). In the same respective categories, buckwheat, oat and kura clover have medium water use, whereas amaranth, barley and lentils have low water use (USDA ARS NGPRL 2018). This information, along with other characteristics that the chart provides (i.e. salinity tolerance, risk when planted into various crop residues), allows farmer to make informed choices that are aligned with their goals and that provide specific responses to obstacles. However, the chart information is for individual species and not mixtures of species. Similar to intercropping cash crops, the intercropping of cover crops can have a variety of outcomes based on their expression of genetic traits, antecedent soil conditions at time of planting, subsequent rainfall frequency and period to accumulate biomass. In the semi-arid Central Great Plains, there was no disparity in water use (216 mm) among a cover crop mix of 10 species (flax, oat, pea, rapeseed, lentil, vetch, berseem clover, barley, phacelia and safflower) or monocultures (i.e. flax, oat, pea or rapeseed) grown for a 1.5- to 2-month period before terminating (Nielsen *et al.* 2015a). In these systems, the cover crops used more water than the evaporative losses (122 mm) from a fallow proso millet residue. However, in the semi-arid and frigid Northern Great Plains, various cover crop mixtures (e.g. millet, triticale, red clover, radish) produced 30–60% more biomass (2.4 Mg ha^{-1}) than monocultures (1.7 Mg ha^{-1}), with the exception of monocropped triticale (3.2 Mg ha^{-1}), when planted into wheat stubble during the spring months after winter thaw and harvest after 1.5–2 months (Sanderson *et al.* 2018). The same cover crop systems produced <0.2 Mg ha^{-1} biomass when planted in late summer after the annual cash crop, which was attributed to low soil water conditions and erratic rainfall (Sanderson *et al.* 2018). Liebig *et al.* (2015) reported similar outcomes in late planting of cover crops in this environment.

Finally, agroforestry systems can increase WUE by as much as 100% in alley systems of trees and crops (Ong *et al.* 2002) by increasing rainfall capture (i.e. reducing run-off losses by increasing infiltration capacity by up to 300%; Ilstedt *et al.* 2007) and reducing evaporative and deep drainage losses. Although WUE and total yields (i.e. combined products from trees and crops) may increase, the individual yields for each species will likely be reduced due to water competition. Efforts to differentiate the spacing of shrub and tree hedgerows as a means of overcoming the water competition appears to be ineffective, unless the shrubs and trees have access to groundwater supplies (Ong *et al.* 2002). For example, Lefroy *et al.* (2001) reported that trees in an alley cropping system of south-western Australia used 68–73% of their annual water from the groundwater. In these situations, the tree roots tend to preferentially accumulate near the groundwater supplies rather than laterally underneath the annual crop (Lehmann *et al.* 1998; Ilstedt *et al.* 2007; Lovell *et al.* 2018). However, the yields of annual crops, within 5 m of the tree hedgerows, are reduced as a function of distance from the trees (Lefroy *et al.* 2001).

Adopting diversity: formulating a strategy

The adoption of crop diversity into fields that have little to no history of such systems will require thoughtful analysis to formulate a good strategy. In his book *Good Strategy, Bad Strategy*, Richard Rumelt (2011) wrote '*A good strategy identifies the critical obstacles and then develops coherent actions to overcome them...The first natural advantage of good strategy arises because others often don't have one*'. In other words, you need more than a list of goals or outcomes to have a successful system; you need to have identified the obstacles you will encounter and have a response to each one. At a minimum, a strategy will include decisions about which new species to introduce (i.e. profitable product to harvest, ecosystem services, nurse crop), the resources they need (i.e. water, light, nutrients, disease suppressants) and their logistics (i.e. timing of planting and harvesting/terminating, placement and sequence in respect to the other crop species and their compatibilities; Bodner *et al.* 2015; Lovell *et al.* 2018; Hatfield and Dold 2019). These choices must be customised to a farmer's local environment, markets and cultural context.

For example, in Brazil, rotations of annual soybean and wheat have been evaluated for their profitability after being double cropped with winter planted maize, forage cover crop or the intercropping of both. The most profitable systems were those that substituted Congo grass for the maize cover every third year and when soybean was annually intercropped with maize (Garbelini *et al.* 2020). In humid regions where farmers want to plant cover crops to remove excess soil water, it is desirable to plant a low-WUE cover crop that can produce >1–3 Mg ha^{-1} of transpiring biomass before the period of frequent and heavy rainfalls. In humid regions where farmers also desire weed control in organic-based systems, then cover crops that can produce >8 Mg ha^{-1} and be effectively terminated and flattened to a >10-cm residue layer with roller-crimpers are desired (Mirsky *et al.* 2012). In semi-arid regions where farmers want to plant rotations or cover crops where the dead residue mulch will subsequently conserve soil water, it is desirable to plant a high-WUE crop that can produce close to 1–3 Mg ha^{-1} biomass before periods of infrequent rainfall. If the cover crop is to be used as forage or harvested hay, then the mulch effect will be reduced and its positive effect on conserving soil water will be negated (Unger and Vigil 1998). In semi-arid and arid regions, too little rainfall and soil water may prevent cover crops from being established or profitable. In these cases, focus should be on the high-WUE crop rotations that can leave enough residues to help capture subsequent rainfall for the next crop. If markets are available for multiple species in the region, then intercropping diverse species may provide a means of managing weather-related risks. Although water competition will exist and there will be lower yields for the individual species, the production and profits are likely similar on an equivalent land basis. In regions where crop failure due to weather conditions is common, the added crop diversity may help alleviate complete loss of cropped fields. If water competition is too great, and the annual temperatures allow for a somewhat long growing season, then relaying crops or implementing alley cropping systems may help alleviate the stress by creating physical separation in the soil depths and corridors where the roots of each species are acquiring water. In existing agroforestry alley crops, pruning trees in hedgerows to reduce their transpiration rates can be effective, in some circumstances, to reduce water competition with the alley annual crop when groundwater resources are not accessible (Lehmann *et al.* 1998). In new agroforestry alley systems, the geometries, spacing and polycultures of hedgerows can be considered for the specific land characteristics to optimise production and minimise water competition (Lefroy *et al.* 2001; Lovell *et al.* 2018). Similar small-scale adjustments to relayed intercropped geometries of annual crops (e.g. maize and soybean; cover crops) and their timings (e.g. planting high-WUE crops before low-WUE crops in rotations) can also be made to improve WUE and resource allocations (Bodner *et al.* 2015; Rahman *et al.* 2017; Sanderson *et al.* 2018; Tetrick 2019; Garbelini *et al.* 2020). Examples of crop species arrangements are shown in Fig. 5.2.

Other obstacles farmers face are the input cost and time for adopting new systems that may have limited or unstable markets. Government cost share programs and incentives can help alleviate monetary burdens where farmers are hesitant about a system's return on investment, particularly during the initial period of transition. Regional market

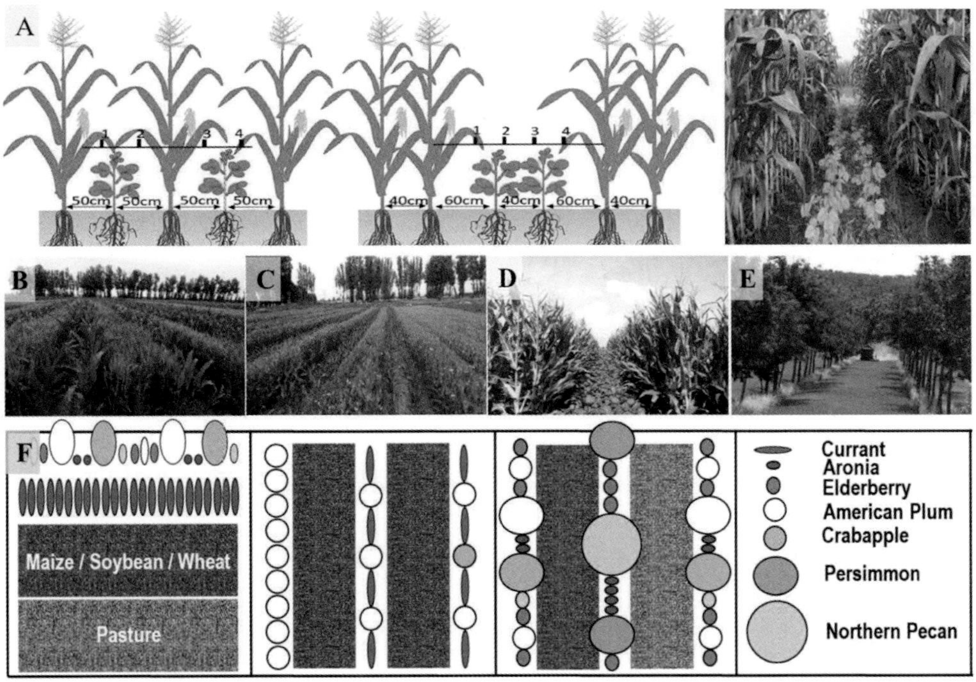

Fig. 5.2. Examples of crop species arrangements for balancing resource use, competition and production. (a) Relay intercropping arrangements for maize and soybean (CC BY 4.0; Fan *et al.* 2018). (b–d) Relay intercropping of maize and wheat (b), maize and pea (c) and maize and soybean with agroforestry among edge of fields (d). (Reproduced with permission from Yin *et al.* (2020).) (e) Traditional alley cropping of cereals and walnut (CC BY 4.0; Udawatta *et al.* 2019). (f) Three arrangements of agroforestry polyculture systems with multilayer structures (i.e. the arrangement and pairing of tall- and medium-height trees, fruit-bearing shrubs and forages/grain crops).

development for products can be bolstered by collaborations among producers, private companies, non-profits, university extension agents and governments. The absence of markets beyond a few crops is a major obstacle for farmers to stabilise their profits. In these situations, community support and buy-in from multiple stakeholders are essential.

Diversifying the near-continuous expanses of large-scale production cropping systems is one of the key features of ecologically and environmentally minded farming systems that are needed for our future global health and prosperity in a changing climate (Muschler 2016; Kaye and Quemada 2017; Renard and Tilman 2019; Franzluebbers *et al.* 2020). Vast landscapes with a single annual crop grown over a relatively short portion of the year have lower evapotranspiration and biomass accumulation than the native lands that preceded them. Therefore, the land conversion to these simple cropping systems has innately lowered WUE and uses less total water than the environment and ecosystem allows (Schilling *et al.* 2008; Nosetto *et al.* 2012). This results in more water leaving those landscapes in the river systems, with a cascade of negative externalities downstream. Diversifying these homogeneous cropping systems is a clear step towards maximising the productive potential of the landscape while taking a key step towards sustaining the world's biodiversity and freshwater resources.

Conclusions

In the past 30 years, crop diversity has continued to decrease in many of the mechanised crop-producing regions of the world. In contrast, the apparent interest in and research investment to sustainably rediversify these systems has been tremendous. In 1989, there were zero papers published on crop diversity and its associations with soil water. However, by 2020, we now add an average of 3.39 new peer-reviewed journal articles every day to the scientific literature. These works have focused on evaluating the coordination and pairing of crop species under numerous spatial arrangements and temporal sequences and how they influence soil water storage, crop water use and WUE, crop survival and productivity, crop species competition and non-productive loses of water (i.e. evaporation, run-off, deep drainage), as well as their effects on soil hydraulic properties. The most common strategies to diversify cropping systems have included the use of cover crops, diversified crop rotations, intercropping, relay cropping and alley cropping. These systems may affect soil water from negligible amounts to ± 0.10 m^3 m^{-3}, increase water infiltration up to 300% and reduce water loss to evaporation and drainage. These diversified cropping systems can result in higher crop productivity and stability, added ecosystem services and fewer consequences to the environment when they are designed strategically to thrive in the local climate, soil, cultural and market conditions and when they have community support and buy-in from multiple stakeholders.

References

Abebe T, Wiersum K, Bongers F (2010) Spatial and temporal variation in crop diversity in agroforestry homegardens of southern Ethiopia. *Agroforestry Systems* **78**, 309–322. doi:10.1007/s10457-009-9246-6

Acharya U, Chatterjee A, Daigh ALM (2019) Effect of subsurface drainage, crop rotation, and tillage on crop yield in Fargo clay soil. *Journal of Soil and Water Conservation* **74**, 456–465. doi:10.2489/jswc.74.5.456

Aguilar J, Gramig GG, Hendrickson JR, Archer DW, Forcella F, Liebig MA (2015) Crop species diversity changes in the United States: 1978–2012. *PLoS One* **10**, e0136580. doi:10.1371/journal.pone.0136580

Alhameid A, Singh J, Sekaran U, Ozlu E, Kumar S, Singh S (2019) Crop rotational diversity impacts soil physical and hydrological properties under long-term no- and conventional-till soils. *Soil Research* **58**, 84–94. doi:10.1071/SR18192

Ali S, Jan A, Manzoor, Sohail A, Khan A, Khan MI, Inamullah, Zhang J, Daur I (2018) Soil amendments strategies to improve water-use efficiency and productivity of maize under different irrigation conditions. *Agricultural Water Management* **210**, 88–95. doi:10.1016/j.agwat.2018.08.009

Alvarez R, Steinbach HS, De Paepe JL (2017) Cover crop effects on soils and subsequent crops in the pampas: a meta-analysis. *Soil & Tillage Research* **170**, 53–65. doi:10.1016/j.still.2017.03.005

Álvaro-Fuentes J, Lampurlanés J, Cantero-Martínez C (2009) Alternative crop rotations under Mediterranean no-tillage conditions: biomass, grain yield, and water-use efficiency. *Agronomy Journal* **101**, 1227–1233. doi:10.2134/agronj2009.0077

Anderson RL (2005) Are some crops synergistic to following crops? *Agronomy Journal* **97**, 7–10.

Anderson RL, Bowman RA, Nielson DC, Vigil DC, Aiken RM, Benjamin JG (1999) Alternative crop rotations for the central Great Plains. *Journal of Production Agriculture* **12**, 95–99. doi:10.2134/jpa1999.0095

Anderson W, Seager R, Baethgen W, Cane M (2017) Crop production variability in North and South America forced by life-cycles of the El Nino Southern Oscillation. *Agricultural and Forest Meteorology* **239**, 151–165. doi:10.1016/j.agrformet.2017.03.008

Anigbogu TU, Taiwo AO, Nwachukwu OF (2016) Performance assessment of farmers multipurpose cooperative societies (FMCS) in marketing of members farm produce in Benue State, of Nigeria. *Cogent Social Sciences* **2**, 1219211. doi:10.1080/23311886.2016.1219211

Barbieri P, Echarte L, Della Maggiora A, Sadras VO, Echeverria H, Andrade FH (2012) Maize evapotranspiration and water-use efficiency in response to row spacing. *Agronomy Journal* **104**, 939–944. doi:10.2134/agronj2012.0014

Bargués Tobella A, Hasselquist NJ, Bazié HR, Nyberg G, Laudon H, Bayala J, Ilstedt U (2017) Strategies trees use to overcome seasonal water limitation in an agroforestry system in semiarid West Africa. *Ecohydrology* **10**, e1808. doi:10.1002/eco.1808

Barker JB, Heeren DM, Koehler-cole K, Shapiro CA, Blanco-Canqui H, Elmore RW, Proctor CA, Irmak S, Francis CA, Shaver TM, Mohammed AT (2018) Cover crops have negligible impact on soil water in Nebraska maize–soybean rotation. *Agronomy Journal* **110**, 1718–1730. doi:10.2134/agronj2017.12.0739

Bartel CA, Archontoulis SV, Lenssen AW, Moore KJ, Huber IL, Laird DA, Dixon PM (2020) Modeling perennial groundcover effects on annual maize grain crop growth with the Agricultural Production Systems sIMulator. *Agronomy Journal* **112**, 1895–1910. doi:10.1002/agj2.20108

Basche A, DeLonge M (2017) The impact of continuous living cover on soil hydrologic properties: a meta-analysis. *Soil Science Society of America Journal* **81**, 1179–1190. doi:10.2136/sssaj2017.03.0077

Basche A, DeLonge M (2019) Comparing infiltration rates in soils managed with conventional and alternative farming methods: a meta-analysis. *PLoS One* **14**, e0215702. doi:10.1371/journal.pone.0215702

Basche AD, Kaspar TC, Archontoulis SV, Jaynes DB, Sauer TJ, Parkin TB, Miguez FE (2016) Soil water improvements with the long-term use of a winter rye cover crop. *Agricultural Water Management* **172**, 40–50. doi:10.1016/j.agwat.2016.04.006

Baumhardt R, Jones O (2002) Residue management and tillage effects on soil-water storage and grain yield of dryland wheat and sorghum for a clay loam in Texas. *Soil & Tillage Research* **68**, 71–82. doi:10.1016/S0167-1987(02)00097-1

Bayala J, Prieto I (2020) Water acquisition, sharing and redistribution by roots: applications to agroforestry systems. *Plant and Soil* **453**, 17–28. doi:10.1007/s11104-019-04173-z

Bengough AG (2012) Water dynamics of the root zone: rhizosphere biophysics and its control on soil hydrology. *Vadose Zone Journal* **11**, vzj2011.0111. doi:10.2136/vzj2011.0111

Bich AD, Reese CL, Kennedy AC, Clay DE, Clay SA (2014) Corn yield is not reduced by mid-season establishment of cover crops in Northern Great Plains environments. *Crop Management* **13**, 1–8. doi:10.2134/CM-2014-0009-RS

Blanco-Canqui H, Mikha MM, Presley DR, Claassen MM (2011) Addition of cover crops enhances no-till potential for improving soil physical properties. *Soil Science Society of America Journal* **75**, 1471–1482. doi:10.2136/sssaj2010.0430

Blanco-Canqui H, Shaver TM, Lindquist JL, Shapiro CA, Elmore RW, Francis CA, Hergert GW (2015) Cover crops and ecosystem services: insights from studies in temperate soils. *Agronomy Journal* **107**, 2449–2474. doi:10.2134/agronj15.0086

Bodner G, Nakhforoosh A, Kaul H (2015) Management of crop water under drought: a review. *Agronomy for Sustainable Development* **35**, 401–442. doi:10.1007/s13593-015-0283-4

Bogie NA, Bayala R, Diedhiou I, Conklin MH, Fogel ML, Dick RP, Ghezzehei TA (2018) Hydraulic redistribution by native Sahelian shrubs: bioirrigation to resist in-season drought. *Frontiers in Environmental Science* **6**, 98–100. doi:10.3389/fenvs.2018.00098

Bogie NA, Bayala R, Diedhiou I, Dick RP, Ghezzehei TA (2019) Intercropping with two native woody shrubs improves water status and development of interplanted groundnut and pearl millet in the Sahel. *Plant and Soil* **435**, 143–159. doi:10.1007/s11104-018-3882-4

Brye KR, Norman JM, Bundy LG, Gower ST (2000) Water-budget evaluation of prairie and maize ecosystems. *Soil Science Society of America Journal* **64**, 715–724. doi:10.2136/sssaj2000.642715x

Chamen T, Alakukku L, Pires S, Sommer C, Spoor G, Tijink F, Weisskopf P (2003) Prevention strategies for field traffic-induced subsoil compaction: a review: Part 2. Equipment and field practices. *Soil & Tillage Research* **73**, 161–174. doi:10.1016/S0167-1987(03)00108-9

Chamen WCT, Moxey AP, Towers W, Balana B, Hallett PD (2015) Mitigating arable soil compaction: a review and analysis of available cost and benefit data. *Soil & Tillage Research* **146**, 10–25. doi:10.1016/j.still.2014.09.011

Chung SO, Horton R (1987) Soil heat and water flow with a partial surface mulch. *Water Resources Research* **23**, 2175–2186. doi:10.1029/WR023i012p02175

Clark AJ, Decker AM, Meisinger JJ, Mclntosh MS (1991) Kill date of vetch, rye, and a vetch-rye mixture: II. Soil moisture and corn yield. *Agronomy Journal* **89**, 434–441. doi:10.2134/agronj1997.00021962008900030011x

Craine JM, Dybzinski R (2013) Mechanisms of plant competition for nutrients, water and light. *Functional Ecology* **27**, 833–840 doi:10.1111/1365-2435.12081.

Craine JM, Fargione J, Sugita S (2005) Supply pre-emption, not concentration reduction, is the mechanism of competition for nutrients. *New Phytologist* **166**, 933–940. doi:10.1111/j.1469-8137.2005.01386.x

D'Odorico P, Chiarelli DD, Rosa L, Bini A, Zilberman D, Rulli MC (2020) The global value of water in agriculture. *Proceedings of the National Academy of Sciences of the United States of America* **117**, 21985–21993. doi:10.1073/pnas.2005835117

Dabney SM (1998) Cover crop impacts on watershed hydrology. *Journal of Soil and Water Conservation* **53**, 207–213.

Daigh ALM, DeJong-Hughes J (2017) Fluffy soil syndrome: when tilled soil does not settle. *Journal of Soil and Water Conservation* **72**, 10A–14A. doi:10.2489/jswc.72.1.10A

Daigh AL, Helmers MJ, Kladivko E, Zhou X, Goeken R, Cavdini J, Barker D, Sawyer J (2014a) Soil water during the drought of 2012 as affected by rye cover crops in fields in Iowa and Indiana. *Journal of Soil and Water Conservation* **69**, 564–573. doi:10.2489/jswc.69.6.564

Daigh AL, Zhou X, Helmers MJ, Pederson CH, Ewing R, Horton R (2014b) Subsurface drainage flow and soil water dynamics of reconstructed prairies and corn rotations for biofuel production. *Vadose Zone Journal* **13**, 1–11. doi:10.2136/vzj2013.10.0177

Daigh ALM, Dick WA, Helmers MJ, Lal R, Lauer JG, Nafziger E, Pederson CH, Strock J, Villamil M, Mukherjee A, Cruse R (2018a) Yields and yield stability of no-till and chisel-plow fields in the Midwestern US Corn Belt. *Field Crops Research* **218**, 243–253. doi:10.1016/j.fcr.2017.04.002

Daigh ALM, Ghosh U, DeJong-Hughes J, Horton R (2018b) Spatial response of near-surface soil water contents to newly imposed soil management. *Agricultural & Environmental Letters* **3**, 180032. doi:10.2134/ael2018.06.0032

Daigh ALM, DeJong-Hughes J, Acharya U (2020) Projections of yield losses and economic costs following deep wheel-traffic compaction during the 2019 harvest. *Agricultural & Environmental Letters* **5**, 1–7. doi:10.1002/ael2.20013

Davis AS, Hill JD, Chase CA, Johanns AM, Liebman M (2012) Increasing cropping system diversity balances productivity, profitability and environmental health. *PLoS One* **7**, e47149. doi:10.1371/journal.pone.0047149

Davis KF, D'Odorico P, Rulli MC (2014) Moderating diets to feed the future. *Earth's Future* **2**, 559–565. doi:10.1002/2014EF000254

DePaula G, Fortes A (2019) The adaptation of soy–corn double-cropping to the Brazilian savanna. *Agricultural Policy Review* **Spring 2019**, 5–6.

Falkenmark M, Rockström J (2004) *Balancing Water for Humans and Nature: The New Approach in Ecohydrology*. Earthscan, Sterling.

Fan Y, Chen J, Cheng Y, Raza MA, Wu X, Wang Z, Liu Q, Wang R, Wang X, Yong T, *et al.* (2018) Effect of shading and light recovery on the growth, leaf structure, and photosynthetic performance of soybean in a maize–soybean relay-strip intercropping system. *PLoS One* **13**, e0198159. doi:10.1371/journal.pone.0198159

Fidantemiz YF, Jia X, Daigh ALM, Hatterman-valenti H, Steele DD, Niaghi AR, Simsek H (2019) Effect of water table depth on soybean water use, growth, and yield parameters. *Water (Basel)* **11**, 931. doi:10.3390/w11050931

Flury M, Mathison JB, Wu JQ, Schillinger WF, Stöckle CO (2009) Water vapor diffusion through wheat straw residue. *Soil Science Society of America Journal* **73**, 37–45. doi:10.2136/sssaj2008.0077

Franzluebbers AJ, Wendroth O, Creamer NG, Feng GG (2020) Focusing the future of farming on agroecology. *Agricultural & Environmental Letters* **5**, e20034. doi:10.1002/ael2.20034

French RJ, Malik RS, Symour M (2015) Crop-sequence effects on productivity in a wheat-based cropping system at Wongan Hills, Western Australia. *Crop and Pasture Science* **66**, 580–593. doi:10.1071/CP14262

Garbelini LG, Franchini JC, Debiasi H, Junior AAB, Junior EB, Telles TS (2020) Profitability of soybean production models with diversified crops in the autumn–winter. *Agronomy Journal* **112**, 4092–4103. doi:10.1002/agj2.20308

Hallett PD, Karim KH, Bengough AG, Otten W (2013) Biophysics of the vadose zone: from reality to model systems and back again. *Vadose Zone Journal* **12**, 1–17. doi:10.2136/vzj2013.05.0090

Hatfield JL, Dold C (2019) Water-use efficiency: advances and challenges in a changing climate. *Frontiers in Plant Science* **10**, 103–117. doi:10.3389/fpls.2019.00103

He Y, Wei Y, Depauw R, Qian B, Lemke R, Singh A, Cuthbert R, McConkey B, Wang H (2013) Spring wheat yield in the semiarid Canadian prairies: effects of precipitation timing and soil texture over recent 30 years. *Field Crops Research* **149**, 329–337. doi:10.1016/j.fcr.2013.05.013

Horton R (1989) Canopy shading effects on soil heat and water flow. *Soil Science Society of America Journal* **53**, 669–679. doi:10.2136/sssaj1989.03615995005300030004x

Horton JL, Hart SC (1998) Hydraulic lift: a potentially important ecosystem process. *Trends in Ecology & Evolution* **13**, 232–235. doi:10.1016/S0169-5347(98)01328-7

Huang M, Shao M, Zhang L, Li Y (2003) Water use efficiency and sustainability of different long-term crop rotation systems in the Loess Plateau of China. *Soil & Tillage Research* **72**, 95–104. doi:10.1016/S0167-1987(03)00065-5

Hunt AG (2016) Spatio-temporal scaling of vegetation growth and soil formation from percolation theory. *Vadose Zone Journal* **15**, 1–15. doi:10.2136/vzj2015.01.0013

Ilstedt U, Malmer A, Verbeeten E, Murdiyarso D (2007) The effect of afforestation on water infiltration in the tropics: a systematic review and meta-analysis. *Forest Ecology and Management* **251**, 45–51. doi:10.1016/j.foreco.2007.06.014

Isbell F, Adler PR, Eisenhauer N, Fornara D, Kimmel K, Kremen C, Letourneau DK, Liebman M, Polley HW, Quijas S, Scherer-Lorenzen M (2017) Benefits of increasing plant diversity in sustainable agroecosystems. *Journal of Ecology* **105**, 871–879. doi:10.1111/1365-2745.12789

Isbell F, Tilman D, Reich P, Clark A (2019) Deficits of biodiversity and productivity linger a century after agricultural abandonment. *Nature Ecology & Evolution* **3**, 1533–1538. doi:10.1038/s41559-019-1012-1

Johnson SE (2009) Crop sequences from expert farmers' fields. In *Crop Rotatation on Organic Farms: A Planning Manual, NRAES 117.* (Eds CL Mohler and SE Johnson) pp. 47–57. Natural Resource, Agriculture, and Engineering Service (NRAES), Sustainable Agriculture Research and Education, Plant and Life Sciences Publishing.

Jose S, Gillespie AR, Seifert JR, Biehle DJ (2000) Defining competition vectors in a temperate alley cropping system in the Midwestern USA: competition for water. *Agroforestry Systems* **48**, 41–59. doi:10.1023/A:1006289322392

Kaihura F, Stocking M (Eds) (2003) *Agricultural Biodiversity in Smallholder Farms of East Africa.* United Nations University Press, New York.

Kassam A, Friedrich T, Derpsch R (2019) Global spread of conservation agriculture. *The International Journal of Environmental Studies* **76**, 29–51. doi:10.1080/00207233.2018.1494927

Katchova A (2010) Structural changes in U.S. agriculture: financial performance of farms in transition. In *114th European Association of Agricultural Economics (EAAE) Seminar,* 15 April, 2010, Berlin. European Association of Agricultural Economics, <https://ageconsearch.umn.edu/record/60965?ln=en>.

Kaye JP, Quemada M (2017) Using cover crops to mitigate and adapt to climate change. A review. *Agronomy for Sustainable Development* **37**, 4–21. doi:10.1007/s13593-016-0410-x

Key N (2019) Farm size and productivity growth in the United States Corn Belt. *Food Policy* **84**, 186–195. doi:10.1016/j.foodpol.2018.03.017

Kool D, Agam N, Lazarovitch N, Heitman JL, Sauer TJ, Ben-Gal A (2014) A review of approaches for evapotranspiration partitioning. *Agricultural and Forest Meteorology* **184**, 56–70. doi:10.1016/j.agrformet.2013.09.003

Krueger ES, Ochsner TE, Porter PM, Baker JM (2011) Winter rye cover crop management influences on soil water, soil nitrate, and corn development. *Agronomy Journal* **103**, 316–323. doi:10.2134/agronj2010.0327

Kumar S, Anderson SH, Bricknell LG, Udawatta RP, Gantzer CJ (2008) Soil hydraulic properties influenced by agroforestry and grass buffers for grazed pasture systems. *Journal of Soil and Water Conservation* **63**, 224–232. doi:10.2489/jswc.63.4.224

Kuyah S, Öborn I, Jonsson M, Dahlin AS, Barrios E, Muthuri C, Malmer A, Nyaga J, Magaju C, Namirembe S, *et al.* (2016) Trees in agricultural landscapes enhance provision of ecosystem services in Sub-Saharan Africa. *The International Journal of Biodiversity Science, Ecosystem Services & Management* **12**, 255–273. doi:10.1080/21513732.2016.1214178

Lefroy EC, Stirzaker RJ, Pate JS (2001) The influence of tagasaste (*Chamaecytisus proliferus* Link.) trees on the water balance of an alley cropping system on deep sand in south-western Australia. *Australian Journal of Agricultural Research* **52**, 235–246. doi:10.1071/AR00035

Lehmann J, Peter I, Steglich C, Gebauer G, Huwe B, Zech W (1998) Below-ground interactions in dryland agroforestry. *Forest Ecology and Management* **111**, 157–169. doi:10.1016/S0378-1127(98)00322-3

Lenssen AW, Sainju UM, Jabro JD, Allen BL, Stevens WB (2018) Dryland pea production and water use responses to tillage, crop rotation, and weed management practice. *Agronomy Journal* **110**, 1843–1853. doi:10.2134/agronj2018.03.0182

Liebig MA, Hendrickson JR, Archer DW, Schmer MA, Nichols KA, Tanaka DL (2015) Short-term soil responses to late-seeded cover crops in a semi- arid environment. *Agronomy Journal* **107**, 2011–2019. doi:10.2134/agronj15.0146

Lovell ST, Dupraz C, Gold M, Jose S, Revord R, Stanek E, Wolz KJ (2018) Temperate agroforestry research: considering multifunctional woody polycultures and the design of long-term field trials. *Agroforestry Systems* **92**, 1397–1415. doi:10.1007/s10457-017-0087-4

Lu J, Zhang Q, Werner AD, Li Y, Jiang S, Tan Z (2020) Root-induced changes of soil hydraulic properties – a review. *Journal of Hydrology* **589**, 125203. doi:10.1016/j.jhydrol.2020.125203

Malézieux E, Crozat Y, Dupraz C, Laurans M, Makowski D, Ozier-Lafontaine H, Rapidel B, de Tourdonnet S, Valantin-Morison M (2009) Mixing plant species in cropping systems: concepts, tools and models. A review. *Agronomy for Sustainable Development* **29**, 43–62. doi:10.1051/agro:2007057

Malik R, Seymour M, French R, Kirkegaard J, Lawes R, Liebig M (2015) Dynamic crop sequencing in Western Australian cropping systems. *Crop and Pasture Science* **66**, 594–609. doi:10.1071/CP14097

Manns HR, Martin RC (2018) Agroecology and sustainable food systems cropping system yield stability in response to plant diversity and soil organic carbon in temperate ecosystems. *Agroecology and Sustainable Food Systems* **42**, 724–750. doi:10.1080/21683565.2017.1423529

Marcillo GS, Miguez FE (2017) Corn yield response to winter cover crops: an updated meta-analysis. *Journal of Soil and Water Conservation* **72**, 226–239. doi:10.2489/jswc.72.3.226

Martin AE, Collins SJ, Crowe S, Girard J, Naujokaitis-Lewis I, Smith AC, Lindsay K, Mitchell S, Fahrig L (2020) Agriculture, ecosystems and environment effects of farmland heterogeneity on biodiversity are similar to – or even larger than – the effects of farming practices. *Agriculture, Ecosystems & Environment* **288**, 106698. doi:10.1016/j.agee.2019.106698

Mavengahama S, Mclachlan M, de Clercq W (2013) The role of wild vegetable species in household food security in maize based subsistence cropping systems. *Food Security* **5**, 227–233. doi:10.1007/s12571-013-0243-2

McMaster GS, Aiken RM, Nielson DC (2000) Optimizing wheat harvest cutting height for harvest efficiency and soil and water conservation. *Agronomy Journal* **92**, 1104–1108. doi:10.2134/agronj2000.9261104x

Meyer N, Bergez JE, Constantin J, Belleville P, Justes E (2020) Cover crops reduce drainage but not always soil water content due to interactions between rainfall distribution and management. *Agricultural Water Management* **231**, 105998. doi:10.1016/j.agwat.2019.105998

Mirsky SB, Ryan MR, Curran WS, Teasdale JR, Maul J, Spargo JT, Moyer J, Grantham AM, Weber D, Way TR, Camargo GG (2012) Conservation tillage issues: cover crop-based organic rotational no-till grain production in the mid-Atlantic region, USA. *Renewable Agriculture and Food Systems* **27**, 31–40. doi:10.1017/S1742170511000457

Mulumba LN, Lal R (2008) Mulching effects on selected soil physical properties. *Soil & Tillage Research* **98**, 106–111. doi:10.1016/j.still.2007.10.011

Muschler RG (2016) Agroforestry: essential for sustainable and climate-smart land use? In *Tropical Forestry Handbook* (Eds L Pancel and M Kohl) pp. 2013–2116. Springer-Verlag, Berlin.

Nichols VA, Ordóñez RA, Wright EE, Castellano MJ, Liebman M, Hatfield JL, Helmers M, Archontoulis SV (2019) Maize root distributions strongly associated with water tables in Iowa, USA. *Plant and Soil* **444**, 225–238. doi:10.1007/s11104-019-04269-6

Nielsen DC, Lyon DJ, Hergert GW, Higgins RK, Calderón FJ, Vigil M (2015a) Cover crop mixtures do not use water differently than single-species plantings. *Agronomy Journal* **107**, 1025–1038. doi:10.2134/agronj14.0504

Nielsen DC, Lyon DJ, Hergert GW, Higgins RK, Holman JD (2015b) Cover crop biomass production and water use in the Central Great Plains. *Agronomy Journal* **107**, 2047–2058. doi:10.2134/agronj15.0186

Nosetto MD, Jobbágy EG, Brizuela AB, Jackson RB (2012) The hydrologic consequences of land cover change in central Argentina. *Agriculture, Ecosystems & Environment* **154**, 2–11. doi:10.1016/j.agee.2011.01.008

O'Brien PL, Daigh ALM (2019) Tillage practices alter the surface energy balance – a review. *Soil & Tillage Research* **195**, 104354. doi:10.1016/j.still.2019.104354

Ong CK, Wilson J, Deans JD, Mulayta J, Raussen T, Wajja-Musukwe N (2002) Tree–crop interactions: manipulation of water use and root function. *Agricultural Water Management* **53**, 171–186. doi:10.1016/S0378-3774(01)00163-9

Or D, Lehmann P, Assouline S (2015) Natural length scales define the range of applicability of the Richards equation for capillary flows. *Water Resources Research* **51**, 7130–7144. doi:10.1002/2015WR017034

Poeplau C, Don A (2015) Carbon sequestration in agricultural soils via cultivation of cover crops – a meta-analysis. *Agriculture, Ecosystems & Environment* **200**, 33–41. doi:10.1016/j.agee.2014.10.024

Prylipko S, Shevchenko N, Hryshchenko O (2016) Efficiency of small farms functioning in Ukraine. *Economic Annals* **158**, 17–21. doi:10.21003/ea.V158-04

Purcell LC, Edwards JT, Brye KR (2007) Soybean yield and biomass responses to cumulative transpiration: questioning widely held beliefs. *Field Crops Research* **101**, 10–18. doi:10.1016/j.fcr.2006.09.002

Qi Z, Helmers MJ, Kaleita AL (2011) Soil water dynamics under various agricultural land covers on a subsurface drained field in north-central Iowa, USA. *Agricultural Water Management* **98**, 665–674. doi:10.1016/j.agwat.2010.11.004

Rahman T, Liu X, Hussain S, Ahmed S, Chen G, Yang F, Chen L, Du J, Liu W, Yang W (2017) Water use efficiency and evapotranspiration in maize–soybean relay strip intercrop systems as affected by planting geometries. *PLoS One* **12**, e0178332. doi:10.1371/journal.pone.0178332

Rankoth LM, Udawatta RP, Gantzer CJ, Jose S, Nelson KA (2019) Cover crop effects on corn plant sap flow rates and soil water dynamics. *Crop Science* **59**, 2227–2236. doi:10.2135/cropsci2019.04.0234

Renard D, Tilman D (2019) National food production stabilized by crop diversity. *Nature* **571**, 257–260. doi:10.1038/s41586-019-1316-y

Rodríguez C, Wiegand K (2009) Evaluating the trade-off between machinery efficiency and loss of biodiversity-friendly habitats in arable landscapes: the role of field size. *Agriculture, Ecosystems & Environment* **129**, 361–366. doi:10.1016/j.agee.2008.10.010

Rumelt RP (2011) *Good Strategy, Bad Strategy: The Difference and Why It Matters.* Currency, New York.

Sanderson MA, Archer D, Hendrickson J, Kronberg S, Liebig M, Nichols K, Schmer M, Tanaka D, Aguilar J (2013) Diversification and ecosystem services for conservation agriculture: outcomes from pastures and integrated crop-livestock systems. *Renewable Agriculture and Food Systems* **28**, 129–144. doi:10.1017/S1742170512000312

Sanderson M, Johnson H, Hendrickson J (2018) Cover crop mixtures grown for annual forage in a semi-arid environment. *Agronomy Journal* **110**, 525–534. doi:10.2134/agronj2017.04.0228

Sauer TJ, Hatfield JL, Prueger JH (1996) Aerodynamic characteristics of standing corn stubble. *Agronomy Journal* **88**, 733–739. doi:10.2134/agronj1996.00021962008800050009x

Sauer TJ, Singer JW, Prueger JH, DeSutter TM, Hatfield JL (2007) Radiation balance and evaporation partitioning in a narrow-row soybean canopy. *Agricultural and Forest Meteorology* **145**, 206–214. doi:10.1016/j.agrformet.2007.04.015

Schilling KE, Jha MK, Zhang Y, Gassman PW, Wolter CF (2008) Impact of land use and land cover change on the water balance of a large agricultural watershed: historical effects and future directions. *Water Resources Research* **44**, W00A09. doi:10.1029/2007WR006644

Shackelford GE, Kelsey R, Dicks LV (2019) Effects of cover crops on multiple ecosystem services: ten meta-analyses of data from arable farmland in California and the Mediterranean. *Land Use Policy* **88**, 104204. doi:10.1016/j.landusepol.2019.104204

Shuttleworth WJ, Wallace JS (1985) Evaporation from sparse crops-an energy combination theory. *Quarterly Journal of the Royal Meteorological Society* **111**, 839–855. doi:10.1002/qj.49711146910

Skaggs TH, Shouse PJ (2008) Roots and root function: introduction. *Vadose Zone Journal* **7**, 1008–1009. doi:10.2136/vzj2008.0076

Steduto P, Albrizio R (2005) Resource use efficiency of field-grown sunflower, sorghum, wheat and chickpea II. Water use efficiency and comparison with radiation use efficiency. *Agricultural and Forest Meteorology* **130**, 269–281. doi:10.1016/j.agrformet.2005.04.003

Steiner JL (1989) Tillage and surface residue effects on evaporation from soils. *Soil Science Society of America Journal* **53**, 911–916. doi:10.2136/sssaj1989.03615995005300030046x

Tan M, Gou F, Stomph TJ, Wang J, Yin W, Zhang L, Chai Q, van der Werf W (2020) Dynamic process-based modelling of crop growth and competitive water extraction in relay strip intercropping: model development and application to wheat-maize intercropping. *Field Crops Research* **246**, 107613. doi:10.1016/j.fcr.2019.107613

Tanveer M, Anjum SA, Hussain S, Cerda A, Ashraf U (2017) Relay cropping as a sustainable approach: problems and opportunities for sustainable crop production. *Environmental Science and Pollution Research International* **24**, 6973–6988. doi:10.1007/s11356-017-8371-4

Teixeira RA, Soares TG, Fernandes AR, Braz AM de S (2014) Grasses and legumes as cover crop in no-tillage system in northeastern Pará Brazil. *Acta Amazonica* **44**, 411–418. doi:10.1590/1809-4392201305364

Tetrick J (2019) Interseeding cover crops in extra-wide corn rows. *Farm Progress*, <https://www.farmprogress.com/cover-crops/interseeding-cover-crops-extra-wide-corn-rows>.

Udawatta RP, Anderson SH (2008) CT-measured pore characteristics of surface and subsurface soils influenced by agroforestry and grass buffers. *Geoderma* **145**, 381–389. doi:10.1016/j.geoderma.2008.04.004

Udawatta RP, Gantzer CJ, Anderson SH, Garrett HE (2008) Agroforestry and grass buffer effects on pore characteristics measured by high-resolution X-ray computed tomography. *Soil Science Society of America Journal* **72**, 295–304. doi:10.2136/sssaj2007.0057

Udawatta RP, Rankoth LM, Jose S (2019) Agroforestry and biodiversity. *Sustainability* **11**, 2879–2901. doi:10.3390/su11102879

Unger PW, Vigil MF (1998) Cover crop effects on soil water relations. *Journal of Soil and Water Conservation* **53**, 200–207.

US Department of Agriculture Agricultural Research Service Northern Great Plains Research Laboratory (USDA ARS NGPRL) (2018) *Cover Crop Chart (v. 3.0)*. USDA ARS NGPRL, Mandan, <https://www.ars.usda.gov/plains-area/mandan-nd/ngprl/docs/cover-crop-chart/>.

Vanlauwe B, Coyne D, Gockowski J, Hauser S, Huising J, Masso C, Nziguheba G, Schut M, Van Asten P (2014) Sustainable intensification and the African smallholder farmer. *Current Opinion in Environmental Sustainability* **8**, 15–22. doi:10.1016/j.cosust.2014.06.001

Vitale P, Vitale J, Epplin F (2019) Factors affecting efficiency measures of Western Great Plains wheat dominant farms. *Journal of Agricultural and Applied Economics* **51**, 69–103. doi:10.1017/aae.2018.24

Ward PR, Flower KC, Cordingley N, Weeks C, Micin SF (2012) Soil water balance with cover crops and conservation agriculture in a Mediterranean climate. *Field Crops Research* **132**, 33–39. doi:10.1016/j.fcr.2011.10.017

Whish JPM, Price L, Castor PA (2009) Do spring cover crops rob water and so reduce wheat yields in the northern grain zone of eastern Australia? *Crop and Pasture Science* **60**, 517–525. doi:10.1071/CP08397

Whitbread AM, Davoren CW, Gupta VVSR, Llewellyn R, Roget D (2015) Long-term cropping system studies support intensive and responsive cropping systems in the low-rainfall Australian Mallee. *Crop and Pasture Science* **66**, 553–565. doi:10.1071/CP14136

Whitney CW (2018) *Agrobiodiversity and Nutrition in Traditional Cropping Systems – Homegardens of the Indigenous Bakiga and Banyakole in Southwestern Uganda*. University of Kassel, Witzenhausen.

Wick AF (2018) *Incorporating Cover Crops*. Publication number SF1901. North Dakota State University Extension, Fargo, <https://www.ag.ndsu.edu/publications/crops/incorporating-cover-crops>.

Wolz KJ, Lovell ST, Branham BE, Eddy WC, Keeley K, Revord RS, Wander MM, Yang WH, Delucia EH (2018) Frontiers in alley cropping: transformative solutions for temperate agriculture. *Global Change Biology* **24**, 883–894. doi:10.1111/gcb.13986

World Bank Open Data (2020a) *Arable Land (% of Land Area)*. World Bank, Washington, DC, <https://data.worldbank.org/indicator/AG.LND.ARBL.ZS?end=2016&start=1961&view=chart>.

World Bank Open Data (2020b) *Population, Total*. World Bank, Washington, DC, https://data.worldbank.org/indicator/SP.POP.TOTL>.

Wunsch EM, Bell LW, Bell MJ (2017) Can legumes provide greater benefits than millet as a spring cover crop in southern Queensland farming systems? *Crop and Pasture Science* **68**, 746–759. doi:10.1071/CP17223

Yin W, Chai Q, Zhao C, Yu A, Fan Z, Hu F, Fan H, Guo Y, Coulter J (2020) Water utilization in intercropping: a review. *Agricultural Water Management* **241**, 106335. doi:10.1016/j.agwat.2020.106335

Zaldivar M, Rocha O, Castro E, Barrantes R (2002) Species diversity of edible plants grown in homegardens of Chibchan Amerindians from Costa Rica. *Human Ecology* **30**, 301–316. doi:10.1023/A:1016516401789

Zegada-Lizarazu W, Iijima M (2005) Deep root water uptake ability and water use efficiency of pearl millet in comparison to other millet species. *Plant Production Science* **8**, 454–460. doi:10.1626/pps.8.454

Zhang ZQ, Evaristo J, Li Z, Si BC, McDonnell JJ (2017a) Tritium analysis shows apple trees may be transpiring water several decades old. *Hydrological Processes* **31**, 1196–1201. doi:10.1002/hyp.11108

Zhang Q, Wang Z, Miao F, Wang G (2017b) Dryland maize yield and water-use efficiency responses to mulching and tillage practices. *Agronomy Journal* **109**, 1196–1209. doi:10.2134/agronj2016.10.0593

Zhang Y, Duan Y, Nie J, Yang J, Ren J, van der Werf W, Evers JB, Zhang J, Su Z, Zhang L (2019) A lack of complementarity for water acquisition limits yield advantage of oats/vetch intercropping in a semi-arid condition. *Agricultural Water Management* **225**, 105778. doi:10.1016/j.agwat.2019.105778

6

Crop residue management and soil water

Sabrina J. Ruis, Jane M. F. Johnson, Humberto Blanco-Canqui

Introduction

Fluctuating climatic conditions, including heavy rainstorms, flooding, heat waves and droughts, are an increasing concern to agriculture. Weather events such as these are predicted to occur with increasing frequency under the fluctuating climate (Feng *et al.* 2016; Wang *et al.* 2017; Tomasek *et al.* 2017). Large rain events or rapid snowmelt, such as the March 2019 'bomb cyclone', which affected much of the US Great Plains and Midwest, resulted in large impacts to agricultural fields due to flooding, including delays in planting or farmers deciding to not plant at all. Further, the frequency of these large, intense rainfall events has increased by 82% in recent decades (Saunders *et al.* 2012). In contrast to large rain events are heat waves and prolonged droughts. Indeed, some reports by scientists cite drier, hotter summers, with a greater intensity of summer drought periods than before (Tomasek *et al.* 2017). In 2012, the US experienced the fifth most severe drought based on the Palmer Z index that encompassed most of the US Great Plains and Midwest (National Centers for Environmental Information, National Oceanic and Atmospheric Administration 2012).

The question is, how do we manage agricultural soils under fluctuating climatic conditions? Ideally, agricultural soils must have rapid water infiltration for heavy rain events yet store and release water during droughty periods. There are many options to manage soils for these extreme weather events, including no-till, cover crops, residue management and others. Cover crops are a support practice for no-till, but are often terminated early, leading to low biomass production and few, if any impacts, on soils (Ruis *et al.* 2017, 2020). Discussions of residue management are not new, but there is renewed interest in using residues for biofuel feedstock, animal feed and bedding. In some instances, the abundant residues (Table 6.1) produced by many of the major cash crops can be challenges for farmers through cooler soil temperatures in spring, the lack of a uniform planting surface, poor seed–soil contact and interference with equipment during planting (Kenney *et al.* 2015). For example, the wetter and cooler soil (due to reduced evaporation and the mulching effect of the residue) can lead to delayed planting and crop germination, which can, in some cases, result in reduced crop yields. Sometimes this leads to the decision to burn or otherwise harvest crop residues.

Residue burning occurs following wheat in the North American Great Plains and is common in rice-growing regions such as India and China (McCarty *et al.* 2009; Chen *et al.* 2019; Bhuvaneshwari *et al.* 2019). It is estimated that China burns approximately

Table 6.1. Aboveground residue production and area harvested in North America and the world for some major cropping systems

Crop	Above-ground residue production (Mg ha^{-1})	Area harvested[A] (million ha)	
		North America	Worldwide
Corn (*Zea mays* L.)	9–11[B]	35	194
Sorghum (*Sorghum* spp.)	11[C]	2.0	42
Wheat (*Triticum aestivum* L.)	3–10[D]	1.3	214
Rice (*Oryza sativa* L.)	6–6.5[E]	1.2	167
Field pea (*Pisum sativum* L.)	3.2[F]	1.8	7.9
Soybean (*Glycine max* L.)	3[G]	38	125
Oat (*Avena sativa* L.)	2.3–2.6[H]	1.3	10
Barley (*Hordeum vulgare* L.)	2–2.5[H]	3.2	48

[A]Data from the Food and Agriculture Organization of the United Nations (http://fao.org/faostat/en/#data).
[B]Ruis *et al.* (2017).
[C]Blanco-Canqui *et al.* (2009).
[D]Dai *et al.* (2016).
[E]Prasad *et al.* (1999).
[F]Stevenson and van Kessel (1997).
[G]Johnson *et al.* (2006).
[H]Sokhansanj *et al.* (2006).

one-quarter of its crop residues (Chen *et al.* 2019), the US burns ~1.2 billion ha cropland (McCarty *et al.* 2009) and India burns the majority of the 360 Mg of sugarcane (*Saccharum* sp.), wheat and rice residues (Bhuvaneshwari *et al.* 2019). In the US, corn is the most common high-residue-producing crop. It is estimated that ~0.81 million ha, or ~0.2% of the corn area, is baled and ~4.06 million ha corn area is grazed in 19 states surveyed (Schmer *et al.* 2017; Obrycki and Karlen 2018). Grazing is more common than baling, particularly in the western US Corn Belt (Schmer *et al.* 2017) and in developing countries (Herrero *et al.* 2010). Thus, these common harvest practices are increasing due to the interest in residues as a biofuel feedstock and the renewed interest in integrating crops and livestock.

One of the recurring questions around residue removal is, what quantity of residues can we safely remove without negative impacts to soils? Further, do we remove residues based on a percentage of the total or based on the mass of residues present? Minimum rates of residue to sustain soil organic matter are a function of the mass returned (Wilhelm *et al.* 2007). However, studies often refer to the percentage of crop residue removed, probably because erosion control is more closely linked to the percentage of soil covered by residue. It should also be noted that the percentage of crop residue is very different from the mass of residue either harvested or returned (Wilhelm *et al.* 2011). Some have suggested residue removal at rates ≤50% may not negatively affect soils in most cases, although these quantities are meant to reduce erosion risks (Graham *et al.* 2007; Blanco-Canqui and Lal 2009). However, a review summarising and integrating the effects of residue removal on soil hydraulic properties and the permissible level of residue removal as it relates to soil hydraulic properties is absent from the literature. Therefore, the questions are: (1) what happens to soil hydrological properties when residues are removed; and (2) how do we best manage residues under fluctuating climatic conditions? The objective of this chapter is to answer these two questions by discussing the impacts of crop residue management strategies including baling, grazing and burning on hydraulic properties and soil hydrology.

Residue management and soil hydraulic properties

In this chapter, we discuss residue management impacts on soil hydraulic properties, namely water infiltration rate, saturated (K_{sat}) and unsaturated (K_{unsat}) hydraulic conductivity and water retention characteristics. These soil hydraulic properties are important for water entry, movement and storage within the soil. Further, as discussed later in this chapter, negative shifts in soil hydraulic properties may indicate negative impacts on other soil properties and crop yields.

Water infiltration rate

The water infiltration rate governs the entry of water into the soil and is driven, in part, by soil aggregation and porosity, particularly macroporosity. Changes to these and other soil properties can alter water infiltration rates. Thus, soil management activities, such as tillage or residue removal, may affect water infiltration rates. Management practices that reduce infiltration rates may increase run-off rates and the loss of sediment and nutrients, and contribute to non-point source pollution.

Residue baling

Crop residue baling generally involves chopping, windrowing and baling the residues. Baling residues occurs in the autumn or spring after harvest and removes 60–70% of crop residues (Fig. 6.1; Rakkar and Blanco-Canqui 2018). Table 6.2 summarises studies

Fig. 6.1. Baled corn residues at a field site in Nebraska. Photograph by H. Blanco-Canqui.

Table 6.2. Impact of residue baling by location, removal rate (by cover), tillage, duration and crop on water infiltration rate

Within a site, values with the same lowercase letters do not differ significantly ($P > 0.05$)

Location	Texture	Residue management	Duration (years)	Crop	Water infiltration rate (cm h^{-1})	Change in infiltration (%)	Reference
No-till							
Nebraska, USA	Silt loam	No removal vs 63% removal	3	Corn	No effect		Blanco-Canqui *et al.* (2014)
Ohio, USA	Silt loam	No removal	3	Corn	70.2a		Blanco-Canqui *et al.* (2007), Blanco-Canqui and Lal (2009)[A]
	10% slope	25% removal			52.5ab		
		50% removal			33.7b	–51.9	
		75% removal			19.9b	–71.7	
		100% removal			18.2b	–74.1	
Ohio, USA	Silt loam 2% slope	No removal vs 25%, 50%, 75% or 100% removal	3	Corn	No effect		Blanco-Canqui *et al.* (2007), Blanco-Canqui and Lal (2009)[A]
Ohio, USA	Clay loam <1% slope	No removal vs 25%, 50%, 75% or 100% removal	3	Corn	No effect		Blanco-Canqui *et al.* (2007), Blanco-Canqui and Lal (2009)[A]
Nebraska, USA	Silt loam	No removal	6	Corn	18.83a		Sindelar *et al.* (2019)[A]
		56% removal			14.88b	–27.0	
South Dakota, USA	Silty loam	No removal vs removal	15	Corn–soybean	No effect		Chalise *et al.* (2018
New York, USA	Silt loam	No removal vs removal	32	Corn	No effect		Moebius-Clune *et al.* (2008)
Tillage							
New Dehli, India	Sandy loam	No removal	3	Pearl millet (*Pennisetum glaucum* L.)–mustard (*Brassica* sp.)	1.52a		Choudhary *et al.* (2019)[B]
		50% removal			1.47ab		
		100% removal			1.38b	–10.1	
New York, USA	Silt loam	No removal vs removal	32	Corn	No effect		Moebius-Clune *et al.* (2008)
Alberta, Canada	Clay loam	No removal vs removal	10	Barley	No effect		Singh *et al.* (1996)

[A]Cumulative infiltration (cm).
[B]50% removal = 2 Mg ha^{-1} residue left; 0% removal = 4 Mg ha^{-1} residue left.

evaluating the water infiltration rate and shows that seven of 10 study locations reported no effect on water infiltration rates due to residue baling. The remaining three studies (30%) reported reduced water infiltration rates due to baling. Table 6.2 and Fig. 6.2 also indicate that residue baling may reduce water infiltration rates by 10–74% (mean 50%). These data suggest that baling often (in more than 50% of cases) may have no effect on water infiltration rates.

The permissible rate of residue removal that reduces water infiltration rates depends on site slope and soil texture (Fig. 6.3). Relatively flat or fine-textured soils may show no effects of residue removal on water infiltration compared with sloping soils (Blanco-Canqui *et al.* 2007; Table 6.2). However, the ability of no-tillage management to protect the soil may be undermined if inadequate residue is returned (Johnson *et al.* 2016). In the sloping soils, residue baling rates <50% appear to have few effects on water infiltration rate compared with residue baling rates ≥50% (Blanco-Canqui *et al.* 2007). Thus, relatively flat and clayey soils may be more resilient to residue removal-induced changes to water infiltration.

Residue baling paired with no-till may have fewer negative effects in the short to medium term than tillage (Table 6.2). Although tillage does result in the destruction of

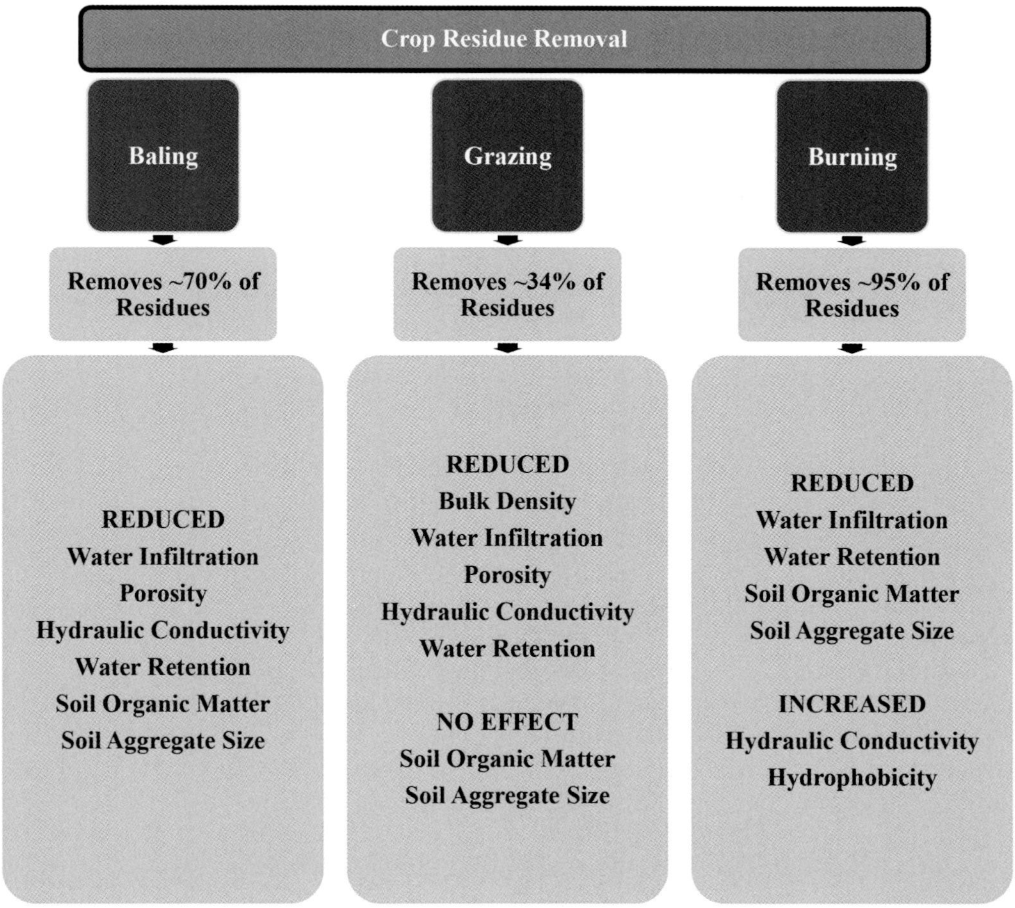

Fig. 6.2. Expected impacts of crop residue removal by baling, grazing and burning on soil hydraulic and related properties.

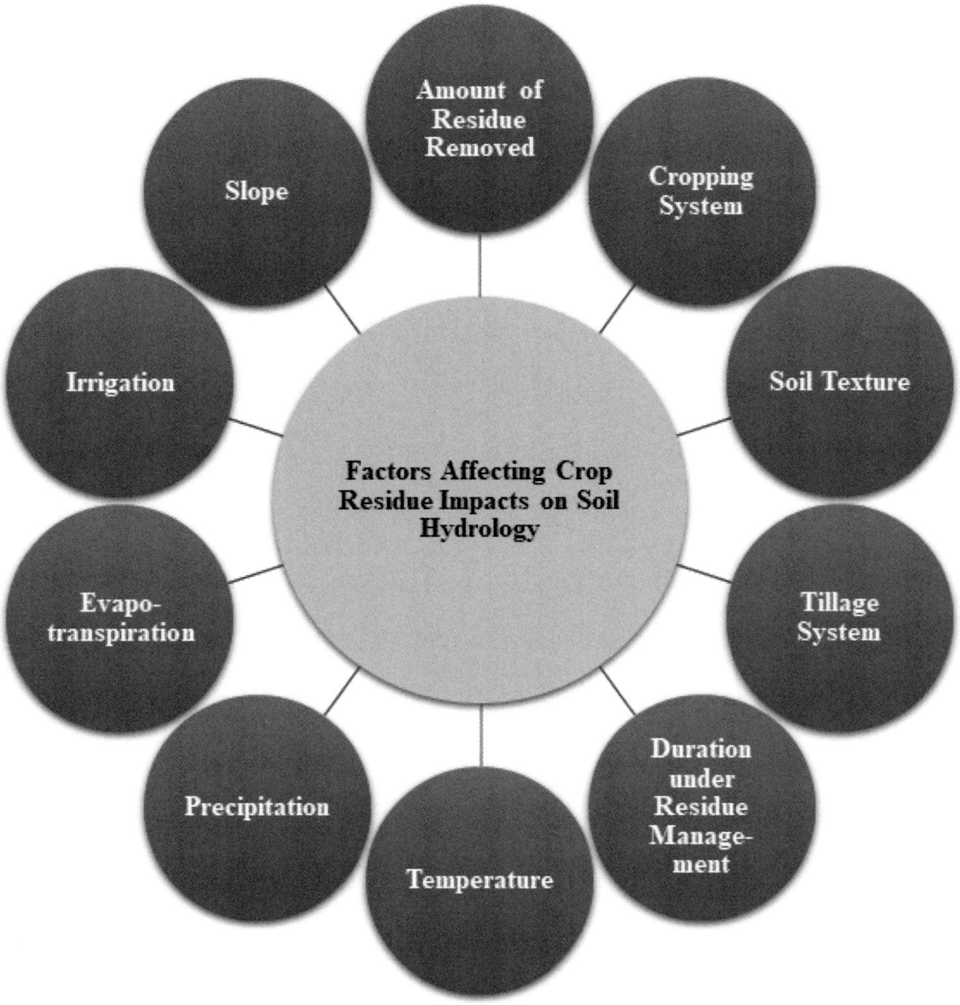

Fig. 6.3. Factors affecting the impacts of crop residue on soil hydrology.

soil aggregates, resulting in losses of soil C and soil structural quality and thereby resulting in reductions in water infiltration compared with no-till (Blanco-Canqui and Ruis 2018), no-till has higher surface soil C and therefore greater capacity for changes in soil properties.

Rain-fed systems are expected to respond more negatively (i.e. greater reductions in water infiltration) to residue baling than irrigated systems. Rain-fed cropping systems typically produce less residue than irrigated systems (Kenney *et al.* 2015). For example, if a corn system produced 3 Mg ha^{-1} residue under low-productivity systems and 6 Mg ha^{-1} under a high-productivity system and 50% of the residue cover was removed from both systems, the amount of remaining residue is less under rain-fed than irrigated systems. Thus, in systems with lower productivity, residue removal may have larger effects on water infiltration than systems with higher productivity. Overall, residue baling may have no effect on water infiltration rates, but impacts may be more negative in sloping soils and rain-fed systems (Table 6.3).

Table 6.3. Summary of residue management by baling, grazing and burning on soil hydraulic properties of water infiltration rate, hydraulic conductivity and water retention characteristics

Hydraulic property	Baling	Grazing	Burning
Water infiltration rate	No effect	No effect	No effect
Hydraulic conductivity	Decrease or no effect	No effect	Increase or no effect
Water retention characteristics	Decrease	No effect	Decrease or no effect

Residue grazing

Crop residue grazing may occur from harvest through the early winter and spring, and returns part of the carbon and nutrients from residues as manure and urine, and removes only ~30% of crop residues (Fig. 6.4; Rakkar and Blanco-Canqui 2018). Crop residue grazing offers a unique opportunity to integrate crops and livestock (Fig. 6.3), as well as a cheap, readily available feedstock. However, does cattle grazing negatively affect water infiltration? In 11 of 13 studies (85%), residue grazing did not reduce the water infiltration rate (Baumhardt *et al.* 2017; Rakkar and Blanco-Canqui 2018). In the remaining two studies in which grazing did reduce water infiltration rates, it was attributed to low crop residue input leading to degradation of soil structure and subsequent reductions in water infiltration rate (Rakkar and Blanco-Canqui 2018). In addition, in two studies in Nebraska, across 15 site-years, residue grazing reduced sorptivity or initial water infiltration rates by 33–63% in two site-years (Rakkar *et al.* 2018; Ruis *et al.* 2018). The general lack of effect of

Fig. 6.4. Cattle grazing corn residue in Nebraska. Photograph by S. J. Ruis.

residue grazing on water infiltration rate indicates that residue grazing, at appropriate stocking rates, does not reduce the water infiltration rate in most cases. If residue grazing does reduce the water infiltration rate, it is likely due to a too-high stocking rate for the residue present.

Most studies evaluating residue grazing are under no-till and reduced-till farming conditions. The effects of combining residue grazing with conventional tillage are less known. However, because grazing at appropriate stocking rates may not reduce soil C (Rakkar *et al.* 2017), the already reduced soil C from conventional tillage (Kibet *et al.* 2016) may not be further reduced. The lack of further reductions in soil C due to grazing suggests a lower likelihood of further reductions in soil aggregation, or the related properties of porosity and water infiltration rates. Overall, the effects of residue grazing on water infiltration rate are minimal unless stocking rates are too high for the amount of residue present (Table 6.3).

Residue burning

Burning crop residues is a quick way to dispose of the residues close to planting time (Smil 1999). Because burning removes most (~95%) of the crop residue C, burning may have unwanted effects on soil hydrological properties. However, very few studies have investigated the impacts of residue burning on water infiltration rates. Three studies reported the impacts of residue burning on water infiltration rates compared with no residue burning. In all three studies, regardless of tillage method (conventional or no-till), residue burning did not affect the water infiltration rate compared with no removal (Wuest *et al.* 2005; Gangwar *et al.* 2006; Desrochers *et al.* 2019). Soil hydraulic properties change slowly in response to management, so observed responses may differ as a function of study duration. However, because the three studies lasted 3, 14 and 70 years (Wuest *et al.* 2005; Gangwar *et al.* 2006; Desrochers *et al.* 2019), the duration of study may not be a factor affecting the effects of burning on water infiltration rate. In summary, based on the few available studies, residue burning may have limited effects on water infiltration rates.

K_{sat} and K_{unsat}

K_{sat} and K_{unsat} are related hydraulic properties that determine water entry and flow through the soil (Hillel 2004). Unlike K_{sat}, which occurs under saturated soil conditions, K_{unsat} refers to the ability of the soil to conduct water under unsaturated conditions, which is more prevalent under field conditions. K_{sat} and K_{unsat} are critical to water storage, deep percolation, chemical and nutrient transport and groundwater recharge. The question is, does residue removal affect K_{sat} and K_{unsat}?

There are few data on the effects of residue removal on K_{sat} and K_{unsat} under baling, grazing and burning. The data show that residue baling can reduce (Blanco-Canqui and Lal 2009) or have no effects on K_{sat} (Singh *et al.* 1996; Moebius-Clune *et al.* 2008). For example, across three no-till sites in Ohio, residue removal by baling at rates >50% reduced K_{sat} by 458–2990% compared with no removal (Blanco-Canqui and Lal 2009). In contrast, Singh *et al.* (1996) showed that under tillage, residue removal had no effect on K_{sat} compared with no removal. Similarly, Moebius-Clune *et al.* (2008) showed no effect of residue removal on K_{sat} under no-till and plough till compared with no removal.

Unlike the negative effects of baling on K_{sat}, residue grazing generally has no effect on K_{sat} according to a review of seven studies (Rakkar and Blanco-Canqui 2018). However, residue burning can increase (Skidmore *et al.* 1986; Valzano *et al.* 1997) or have no effect (Skidmore *et al.* 1986; Celik *et al.* 2012) on K_{sat} (Table 6.4). When residue burning increased

Table 6.4. Impact of residue burning on saturated hydraulic conductivity (Ksat) by tillage, duration of study and cropping system

Location	Residue management	Tillage	Duration (years)	Crop	K_{sat} (mm h^{-1})	Reference
Turkey	No removal vs burned	Tilled	3	Continuous wheat	No effect	Celik *et al.* (2012)
Kansas, USA	No removal vs burned	Tilled.	14	Wheat	No effect	Skidmore *et al.* (1986)
Kansas, USA	No removal vs burned	Tilled	14	Sorghum	5.4-fold increase	Skidmore *et al.* (1986)
Australia	No removal vs burned	No-till to minimum till	15	Wheat rotation	2.2-fold increase	Valzano *et al.* (1997)

K_{sat}, it increased it by 2.2- to 5.4-fold compared with no removal. Cropping system under burning may drive the differences among studies. For example, as indicated in Table 6.4, sites that were under wheat showed no effect of residue burning on K_{sat} compared with sites where wheat was rotated to another crop or to sorghum.

K_{unsat} is a less assessed soil property than K_{sat} under crop residue removal. A study on two loam soils in Minnesota in the US evaluated the effects of 78% residue removal and no residue removal under chisel plough and no-till (Johnson *et al.* 2016). In that study, residue removal reduced K_{unsat} at a tension infiltrometer pressure head of –15 mm from 57.1 to 30.4 mm h^{-1} under no-till, but not under chisel plough (Johnson *et al.* 2016). The lack of difference in residue removal under chisel ploughing was attributed to chisel ploughing disturbing the soil and soil consolidation blocking water transmission pores. Under no-till, the lower K_{unsat} with residue removal was potentially due to raindrop impact on the unprotected soil surface breaking soil aggregates and subsequent soil consolidation. Under residue removal, the lower K_{unsat} and reduced water transmission can lead to run-off, and thus less water available for crop production (Johnson *et al.* 2017). Studies directly evaluating the effects of residue grazing and burning on K_{unsat} are scarce. However, the effects of grazing and burning on K_{unsat} are likely similar to those on K_{sat} and water infiltration. One study that assessed K_{unsat} under burned and unburned residue showed that K_{unsat} was 16.1 cm h^{-1} under no residue burning and 5.3 cm h^{-1} under residue burned, the rate being 2.1-fold greater under no residue burning (Desrochers *et al.* 2019). Overall, residue baling can reduce hydraulic conductivity, grazing may have no effect and burning may increase or have no effect on hydraulic conductivity (Table 6.3).

Soil water retention

Crop residue management can also affect the capacity of the soil to retain plant-available water. Plant-available water may be defined as the amount of stored water between field capacity (–33 kPa) and permanent wilting point (–1500 kPa; Hillel 2004). Note that the –33 kPa threshold for field capacity is an average and varies among soils. Water held at field capacity is driven, in part, by macro- and mesoporosity, soil organic matter and soil aggregation, whereas water held at permanent wilting point is driven by a combination of the clay and soil organic matter content. In general, clay content is not changed by soil management practices, unless severe erosion exposes clayey subsoil, but soil organic matter does change with management. Thus, any soil management practice that changes soil organic matter has the potential to also change water retention.

Residue baling

Water retention or storage, like water infiltration rate and K_{sat}, may also be affected by residue baling (Fig. 6.2). Based on six studies worldwide, water content at field capacity may decrease by 3–79% with residue removal (Blanco-Canqui and Lal 2009). In general, high rates (≥50%) of removal reduce field capacity water content compared with no removal. Studies from Lithuania (Feiziene *et al.* 2018), New York (Moebius-Clune *et al.* 2008), Wisconsin, Iowa (Tormena *et al.* 2017) and Nebraska (Sindelar *et al.* 2019) show residue removal can also reduce plant available water by 0.03–0.35 cm for the upper 10 cm of soil (Table 6.5). Similar to the water infiltration rate, the critical removal rate that results in reductions to plant-available water is likely around 50% (Blanco-Canqui and Lal 2009). Crop residue removal by baling significantly reduced field capacity and plant-available water, but the degree of reduction varied depending on the age of the site and other factors. For example, the percentage decrease in plant-available water can increase with increasing duration of residue removal. Tormena *et al.* (2017), on a loam to clay loam in Iowa, showed that residue removal for 5 years reduced plant-available water by 3% compared with no removal, whereas Moebius-Clune *et al.* (2008), on a silt loam in New York, showed that 32 years of residue removal reduced plant-available water by 16%. Therefore, residue baling may reduce plant-available water through reductions in field capacity (Table 6.5).

Residue grazing and burning

There are limited studies in the literature evaluating the effects of crop residue grazing and burning on soil water retention. Therefore, we draw upon studies discussing water retention under different rates of residue baling, particularly low rates similar to what would be removed by grazing. As discussed earlier, residue baling can reduce plant-available water. However, a study with three sites in Ohio showed that residue removal at rates of ≤25% on silt loams and ≤50% on clay loam may not affect plant-available water (Blanco-Canqui *et al.* 2007). Because removing <50% of residues by baling does not affect water retention, we would expect grazing would have a limited effect on water retention because it only removes ~34% of residues.

Residue burning, in contrast to grazing or baling, can reduce field capacity water content by 0.3–4.3% (Bescansa *et al.* 2006; Castellini *et al.* 2014), but have no effect on permanent wilting point (Bescansa *et al.* 2006). However, plant-available water may also decrease (Table 6.3). Studies have shown that residue burning may reduce or tend to reduce plant-available water by 2.8–3.2 mm, depending on study duration (Bescansa *et al.* 2006; Castellini *et al.* 2014). These data suggest reductions in water content at field capacity and variable effects on plant-available water from residue burning are probably due to hydrophobicity-induced changes to organic matter and soil particles. In summary, residue grazing may have no effect on water retention, whereas burning may reduce water retention.

Mechanisms of crop residue management effects on hydraulic properties

Some of the mechanisms by which crop residue removal may affect soil hydraulic properties include: (1) removal of surface cover and subsequent raindrop impacts; (2) the creation of surface seals and crusts; (3) a reduction in soil organic matter; and (4) reduced aggregation, and thus reduced porosity (Fig. 6.5). Baling removes large portions (up to 70%; Rakkar and Blanco-Canqui 2018) of the residue present on the soil surface. In contrast, grazing removes <35% of the residue cover (Rakkar and Blanco-Canqui

Table 6.5. Effect of crop residue removal on plant-available water according to tillage, duration of management, cropping system and soil texture

For a given location, different letters after plant-available water values indicate significant differences ($P < 0.05$)

Location	Texture	Tillage	Residue management	Duration (years)	Crop	Soil depth (cm)	Plant-available water ($cm^3\ cm^{-3}$)	% Change in retention	Reference
Lithuania	Loam	Conventional till	No removal	17	Winter wheat–oilseed rape (*Brassica napus*)–spring wheat–spring barley–pea	0–10	0.183c		Feiziene *et al.* (2018)
			Residue removed				0.173d	–5	
		No-till	No removal				0.209a		
			Residue removed				0.199b	–5	
Lithuania	Sandy loam	Conventional till	No removal	17	Winter wheat–oilseed rape–spring wheat–spring barley–pea	0–10	No effect		Feiziene *et al.* (2018)
			Residue removed						
		No-till	No removal				No effect		
			Residue removed						
Iowa	Loam to clay loam	Chisel plough	No removal	7	Corn	0–7.5	0.159a		Tormena *et al.* (2017)
			50% removal				0.156b	–2	
			90% removal				0.162a		
		No-till	No removal				0.159a		
			50% removal				0.155b	–3	
			90% removal				0.155b	–3	
Nebraska	Silt loam	No-till	No removal	6	Corn	0–5	0.215a		Sindelar *et al.* (2019)
			56% removal				0.146b	–32	
New York	Silt loam	No-till	No removal	31	Corn	5–6.1	0.357a		Moebius-Clune *et al.* (2008)
			100% removal				0.308b	–14	
		Plough	No removal				No effect		
			100% removal						

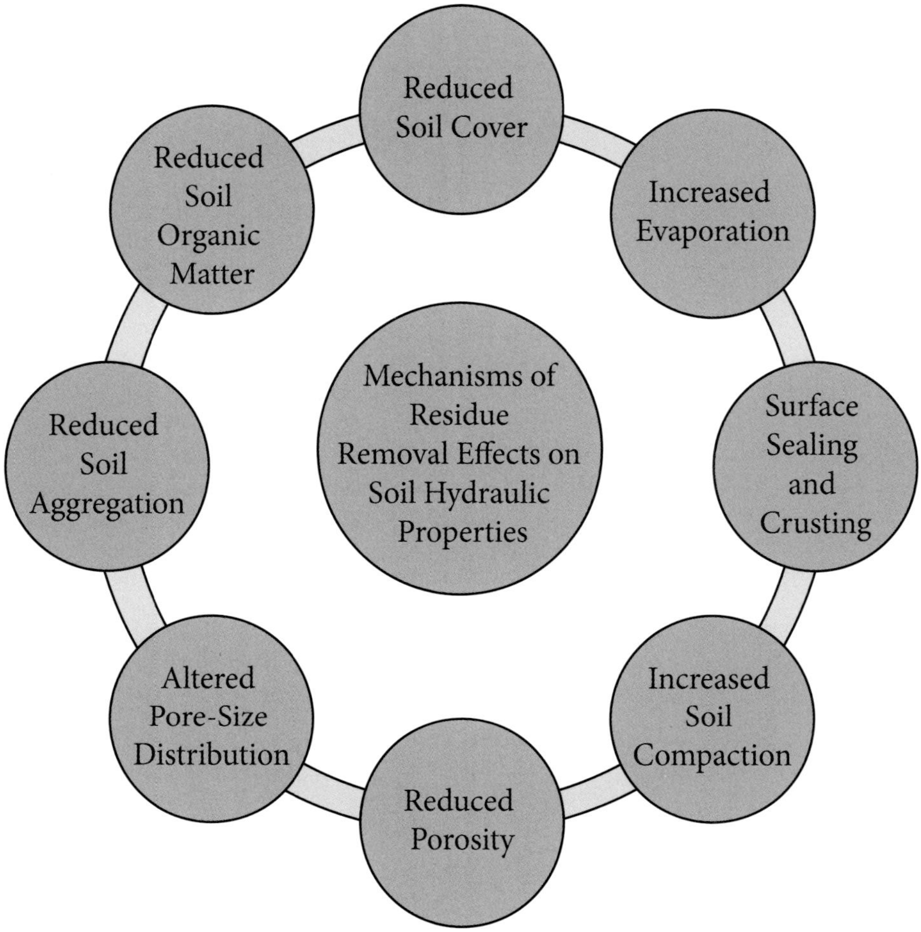

Fig. 6.5. Mechanisms underlying the effects of residue removal on soil hydraulic properties.

2018). Burning, although removing much of the residue cover, generally occurs only a short time before planting. Thus, the high removal rate by baling may lead to greater negative effects on soil hydraulic properties, whereas grazing, as long as it is at appropriate stocking rates, may have minimal effects on soil hydraulic properties.

Residues protect the soil by blocking the highly erosive impact of raindrops. Raindrop impact results in splash erosion, dislodging and disintegrating soil aggregates, and the creation of craters. The dislodgment and disintegration of aggregates then lead to the formation of surface seals and the clogging of soil pores (Kumar *et al.* 2012; Fernández *et al.* 2015; Gómez-Paccard *et al.* 2015), thus physically blocking water infiltration. The disintegration of soil aggregates releases soil organic matter from its protective place inside aggregates, exposing it to microbial and macrofaunal activity, causing the oxidation of C contained in organic matter.

In addition to erosion, reduction in soil C involves the direct removal of organic matter input via baling. For example, residue removal at increasing rates can reduce organic matter at a rate of 0.05 g kg^{-1} per percentage increase in residue removal (Fig. 6.6). Organic matter is critical for soil aggregation, which, in turn, is essential for the

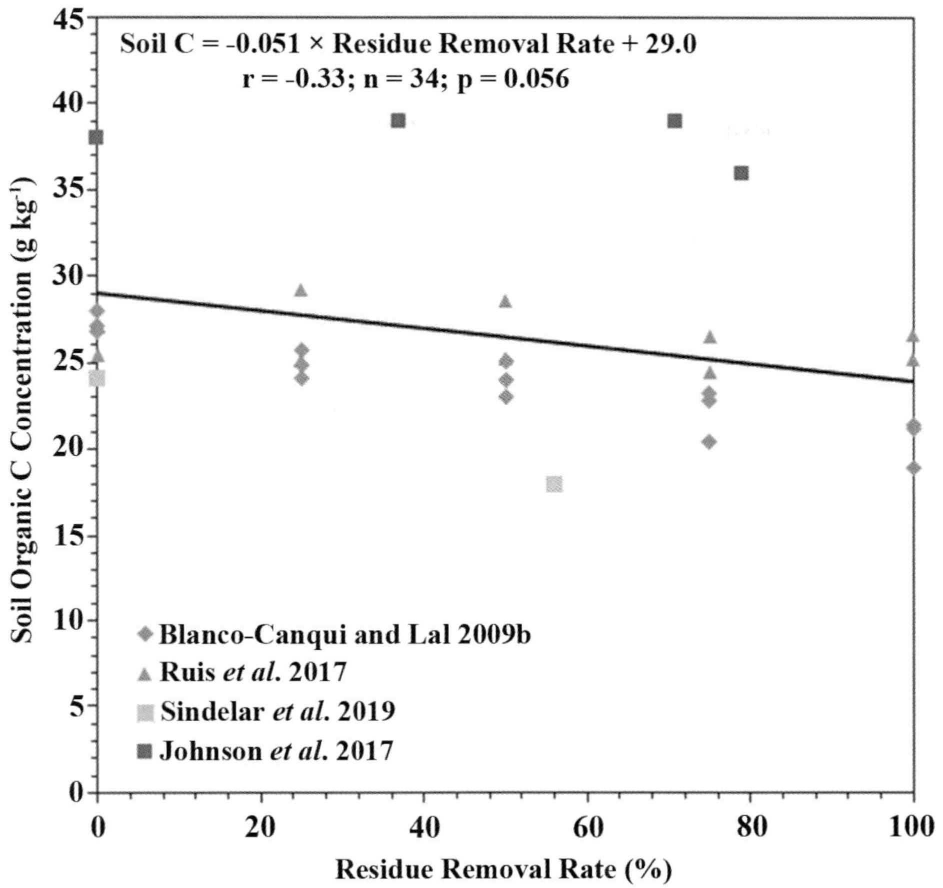

Fig. 6.6. Impact of crop residue removal rate on soil organic C concentration across eight sites. Drawn from data in Blanco-Canqui and Lal (2009), Johnson *et al.* (2017), Ruis *et al.* (2017) and Sindelar *et al.* (2019).

protection of organic matter (Tisdall and Oades 1982). Ojekanmi and Johnson (2021) reported that the percentage of water-stable aggregates increased by 0.85% and 2.25% per Mg stover returned in field managed with chisel ploughing and no-tillage respectively, which can then improve water infiltration. The activity of microbes and macrofauna creates glue-like binding agents that stick soil particles, organic matter and smaller aggregates together. Further, soils high in silt and clay can store more organic matter than coarser-textured soils due to the association of organic matter with soil minerals (Six *et al.* 2002; Kleber *et al.* 2007) particularly with smectite surfaces (Gonzalez and Laird 2003). However, without the input from residues, aggregate size decreases (Sindelar *et al.* 2019; Ruis *et al.* 2017), thereby decreasing porosity (Sindelar *et al.* 2019) and ultimately water infiltration.

Residue removal may be more detrimental under tillage than no-till. No-till can result in larger, more stable aggregates than conventional tillage because conventional tillage mixes crop residues with the soil, loosens the soil and can reduce soil C (Kibet *et al.* 2016) due to the breaking of soil aggregates, which exposes soil organic matter to microbial decay. However, because no-till soils have higher soil C and overall better soil structure, residue removal may result in larger changes to soil properties under no-till than tilled conditions.

The mechanisms for the reduction in hydraulic conductivity are likely similar to those for water infiltration. The removal of crop residues reduces soil organic C, the binding agent for soil aggregation, which reduces soil aggregation, reducing soil porosity (Sindelar *et al.* 2019). The increase in K_{sat} under burning could be due to hydrophobicity (water repellency) of the charred residues. The disintegration of soil aggregates and subsequent soil consolidation can physically block macropores, further reducing the ability of soil to conduct water. The reduction in plant-available water due to residue baling is likely due to multiple mechanisms. First, the reduction in plant-available water can be attributed to a loss of soil organic matter through the physical removal of residues, although this is unlikely because large increases (or decreases) in soil organic C (>1%) may not have large effects on available water (Minasny and McBratney 2018). For example, an analysis of 60 studies showed that an increase of 1% in soil organic C may lead to only a 1.16% increase in plant-available water (Minasny and McBratney 2018). Similarly, Libohova *et al.* (2018) reported only a 1% increase in available water-holding capacity, which was variable, interacting with soil texture and clay mineralogy. Similarly, a reduction in soil organic matter through the removal of crop residues can degrade soil structure, reducing porosity (Bescansa *et al.* 2006). Reductions in aggregate stability and porosity affect the amount of available water retained, all of which can occur with residue baling at high rates. The lower water content at field capacity with residue burning was similar to the water infiltration rate.

Crop residue management and soil hydrology

Soil hydrology or water balance is composed of six terms: precipitation (*P*), irrigation (*I*), run-off (*RO*), evapotranspiration (*ET*), deep percolation (*D*) and the change in water storage (ΔW). These terms are arranged in Eqn 1 to form the water balance equation:

$$P + I - RO - ET = D + \Delta W \tag{1}$$

In this section, we discuss the effects of crop residue management on several components of the water balance.

Precipitation capture

Precipitation capture includes the capture of rainfall, snow, snowmelt and irrigation. These three components of the soil hydrological cycle are key to the interception and infiltration of water into the soil for other processes, including hydraulic properties such as hydraulic conductivity and retention. In this section, we discuss how residue management affects these soil hydrological parameters.

Rainfall and irrigation

Crop residue management can affect rainfall and irrigation capture and irrigation needs depending on the level of removal. Residue removal at high rates may reduce the water infiltration rate and transmission, as discussed in previous sections of this chapter. High rates of residue removal expose large amounts of the soil surface to high-energy impact rain or irrigation drops. The dislodged soil particles result in smaller soil aggregates, leading to smaller pore sizes, and physical blocking of soil pores. Residue removal also leads to soil surface sealing and crusting (Turmel *et al.* 2015). Soil crusts can significantly reduce the water infiltration rate due to lower porosity and reduced macroporosity (Hardie and Almajmaie 2019). All these processes then lead to reductions in the water infiltration rate, and thus run-off, which is discussed in a following section.

Residue grazing at low to moderate rates results in fewer of these processes occurring due to the loss of small amounts of residues. However, residue burning can result in some of the same mechanisms as those responsible for the reduction in water infiltration rate, plus one additional process, which is the creation of hydrophobicity (water repellency). Through the combustion process, the residue and other organic matter can become hydrophobic or water repellent, which means they will not rapidly absorb water. Further, during the combustion process, the soil particles and organic matter can interact to create coatings on the soil particles or small aggregates that are also water repellent. Actions that chop and cut residue much shorter than harvest, such as baling or heavy grazing, can negatively affect rainfall or irrigation capture. For example, residue rates of 1–4 Mg ha^{-1} intercepted 6–18% of rainfall, an amount that increased as the cutting height increased (Swella *et al.* 2015). Further, in locations with small and frequent rainfall, heavy residue can intercept rainfall and prevent the transfer of rainwater to the soil (Turmel *et al.* 2015). In contrast, in wet environments, residue can lead to water logging and other signs of excessive soil moisture (Turmel *et al.* 2015).

Snow

Crop residues are important for the capture and retention of snowfall (Fig. 6.7). The standing stems of small grains and stalks of corn capture snowfall and prevent it from blowing off-site during blizzards or high winds. Standing stalks are key to snowfall retention and subsequent snowmelt. For example, stalks 68–74 cm tall captured more snowfall than stalks 43–46 cm tall, which captured more snowfall than flat stalks in winters with high wind speed and normal to above normal snowfall (Nielsen 1998). Not only is the height of the stalk important, but so is stalk diameter. Stalks with greater diameter and height were more effective in snow capture and water storage in spring (Nielsen 1998). The

Fig. 6.7. Snow capture and retention in an unbaled and baled field. Photograph by H. Blanco-Canqui.

retention of snowfall leads to snowmelt during warm weather or spring, thereby increasing soil water. The presence of residue also leads to shallower frost depths, with deeper snow cover, warmer temperatures and potentially faster warming in spring (Sharratt *et al.* 1998).

When crop residues are shredded and baled, the standing stems and stalks are no longer present to capture and retain snowfall. Even if the amount of residue removed through baling is less than the typical 60% of soil coverage, standing stalks are no longer present unless select rows are skipped. The residue remaining after a lower baling rate will capture snow, but to a lesser degree than standing stems and stalks. However, the baled residues could be stored on-site as a windbreak, increasing snow retention on a portion of the field. When crop residues are grazed, many of the stems and stalks remain in place, so grazing has fewer effects on snow retention than residue baling. Residue burning typically happens in spring in the US Great Plains, leaving residues in place during the winter to retain the snowfall and spring snowmelt. Therefore, residue management affects snow capture and retention in this order: baling > grazing > burning > no removal.

Run-off

Crop residue removal can increase the amount of run-off through a combination of factors. The first is the loss of protective residue cover over the soil, opening the soil to high-impact raindrops. Second, the impact of the raindrops leads to the disintegration of soil aggregates. Third, the disintegration of soil aggregates through rainfall can lead to surface sealing and crusting, lowering the water infiltration rate. Fourth, the reduction of soil organic matter through crop residue removal leads to lower aggregate stability (Sindelar *et al.* 2019), leading to further reductions in infiltration. Finally, the removal of residues, which would otherwise reduce the flow rate, reduces resistance to run-off. The combination of these forces leads to run-off or loss of water overland and reduces the amount of water entering the soil profile. High rainfall rates alone that exceed infiltration rates can result in run-off.

Residue baling at high rates can increase run-off depth, whereas light grazing may have little to no effect (Kenney *et al.* 2015; Blanco-Canqui *et al.* 2016). For example, at two of three sites in Kansas, which were no-till before and during the experiment, residue removal at any rate had no effect on run-off (Kenney *et al.* 2015). However, at the third site, which was conventionally tilled in the 3 years before and under no-till during the study, residue removal at 50% and 100% of soil coverage increased run-off depth by 11 and 17 mm respectively (Kenney *et al.* 2015).

The effects of residue baling and grazing on run-off depth may differ among rainfall patterns and soil textures (Blanco-Canqui *et al.* 2016). For example, on a clay loam with 816 mm annual precipitation, the run-off amount increased by 14 mm under baling compared with no removal and grazing (Blanco-Canqui *et al.* 2016). However, on a loam with 475 mm annual precipitation, baling and light grazing increased run-off depth by 4 mm compared with no removal, whereas heavy grazing increased run-off depth by 7 mm (Blanco-Canqui *et al.* 2016). These data suggest that higher-rainfall areas or soils with a higher clay content are less susceptible to erosion under low rates of residue removal under the conditions of the studies described.

One may expect that the effects of residue removal on sediment or soil loss would be similar to those on run-off depth. Sediment loss and run-off depth may follow similar rates of loss with increasing residue removal (Fig. 6.8). As shown in Fig. 6.8, the relationships between sediment loss and residue removal rate, as well as between sediment loss and run-off depth, differ by site and management, likely due to soil texture and management history. Note that in some sites, sediment loss increases exponentially with increases in crop residue removal. This means that as the rate of residue removal increases, the rate of sediment loss increases faster after removal of 50% residue coverage.

Residue burning may potentially increase run-off risks. Removal of the protective layer of residue, which intercepts and reduces raindrop force on soils, can lead to soil aggregate degradation via freeze–thaw and wet–dry cycles that are more extreme because crop residues moderate soil temperature fluctuations and reduce evaporation. Run-off losses may depend on year (i.e. rainfall amount and intensity) and may be site specific. For example, in semi-arid Washington state, burning winter wheat residues led to only tendencies for higher soil loss in 2 years compared with conventionally managed wheat residue due to high data variability (McCool *et al.* 2008). In contrast, in subtropical Brazil, residue burning with no-till soybean led to 89% higher phosphorus losses and 5% higher soil organic C losses than no-till without burning (Bertol *et al.* 2007). These data suggest that erosion losses may vary depending on site-specific factors.

Evaporation

Crop residue removal can significantly affect evaporation due to the increasing proportion of bare soil. Evaporation is related to albedo and the cover of residues. Albedo is the reflectivity or proportion of shortwave radiation reflected by a surface (Campbell and Norman

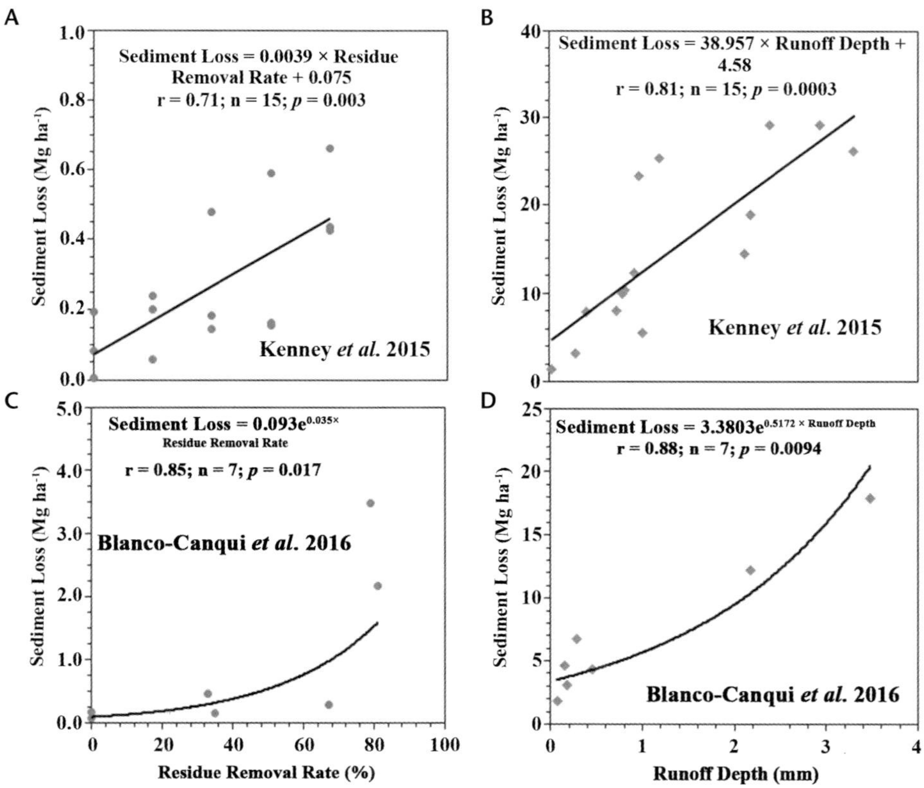

Fig. 6.8. Impact of residue removal rate on sediment loss across (a) three sites in Kansas and (b) two sites in Nebraska, and the relationship between run-off depth and sediment loss for (c) the three sites in Kansas and (d) two sites in Nebraska. The sites in Kansas had residue baling at rates of 0%, 25%, 50%, 75% and 100%, whereas the sites in Nebraska had treatments of no removal, bale or graze.

1998). The albedo of growing corn ranges from 0.18 to 0.22, whereas that of old snow is 0.40–0.70 and that of soil is 0.08–0.18. This means that growing corn reflects ~20% of shortwave radiation, whereas soil only reflects ~13% (Campbell and Norman 1998). Corn residue likely reflects radiation within the range of old snow (40–70%). This suggests that residues can reflect more radiation than soil, which reduces evaporation losses, although this can be complicated by characteristics of photosynthetic and non-photosynthetic vegetation. Residues also act as mulch, preventing the soil from absorbing shortwave radiation and subsequently warming.

Studies of evaporation usually measure the combination of evaporation and transpiration (plant water loss). In a 3-year study in Kansas, removal of 25–75% and 100% of corn residue cover increased cumulative evaporation by 12% and 19% respectively compared with no removal (Klocke *et al.* 2009). In a 2-year experiment in Nebraska with and without complete residue removal in a corn–soybean system, residue removal increased evapotranspiration in Year 1 by 0.66 mm day^{-1} and decreased it in Year 2 by 0.53 mm day^{-1} when averaged across the growing season (van Donk *et al.* 2012). Higher rates of evaporation and greater transpiration early in the season due to higher temperatures in the complete residue removal plots were attributed to the differences in evaporation (van Donk *et al.* 2012). Indeed, temperatures with 100% removal can be warmer than no removal (Kenney *et al.* 2015).

Drainage

Drainage is an important part of the hydrological cycle in the long term. Drainage refers to water movement down through the soil profile and the subsequent addition of said water to groundwater resources. Drainage that occurs too slowly lengthens the return to groundwater, whereas drainage that occurs too quickly can introduce pollutants and nutrients to the groundwater. In order for water to drain or percolate to groundwater it must first enter or infiltrate the soil and then be conducted or transmitted through the soil profile. Thus, any management practice that interferes with soil hydrology can affect drainage. A study using sandy clay loam soil from sugarcane fields with lysimeters under 0%, 25%, 50%, 75% and 100% of sugarcane straw removal showed that 75% and 100% removal had the lowest water storage and least drainage compared with 50% and no removal (Gmach *et al.* 2019). Gmach *et al.* (2019) attributed the dryness of the bare soil needing more water to bring the soil to a point where drainage can occur. They also attributed lower drainage due to surface sealing, degradation of surface soil aggregates and blocked macropores (Gmach *et al.* 2019). Remember from earlier in the chapter that residue removal at high rates reduces hydraulic conductivity or water transmission, thus resulting in reduced drainage. Further, high rates of residue removal may also reduce the water infiltration rate, which further impedes sufficient water entering the soil. Because residue grazing at low to moderate rates has minimal effects on infiltration and hydraulic conductivity, we expect that drainage would not be impeded.

Crop residue management under fluctuating climates

To withstand extreme weather events, soils must be able to drain quickly when rainfall is abundant and retain and slowly release water when rainfall is scarce. Based on the discussion in this chapter, in general, baling and grazing may not affect water infiltration, but baling may reduce water retention. In addition, burning may not affect water infiltration but may reduce water retention. Residue baling can occur with minimal impacts on hydraulic properties, but it needs to be done at low rates (<50%). Further, threshold levels of residue removal for different soils and climates need to be established based on the amount of residue produced.

Ensuring sufficient residue retention (at least 50% residue coverage) can retain soil organic C and thereby maintain soil hydraulic properties. Research in the US Corn Belt showed that the amount of residue needed to maintain soil C is greater than that required to reduce water and wind erosion (Wilhelm *et al.* 2007). Thus, if we work to retain soil C, we preserve soil properties and, at the same time, reduce water erosion losses. The question is, how much residue can be safely removed given a specific crop yield? Relationships among continuous corn or corn–soybean, different tillage systems, and the amount of harvestable residue given a specific grain yield, based on research on corn residue removal in the US Corn Belt by Wilhelm *et al.* (2007), are shown in Fig. 6.9. For example, a grain yield of 10 Mg ha^{-1} in continuous corn with no-till or conservation tillage would allow for ~2.5 Mg ha^{-1} to be harvested (Fig. 6.9). It should be noted that this is a broad average and the amount of residue harvested varies among regions due to climatic conditions at the field or even subfield level depending on soil texture, slope and management (Johnson *et al.* 2010, 2014). Similar principles could be used in other cropping and tillage systems. Other baling options may include occasional baling, such as in years when residue is abundant or baling in spring just before planting to avoid erosion events.

Because grazing has few impacts on soil hydraulic properties, it is important to use best management practices, such as avoiding wet fields and using appropriate stocking rates for

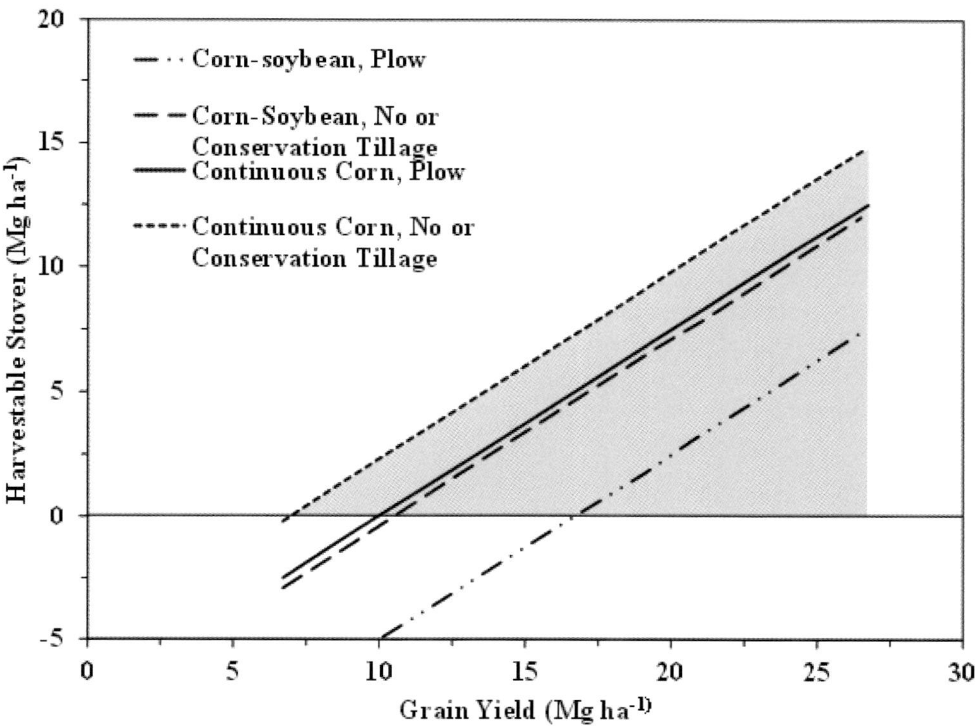

Fig. 6.9. Estimated levels of harvestable corn stover (residue) based on cropping and tillage systems and grain yield. The shaded area denotes what may be sustainably harvested. Redrawn from Wilhelm *et al.* (2007).

the amount of residue present. Regarding residue burning, waiting until close to planting time to burn residues may reduce the risk of water erosion events, but it also reduces water retention and has many other negative effects on air quality, so it would be better to bale at low to moderate rates or graze the residues.

There are some best management and support practices that can offset some of the negative effects of residue removal. Tillage can exacerbate negative effects of residue removal on soil hydraulic properties (Johnson *et al.* 2016; Wilhelm *et al.* 2007). However, no-till is often cited as the first management practice to begin the restoration of soil C lost due to agricultural activity. Thus, no-till is one support practice that may mitigate some of the effects of residue removal because no-till soils are said to be more resilient to management changes. However, a recent study across three sites in the Central and Western Great Plains showed no-till alone is not enough to offset the decreased soil C and aggregation observed with high rates (>60%) of residue removal (Stewart *et al.* 2019).

One support practice, when combined with no-till, that may have the potential to offset residue removal effects on soil hydraulic properties is the use of cover crops. Cover crops grow during the off-season and can add more organic matter to the soil. Indeed, recent reviews have shown cover crops increase soil C by 0.32–0.56 Mg ha^{-1} $year^{-1}$ in the surface 30 cm of the soil (Poeplau and Don 2015; Ruis and Blanco-Canqui 2017; Jian *et al.* 2020). However, the ability of cover crops to induce changes to soil C and soil hydraulic properties is dictated by the amount of biomass produced. Too often cover crops are terminated months or weeks before planting the main crop, giving the cover crop very little time to accumulate biomass (Ruis *et al.* 2017). However, waiting to terminate the cover crops late, such as at corn planting, may be one method to improve biomass production and improve soil properties related to soil hydrological properties, such as aggregation (Ruis *et al.* 2017). Another method to improve cover crop biomass production and thereby improve input to soil organic matter and hydraulic properties is through planting cover crops after a short-season crop such as corn silage or winter wheat.

Other organic amendments, such as animal manure, are options to offset the effects of residue removal on soil hydraulic properties. For example, studies have shown that animal manure can increase soil organic C, thereby increasing water infiltration, K_{sat} and water retention characteristics (Adesanya *et al.* 2016; Ozlu *et al.* 2019). However, the changes to these soil hydraulic properties depend on the manure application rate and site characteristics. One study evaluating the ability of animal manure to offset the impacts of corn residue removal showed small effects of animal manure on soil properties related to soil hydraulic properties after 3 years (Blanco-Canqui *et al.* 2014). These data indicate that manure may need to be applied at high rates or for many years to offset the impacts of residue removal on soil hydraulic properties.

Summary

Crop residues are important to maintain soil organic matter and generate an income source from grazing or residue bales. Crop residues also present a challenge through non-uniform soil surface, planting equipment issues and poor seed–soil contact. The effects of residue baling, grazing and burning on soil hydraulic properties are site specific, but nonetheless some general conclusions can be drawn. Crop residue baling may not always reduce water infiltration rates but can reduce water retention. The minimal changes in water infiltration rates were related to the small changes in macroporosity, whereas the changes in water retention were related to changes in soil organic matter and soil aggregation. Crop residue grazing at light to moderate rates had no effect on soil hydraulic properties, probably attributable to the low residue removal rate. Unlike residue grazing, residue burning had mixed or no effect on water infiltration, hydraulic conductivity and water retention. The variable effects on some hydraulic properties with residue burning were probably the result of hydrophobicity from the charred residues. However, residue removal can increase run-off losses of soils and nutrients, reduce snow capture and increase evapotranspiration depending on the level of residue removal. Thus, grazing, or low rates of baling, may preserve soil hydraulic properties, particularly under fluctuating climatic conditions where we must plan for the unpredictable.

References

Adesanya T, Akinremi O, Zvomuya F (2016) Physical properties of an Orthic Black Chernozem after 5 years of liquid and solid pig manure application to annual and perennial crops. *Canadian Journal of Soil Science* **96**, 145–153. doi:10.1139/cjss-2015-0086

Baumhardt RL, Johnson GL, Schwartz RC, Brauer DK (2017) Grazing and tillage effects on soil properties, rain infiltration, and sediment transport during fallow. *Soil Science Society of America Journal* **81**, 1548–1556. doi:10.2136/sssaj2017.04.0133

Bertol I, Engel FL, Mafra AL, Bertol OJ, Ritter SR (2007) Phosphorus, potassium and organic carbon concentrations in runoff water and sediments under different soil tillage systems during soybean growth. *Soil & Tillage Research* **94**, 142–150. doi:10.1016/j.still.2006.07.008

Bescansa P, Imaz MJ, Virto I, Enrique A, Hoogmoed WB (2006) Soil retention as affected by tillage and residue management in semiarid Spain. *Soil & Tillage Research* **87**, 19–27. doi:10.1016/j.still.2005.02.028

Bhuvaneshwari S, Hettiarachchi H, Meegoda JN (2019) Crop residue burning in India: policy challenges and potential solutions. *International Journal of Environmental Research and Public Health* **16**, 832. doi:10.3390/ijerph16050832

Blanco-Canqui H, Lal R (2009) Crop residue removal impacts on soil productivity and environmental quality. *Critical Reviews in Plant Sciences* **28**, 139–163. doi:10.1080/07352680902776507

Blanco-Canqui H, Ruis SJ (2018) No-till and the soil physical environment. *Geoderma* **326**, 164–200. doi:10.1016/j.geoderma.2018.03.011

Blanco-Canqui H, Lal R, Post WM, Izaurralde RC, Shipitalo MJ (2007) Soil hydraulic properties influenced by corn stover removal from no-till corn in Ohio. *Soil & Tillage Research* **92**, 144–155. doi:10.1016/j.still.2006.02.002

Blanco-Canqui H, Stephenson RJ, Nelson NO, Presley DR (2009) Wheat and sorghum residue removal for expanded uses increases sediment and nutrient loss in runoff. *Journal of Environmental Quality* **38**, 2365–2372. doi:10.2134/jeq2009.0021

Blanco-Canqui H, Ferguson RB, Jin VL, Schmer MR, Weinhold BJ, Tatarko J (2014) Can cover crop and manure maintain soil properties after stover removal from irrigated no-till corn? *Soil Science Society of America Journal* **78**, 1368–1377. doi:10.2136/sssaj2013.12.0550

Blanco-Canqui H, Stalker AL, Rasby R, Shaver TM, Drewnoski ME, van Donk S, Kibet L (2016) Does cattle grazing and baling of corn residue increase water erosion? *Soil Science Society of America Journal* **80**, 168–177. doi:10.2136/sssaj2015.07.0254

Campbell GS, Norman J (1998) *An Introduction to Environmental Biophysics.* Springer-Verlag, New York.

Castellini M, Niedda M, Pirastru M, Ventrella D (2014) Temporal changes of soil physical quality under two residue management systems. *Soil Use and Management* **30**, 423–434. doi:10.1111/sum.12137

Celik I, Turgit MM, Acir N (2012) Crop rotation and tillage effects on selected soil physical properties of a Typic Haploxerert in an irrigated semi-arid Mediterranean region. *International Journal of Plant Production* **6**, 457–480.

Chalise KS, Singh S, Wegner BR, Kumar S, Perez-Gutierrez JD, Osborne SL, Nleya T, Guzman J, Rohila JS (2018) Cover crops and returning residue impact on soil organic carbon, bulk density, penetration resistance, water retention, infiltration, and soybean yield. *Agronomy Journal* **110**, 99–108.

Chen J, Gong Y, Wang S, Guan B, Balkovic J, Kraxner F (2019) To burn or retain crop residues on croplands? An integrated analysis of crop residue management in China. *The Science of the Total Environment* **662**, 141–150. doi:10.1016/j.scitotenv.2019.01.150

Choudhary M, Rana KS, Meena MC, Bana RS, Jakhar P, Ghasal PC, Verma RK (2019) Changes in physico-chemical and biological properties of soil under conservation agriculture based pearl millet-mustard cropping system in rainfed semi-arid region. *Archives of Agronomy and Soil Science* **65**, 911–927. doi:10.1080/03650340.2018.1538556

Dai J, Bean B, Brown B, Bruening W, Edwards J, Flowers M, Karow R, Lee C, Morgan G, Ottman M, et al. (2016) Harvest index and straw yield of five classes of wheat. *Biomass and Bioenergy* **85**, 223–227. doi:10.1016/j.biombioe.2015.12.023

Desrochers J, Brye KR, Gbur E, Mason RE (2019) Infiltration as affected by long-term residue and water management on loess-derived soil in eastern Arkansas, USA. *Geoderma Regional* **16**, e00203. doi:10.1016/j.geodrs.2019.e00203

Feng Z, Leung R, Hagos S, Houze RA, Burleyson CD, Balaguru K (2016) More frequent intense and long-lived storms dominate the springtime trend in central US rainfall. *Nature Communications* **7**, 13429. doi:10.1038/ncomms13429

Fernández FG, Sorensen BA, Villamil MB (2015) A comparison of soil properties after five years of no-till and strip-till. *Agronomy Journal* **107**, 1339–1346. doi:10.2134/agronj14.0549

Feiziene D, Feiz V, Karklins A, Versuliene A, Janusauskaite D, Antanaitis S (2018) After-effects of long-term tillage and residue management on topsoil state in Boreal conditions. *European Journal of Agronomy* **94**, 12–24. doi:10.1016/j.eja.2018.01.003

Gangwar KS, Singh KK, Sharma SK, Tomar OK (2006) Alternative tillage and crop residue management in wheat after rice in sandy loam soils of Indo-Gangetic Plains. *Soil & Tillage Research* **88**, 242–252. doi:10.1016/j.still.2005.06.015

Gmach MR, Scarpare FB, Cherubin MR, Lisboa IP, Belarmino dos Santos AK, Pellegrino Cerri CE, Cerri CC (2019) Sugarcane straw removal effects on soil water storage and drainage in southeastern Brazil. *Journal of Soil and Water Conservation* **74**, 466–476. doi:10.2489/jswc.74.5.466

Gómez-Paccard C, Hontoria C, Mariscal-Sancho I, Perez J, Leon P, Gonzalez P, Espejo R (2015) Soil-water relationships in the upper soil layer in a Mediterranean Palexerult as affected by no-till under excess water conditions – influence on crop yield. *Soil & Tillage Research* **146**, 303–312. doi:10.1016/j.still.2014.09.012

Gonzalez JM, Laird DA (2003) Carbon sequestration in clay mineral fractions from ^{14}C-labeled plant residues. *Soil Science Society of America Journal* **67**, 1715–1720. doi:10.2136/sssaj2003.1715

Graham RL, Nelson R, Sheehan J, Perlack RD, Wright LL (2007) Current and potential US corn stover supplies. *Agronomy Journal* **99**, 1–11. doi:10.2134/agronj2005.0222

Hardie M, Almajmaie A (2019) Measuring and estimating the hydrological properties of a soil crust. *Journal of Hydrology* **574**, 12–22. doi:10.1016/j.jhydrol.2019.04.031

Herrero M, Thornton PK, Notenbaert AM, Wood S, Msangi S, Freeman HA, Bossio D, Dixon J, Peters M, Van De Steeg J, Lynam J, Parthasarathy Rao P, Macmillan S, Gerard B, Mcdermott J, Seré C, Rosegrant M (2010) Smart investments in sustainable food production, revisiting mixed crop–livestock systems. *Science* **327**, 822–825. doi:10.1126/science.1183725

Hillel D (2004) *Environmental Soil Physics*. Academic Press, San Diego.

Jian J, Du X, Reiter MS, Stewart RD (2020) A meta-analysis of global cropland soil carbon changes due to cover cropping. *Soil Biology & Biochemistry* **143**, 107735. doi:10.1016/j.soilbio.2020.107735

Johnson JMF, Allmaras RR, Reicosky DC (2006) Estimating source carbon from crop residues, roots and rhizodeposits using the national grain-yield database. *Agronomy Journal* **98**, 622–636. doi:10.2134/agronj2005.0179

Johnson JMF, Karlen DL, Andrews SS (2010) Conservation considerations for sustainable bioenergy feedstock production: if, what, where, and how much? *Journal of Soil and Water Conservation* **65**, 88A–91A. doi:10.2489/jswc.65.4.88A

Johnson JMF, Novak JM, Varvel GE, Stott DE, Osborne SL, Karlen DL, Lamb JA, Baker J, Adler PR (2014) Crop residue mass needed to maintain soil organic carbon levels: can it be determined? *BioEnergy Research* **7**, 481–490. doi:10.1007/s12155-013-9402-8

Johnson JMF, Strock JS, Tallaksen JE, Reese M (2016) Corn stover harvest changes soil hydrology and soil aggregation. *Soil & Tillage Research* **161**, 106–115. doi:10.1016/j.still.2016.04.004

Johnson JMF, Strock JS, Barbour NW, Tallaksen JE, Reese M (2017) Stover harvest impacts soil and hydrologic properties on three Minnesota farms. *Soil Science Society of America Journal* **81**, 932–944. doi:10.2136/sssaj2017.01.0007

Kenney I, Blanco-Canqui H, Presley DR, Rice CW, Janssen K, Olson B (2015) Soil and crop response to stover removal from rainfed and irrigated corn. *Global Change Biology. Bioenergy* **7**, 219–230. doi:10.1111/gcbb.12128

Kibet LC, Blanco-Canqui H, Jasa P (2016) Long-term tillage impacts on soil organic matter components and related properties on a Typic Argiudoll. *Soil & Tillage Research* **155**, 78–84. doi:10.1016/j.still.2015.05.006

Kleber M, Sollins P, Sutton R (2007) A conceptual model of organo-mineral interactions in soils: self-assembly of organic molecular fragments into zonal structures on mineral surfaces. *Biogeochemistry* **85**, 9–24. doi:10.1007/s10533-007-9103-5

Klocke NL, Currie RS, Aiken RM (2009) Soil water evaporation and crop residues. *Transactions of the ASABE* **52**, 103–110. doi:10.13031/2013.25951

Kumar S, Kadono A, Lal R, Dick W (2012) Long-term tillage and crop rotations for 47–49 years influences hydrological properties of two soils in Ohio. *Soil Science Society of America Journal* **76**, 2195–2207. doi:10.2136/sssaj2012.0098

Libohova Z, Seybold C, Wysocki D, Wills S, Schoeneberger P, Williams C, Lindbo D, Stott D, Owens PR (2018) Reevaluating the effects of soil organic matter and other properties on available water-holding capacity using the National Cooperative Soil Survey Characterization Database. *Journal of Soil and Water Conservation* **73**, 411–421. doi:10.2489/jswc.73.4.411

McCarty JL, Korontzi S, Justice CO, Loboda T (2009) The spatial and temporal distribution of crop residue burning in the contiguous United States. *The Science of the Total Environment* **407**, 5701–5712. doi:10.1016/j.scitotenv.2009.07.009

McCool DK, Pannkuk CD, Kennedy AC, Fletcher PS (2008) Effects of burn/low-till on erosion and soil quality. *Soil & Tillage Research* **101**, 2–9. doi:10.1016/j.still.2008.05.007

Minasny B, McBratney AB (2018) Limited effect of organic matter on soil available water capacity. *European Journal of Soil Science* **69**, 39–47. doi:10.1111/ejss.12475

Moebius-Clune BN, van Es HM, Idowu OJ, Schindelbeck RR, Moebius-Clune DJ, Wolfe DW, Abawi GS, Theis JE, Gugino BK, Lucey R (2008) Long-term effects of harvesting maize stover and tillage on soil quality. *Soil Science Society of America Journal* **72**, 960–969. doi:10.2136/sssaj2007.0248

Nielsen DC (1998) Snow catch and soil water recharge in standing sunflower residue. *Journal of Production Agriculture* **11**, 476–480. doi:10.2134/jpa1998.0476

National Centers for Environmental Information, National Oceanic and Atmospheric Administration (NOAA) (2012) State of the climate: drought for June 2012. NOAA, <https://www.ncdc.noaa.gov/sotc/drought/201206>.

Obrycki JF, Karlen DL (2018) Is corn stove harvest predictable using farm operation, technology, and management variables? *Agronomy Journal* **110**, 749–757. doi:10.2134/agronj2017.08.0504

Ojekanmi A, Johnson JMF (2021) Quantifying improvements in water stable aggregation due to corn stover retention. *Soil Science Society of America Journal* **85**, 776–788. doi:10.1002/saj2.20198

Ozlu E, Kumar S, Arriaga FJ (2019) Responses of long-term cattle manure on soil physical and hydraulic properties under a corn–soybean rotation at two locations in Eastern South Dakota. *Soil Science Society of America Journal* **83**, 1459–1467. doi:10.2136/sssaj2019.03.0077

Poeplau C, Don A (2015) Carbon sequestration in agricultural soils via cultivation of cover crops – a meta-analysis. *Agriculture, Ecosystems & Environment* **200**, 33–41. doi:10.1016/j.agee.2014.10.024

Prasad R, Gangaiah B, Aipe KC (1999) Effect of crop residue management in a rice–wheat cropping system on growth and yield of crops and on soil fertility. *Experimental Agriculture* **35**, 427–435. doi:10.1017/S001447979935403X

Rakkar MK, Blanco-Canqui H (2018) Grazing of crop residues: impacts on soils and crop production. *Agriculture, Ecosystems & Environment* **258**, 71–90. doi:10.1016/j.agee.2017.11.018

Rakkar MK, Blanco-Canqui H, Drijber RA, Drewnoski ME, MacDonald JC, Klopfenstein T (2017) Impacts of cattle grazing of corn residues on soil properties after 16 years. *Soil Science Society of America Journal* **81**, 414–424.

Rakkar MK, Blanco-Canqui H, Rasby RJ, Ulmer K, Cox-O'Neill J, Drewnoski ME, Drijber RA, Jenkins K, MacDonald JC (2018) Grazing crop residues has less impact in the short-term on soil properties than baling in the Central Great Plans. *Agronomy Journal* **110**, 109–121.

Ruis SJ, Blanco-Canqui H (2017) Cover crops could offset crop residue removal effects on soil carbon and other soil properties: a review. *Agronomy Journal* **109**, 1785–1805. doi:10.2134/agronj2016.12.0735

Ruis SJ, Blanco-Canqui H, Jasa PJ, Ferguson RB, Slater G (2017) Can cover crop use allow increased levels of corn residue removal for biofuel in irrigated and rainfed systems? *BioEnergy Research* **10**, 992–1004. doi:10.1007/s12155-017-9858-z

Ruis S, Blanco-Canqui H, Burr C, Olson B, Reiman M, Rudnick D, Drijber R, Shaver T (2018) How does corn residue baling and grazing under different irrigation and tillage levels affect soil properties? *Soil Science Society of America Journal* **82**, 202–213. doi:10.2136/sssaj2017.05.0177

Ruis SJ, Blanco-Canqui H, Elmore RW, Proctor C, Koehler-Cole K, Shapiro CA, Francis CA, Ferguson RB (2020) Preharvest-planted cover crops and mixes did not change most soil properties after four years. *Agronomy Journal* **112**, 1649–1665. doi:10.1002/agj2.20143

Saunders S, Findlay D, Easley T, Spencer T (2012) Doubled trouble: more Midwestern extreme storms. The Rocky Mountain Climate Organization and Natural Resources Defense Council, <http://www.rockymountainclimate.org/images/DoubledTroubleHigh.pdf>.

Schmer MR, Brown RM, Jin VL, Mitchell RB, Redfearn DD (2017) Corn residue use by livestock in the United States. *Agricultural & Environmental Letters* **2**, 160043. doi:10.2134/ael2016.10.0043

Sharratt BS, Benoit GR, Voorhees WB (1998) Winter soil microclimate altered by corn residue management in the northern Corn Belt of the USA. *Soil & Tillage Research* **49**, 243–248. doi:10.1016/S0167-1987(98)00181-0

Sindelar M, Blanco-Canqui H, Jin VL, Ferguson R (2019) Cover crops and corn residue removal: impacts on soil properties and their relationships with carbon. *Soil Science Society of America Journal* **83**, 221–231. doi:10.2136/sssaj2018.06.0225

Singh D, Chanasyk DS, McGill WB (1996) Soil hydraulic properties of an Orthic Black Chernozem under long-term tillage and residue management. *Canadian Journal of Soil Science* **76**, 63–71. doi:10.4141/cjss96-010

Six J, Conant RT, Paul EA, Paustian K (2002) Stabilization mechanisms of soil organic matter: implications for C-saturation of soils. *Plant and Soil* **241**, 155–176. doi:10.1023/A:1016125726789

Skidmore EL, Layton JB, Armbrust DV, Hooker ML (1986) Soil physical properties as influenced by cropping and residue management. *Soil Science Society of America Journal* **50**, 415–419. doi:10.2136/sssaj1986.03615995005000020032x

Smil V (1999) Crop residues: agriculture's largest harvest. *Bioscience* **49**, 299–308. doi:10.2307/1313613

Sokhansanj S, Mani S, Stumborg M, Samson R, Fenton J (2006) Production and distribution of cereal straw on the Canadian prairies. *Canadian Biosystems Engineering* **48**, 39–46.

Stevenson FC, van Kessel C (1997) Nitrogen contribution of pea residue in a hummocky terrain. *Soil Science Society of America Journal* **61**, 494–503. doi:10.2136/sssaj1997.03615995006100020018x

Stewart CE, Roosendaal DL, Sindelar A, Pruessner E, Jin VL, Schme MR (2019) Does no-tillage mitigate stover removal in irrigated continuous corn? A multi-location assessment. *Soil Science Society of America Journal* **83**, 733–742. doi:10.2136/sssaj2018.09.0352

Swella GB, Ward PR, Siddique KHM, Flower KC (2015) Combinations of tall standing and horizontal residue affect soil water dynamics in rainfed conservation agriculture systems. *Soil & Tillage Research* **147**, 30–38. doi:10.1016/j.still.2014.11.004

Tisdall JM, Oades JM (1982) Organic matter and water stable aggregates in soils. *Journal of Soil Science* **33**, 141–163. doi:10.1111/j.1365-2389.1982.tb01755.x

Tomasek BJ, Williams MM, Davis AS (2017) Changes in field workability and dorught risk from projected climate change drive spatially variable risks in Illinois cropping systems. *PLoS One* **12**, e0172301. doi:10.1371/journal.pone.0172301

Tormena CA, Karlen DL, Logsdon S, Cherubin MR (2017) Corn stover harvest and tillage impacts on near-surface soil physical quality. *Soil & Tillage Research* **166**, 122–130. doi:10.1016/j.still.2016.09.015

Turmel MS, Siperatti A, Baudron F, Verhulst N, Govaerts B (2015) Crop residue management and soil health: a systems analysis. *Agricultural Systems* **134**, 6–16. doi:10.1016/j.agsy.2014.05.009

Valzano FP, Greene RSB, Murphy BW (1997) Direct effects of stubble burning on soil hydraulic and physical properties in direct tillage system. *Soil & Tillage Research* **42**, 209–219. doi:10.1016/S0167-1987(96)01101-4

van Donk SJ, Shaver TM, Peterson JL, Davidson DR (2012) Effects of crop residue removal on soil water content and yield of deficit-irrigated soybean. *Transactions of the ASABE* **55**, 149–157. doi:10.13031/2013.41266

Wang G, Wang D, Trenberth KE, Erfanian A, Yu M, Bosilovich MG, Parr DT (2017) The peak structure and fututre changes of the relationships between exstreme precipitation and temeprature. *Nature Climate Change* **7**, 268–274. doi:10.1038/nclimate3239

Wilhelm WW, Johnson JMF, Karlen DL, Lightle DT (2007) Corn stover to sustain soil organic carbon further constrains biomass supply. *Agronomy Journal* **99**, 1665–1667. doi:10.2134/agronj2007.0150

Wilhelm WW, Johnson JMF, Lightle DT, Karlen DL, Novak JM, Barbour NW, Laird DA, Baker J, Ochsner TE, Halvorson AD, Archer DW, Arriaga F (2011) Vertical distribution of corn stover dry mass grown at several US locations. *BioEnergy Research* **4**, 11–21. doi:10.1007/s12155-010-9097-z

Wuest SB, Caesar-TonThat TC, Wright SF, Williams JD (2005) Organic matter addition, N, and residue burning effects on infiltration, biological, and physical properties of an intensively tilled silt-loam soil. *Soil & Tillage Research* **84**, 154–167. doi:10.1016/j.still.2004.11.008

7

Conservation buffers and soil water

Salah M. Alagele, Stephen H. Anderson, Ranjith P. Udawatta

Introduction

Increasing population growth over the past four decades, from 4.5 billion people in 1980 to 7.8 billion in 2020, and expected to increase to 9.1 billion by 2050, has associated increases in demand for food, fibre and fuel (United Nations Department of Economic and Social Affairs 2009). In addition, higher world population growth rates have contributed to an unsustainable consumption of natural resources (soil and water) and accelerated the climate change (Nema *et al.* 2012). Soil and essential nutrients are continuously being depleted due to climate change, as well as modernised intensive agricultural operations and reduced return of organic matter to the soil. Degraded soil and impaired water quality can be attributed to a misuse of chemical fertilisers, pesticides and tillage practices. Depletion of fresh water in aquifers and rivers has been increased to target higher crop production and meet food demands (Spiertz and Ewert 2009; Alagele *et al.* 2019a). Soil and water management for climate change adaptation are important approaches for environmental conservation, sustainability of cropping systems, soil and water quality and food security (Lal *et al.* 2011). Thus, sustainable maintenance, rehabilitation and improvement of soil and water resources are required for land use practices to minimise further deterioration of these natural resources (Bossio *et al.* 2010). Implementation of effective sustainable agricultural approaches is needed to conserve soil and water and to improve agricultural production to meet the demands of increasing population. A conservation buffer (CB) system is a strategy that provides many benefits for soil and water and may enhance adaptation to climate change.

CBs are conservation vegetation perennial management systems. These systems use trees, grasses, shrubs or a combination of any two or three vegetation types that are established as narrow strips of vegetation widely prescribed to reduce sediment and chemical transport and improve landscape diversity (Jose 2012; Kumar *et al.* 2012; Fig. 7.1). Studies have confirmed the practical importance of the integration of multispecies CBs within continuous monocrop management in improving soil quality by increasing soil organic carbon (SOC; Veum *et al.* 2012; Alagele *et al.* 2019a). In addition, CBs such as agroforestry (ABs) and grass (GBs) buffers help reduce non-point source pollution (NPSP) from row crop (RC) areas by improving soil hydraulic properties, increasing water storage and minimising run-off (Udawatta *et al.* 2002; Seobi *et al.* 2005; Blanco-Canqui *et al.* 2014).

Fig. 7.1. Agroforestry and grass–legume integrated with switchgrass biomass in north-eastern Missouri, USA. (Photographs taken in the spring and summer of 2017 by S. Alagele.)

Benefits of CBs on soil hydraulic and physical properties, including saturated hydraulic conductivity (K_{sat}), available water capacity, bulk density (Db), soil aggregate stability and pore size distribution (PSD), as well as on soil hydrology (e.g. infiltration, water storage, evapotranspiration (ET) and redistribution), have been widely and well documented (Rachman *et al.* 2004a; Seobi *et al.* 2005; Kumar *et al.* 2008; Mudgal *et al.* 2010; Zaibon *et al.* 2016; Alagele *et al.* 2019b). These articles report that the combination of various CBs with cropping systems enhances soil physical and hydraulic properties and contributes higher amounts of SOC. Knowledge of soil physical properties and behaviours is crucial for the successful and sustainable application of CBs to achieve many beneficial effects on soil and crop production and ecosystem functions. The link between applying CBs and improving soil physical and hydraulic properties is SOC. Increased SOC contributes to increased K_{sat}, aggregation, soil available water capacity and other soil properties.

Soil hydrology concerns the retention and transport of water in the soil. The movement of water between the soil and atmosphere, in the soil and in the groundwater can be divided into run-off, infiltration, soil water storage and ET (Novák and Hlaváčiková 2019). Water movement and PSD through the soil profile can also be altered by applying CBs. Researchers have shown that the establishment of CBs positively affects soil hydrology and increases crop yields, and may provide economic incentives for farmers near cities and towns (Anderson *et al.* 2009; Kumar *et al.* 2012; Zaibon *et al.* 2017). Therefore, the use of CBs in the long term may improve soil hydrology.

This chapter reviews the importance of applying CBs (trees, grasses, shrubs), particularly integrated with a continuous monocrop agricultural system, and that their use in various soils and under different climatic conditions may improve soil hydraulic, physical and hydrological properties, concomitantly improving agricultural sustainability. The specific objective of this chapter is to determine the effects of CBs on selected soil hydraulic and physical properties and soil hydrology using published studies from selected climatic regions.

CB systems

The continuous use of intensive agricultural monocropping systems causes soil erosion, returns less SOC and wastes scarce natural resources (Pramanik *et al.* 2019). The selection of better varieties of trees and the potential use of drought-tolerant crops are considered the best strategies to enhance SOC levels and water use efficiency to reduce soil erosion (Montgomery 2007; Evans and Sadler 2008). The most important types of CBs are ABs, biomass crops and grass hedges and riparian buffers, as discussed in detail below.

Conservation buffers

CBs have become a promising production approach that can decrease soil disturbance, preserve permanent vegetation cover on most of the soil surface for the entire year and achieve higher crop productivity (Lovell and Sullivan 2006; Hobbs 2007). CBs include any species of trees or grasses (e.g. ABs, GBs, riparian buffers, grass waterways (GWWs) and biomass crops) grown in narrow strips of permanent vegetation to enhance many ecosystem services (Rachman *et al.* 2004b; Kumar 2009; Blanco-Canqui 2010). Multipurpose CBs can be installed where they may be considered a sustainable, cost-effective method for crop production and climate change mitigation (Derpsch 2008). The use of CBs under various climate change settings in different landscapes can help reduce production costs and mitigate threats to soil health (Matocha *et al.* 2012). CBs represent an approach for maintaining and protecting soil and water resources, and deserve wider application to promote agricultural production (Alagele *et al.* 2019a). CBs protect the ground with living, and dead, vegetation and root materials and increase the quantity of plant residues that can be readily utilised and decomposed by soil organisms, and thus may increase SOC (Brady and Weil 2008; Han *et al.* 2018). Thus, CBs play an important role in enhancing soil microbial biomass and provide more appropriate habitat for soil microbial species relative to RC management (Alagele *et al.* 2020a).

Agroforestry

Agroforestry, which may be established as filter strips, is a land management system that incorporates trees in the landscape and traditional plants simultaneously in the same area for environmental and economic benefits (Ramachandran Nair 1993, 2007). Agroforestry combines trees, grasses or shrubs with crops and/or livestock systems (Gold and Garrett 2009; Ramachandran Nair *et al.* 2009). A principal benefit of agroforestry management is that the combination of trees with crops has a greater potential for the production of food, forage and fibre than any one element alone (Sauer and Hernandez-Ramirez 2011). ABs have been used to enhance soil quality by increasing the addition of SOC and enhancing soil microbial biomass (Paudel *et al.* 2011; Lorenz and Lal 2014; Weerasekara *et al.* 2016). Agroforestry is being promoted as an alternative management practice that can provide additional income to farms (Garrity 2004; Dhyani *et al.* 2009) and can be important in improving soil health for future sustainable production (Dollinger and Jose 2018; Anderson and Udawatta 2019). ABs enhance soil conservation and water resources in temperate areas, especially on claypan landscapes (Jose 2009). A strategically located AB can provide positive environmental benefits, including improvements in soil hydraulic properties, reductions in run-off and mitigation of climate change (Udawatta *et al.* 2002; Mbow *et al.* 2014).

Biomass crops and grass hedges

Biomass crops and grass hedges are considered effective types of CBs and have been used to promote soil conservation (Blanco-Canqui *et al.* 2017). Perennial warm-season grasses, such as switchgrass (*Panicum virgatum* L.), have been widely examined as a biofuel feedstock that has a broad range of benefits on soil properties and the environment (Monti 2012; Blanco-Canqui 2016). The advantages of perennial grasses (e.g. grass hedges) in conserving soil have gained the attention of soil scientists because they provide numerous benefits (Ma *et al.* 2000; Zaibon *et al.* 2016). Perennial grasses enhance soil and water quality while increasing the production of biofuel feedstocks on degraded lands (Bonin *et al.* 2012; Zaibon *et al.* 2017). The long-term use of switchgrass hedges can provide numerous benefits, including increased water use efficiency, reduced chemical leaching, enhanced soil microbial activity and augmented SOC (Zeri *et al.* 2013). The integration of switchgrass and trees can have multiple environmental benefits and help mitigate or adapt to climate change.

Riparian buffers

Riparian buffers are another successful system in which trees and grasses are grown for soil and water conservation, and they have long been a feature of agricultural landscapes (McKergow *et al.* 2003). Riparian buffers play a major role in many ecosystem functions, such as trapping or reducing NPSP transported from upslope areas (Stutter *et al.* 2012). Riparian buffers are broad strips of a permanent combination of woody and herbaceous vegetation established near a stream along edges of agricultural fields that have been designed to reduce chemical transport to water sources (Blanco-Canqui and Lal 2008). Riparian buffers are used in both agricultural and urban soils along streams to control sediment transport (Hickey and Doran 2004). The use of riparian buffers near streams can improve surface water quality run-off from agricultural areas by reducing off-site sediment and other chemicals (Liu *et al.* 2008). Along with these benefits, riparian buffers can improve SOC levels and soil biodiversity (Kremen and Miles 2012).

Improving inputs of SOC is the key for the various soil services provided by riparian buffers because soil carbon has benefits on many soil properties. The growth and decay of large and deep tree roots in riparian buffers increase SOC and create numerous different-sized channels and pores, especially macropores, which greatly enhance the ability of soil to percolate and conduct water compared with cropping systems (Alagele *et al.* 2019a, 2020b). Macropores created by decayed roots in riparian buffers allow surface water to easily enter into the soil and increase soil water storage compared with soils that are managed without buffers (Rachman *et al.* 2005). These changes help improve soil quality properties and soil productivity. For example, the extensive deep root systems of these trees, grasses and shrubs reduce run-off from watersheds in diverse landscapes in various climate regions (Udawatta *et al.* 2002, 2004).

Effects of CBs on soil hydraulic and physical properties

The CBs provide many potential benefits that build and sustain good levels of SOC. Because there is a mutual relationship between SOC and soil hydraulic and physical properties, establishing CBs can enhance soil properties. Increases in SOC inputs may enhance soil structure, K_{sat}, PSD and other properties and improve the quantity, diversity and activity of soil organisms (Alagele *et al.* 2019a, 2019b, 2020a). Inputs of SOC produce organic binding agents (e.g. polysaccharides and glycoproteins) that enhance the development of soil structure by stabilising soil aggregates, which are strongly related to other soil properties and soil moisture conditions. The effects of CBs and their benefits on soil hydraulic and physical properties are discussed below.

Saturated hydraulic conductivity

K_{sat} is an indicator of water movement in soil layers, affects solute transport and influences patterns of infiltration and run-off (Becker *et al.* 2018). K_{sat} depends on the PSD and pore continuity, especially macropores, which are created by roots of perennial plants and macrofauna that form channels that facilitate water movement (Brady and Weil 2008). K_{sat} reflects how fast water flows through the soil profile when soil pores are filled with water (Anderson and Udawatta 2019). Knowledge of K_{sat} is beneficial in determining environmental management for water conservation and chemical movement into groundwater and water recharge (Hillel 2004). K_{sat} is significantly affected by land management. Significantly greater K_{sat} has been observed with CBs such as AB and GB, restored prairie, native grass, biomass crops and soil without tillage in agriculture (Kumar 2009). Several researchers have studied the effects of various CBs on K_{sat} in humid and temperate regions, finding that K_{sat} was higher with CBs than annual RC management (Rachman *et al.* 2004a; Seobi *et al.* 2005; Kumar *et al.* 2008; Mudgal *et al.* 2010; Zaibon *et al.* 2016; Alagele *et al.* 2019b).

Seobi *et al.* (2005) investigated the effects of ABs and GBs 6 years after establishment on K_{sat} for a Vertic Albaqualf (claypan soil) compared with RC management. K_{sat} in the GBs and ABs were 3- and 14-fold greater respectively than in RC. In the same watersheds, Alagele *et al.* (2019b) evaluated the effects of AB and GBs 20 years after their establishment, as well as that of biomass crops 5 years after establishment, on K_{sat}. The mean K_{sat} values for the ABs, biomass crops and GBs were 143%, 240% and 23% higher than for the RC respectively, especially at the soil surface (Fig. 7.2a). Both studies indicate that K_{sat} is higher under long-term CBs than with RC management because these systems may create deep root systems, enhance soil structure, add more SOC and increase macropores (Seobi *et al.* 2005; Alagele *et al.* 2019b). CB enhancement of K_{sat} may contribute to the control of NPSP from RC areas.

Kumar *et al.* (2008) observed that K_{sat} values were significantly higher in both ABs and GBs (61.3 and 57.0 mm h^{-1} respectively) than in rotationally and continuously grazed pasture (3.1 and 4.0 mm h^{-1} respectively). Similarly, Zaibon *et al.* (2016) reported that the K_{sat} under switchgrass (122.6 mm h^{-1}) was 74% higher than that under RC (32.5 mm h^{-1}) for a Mexico silt loam soil (Fig. 7.2b). Rachman *et al.* (2004a) evaluated the effect of stiff-stemmed grass hedge systems on K_{sat} for a Monona silt loam soil in Iowa, USA, finding that K_{sat} values were sixfold greater under a switchgrass hedge than under cropped areas. Other studies have also assessed the effects of various CBs on K_{sat}, finding that K_{sat} is higher in remnant uncultivated prairie than under long-term RC management (Fuentes *et al.* 2004; Mudgal *et al.* 2010). These differences are due to diverse long-term perennial CBs, the roots of which may produce root exudate materials that stimulate greater microbial abundance and more diverse microbial communities (e.g. bacteria, fungi and many other species; Lupwayi *et al.* 1998; Udawatta *et al.* 2008a). These microbial activities increase pronounced perennial roots of trees and grass, promote an abundance of connected pores and enhance macropores (Schoonover and Crim 2015), which may increase water movement downward into the soil during intense rainstorms and improve K_{sat} (Zaibon *et al.* 2016).

In general, the root systems in CBs are more extensive and deeper than annual RC, which increases porosity and SOC, and so improves soil structure (Blanco-Canqui *et al.*

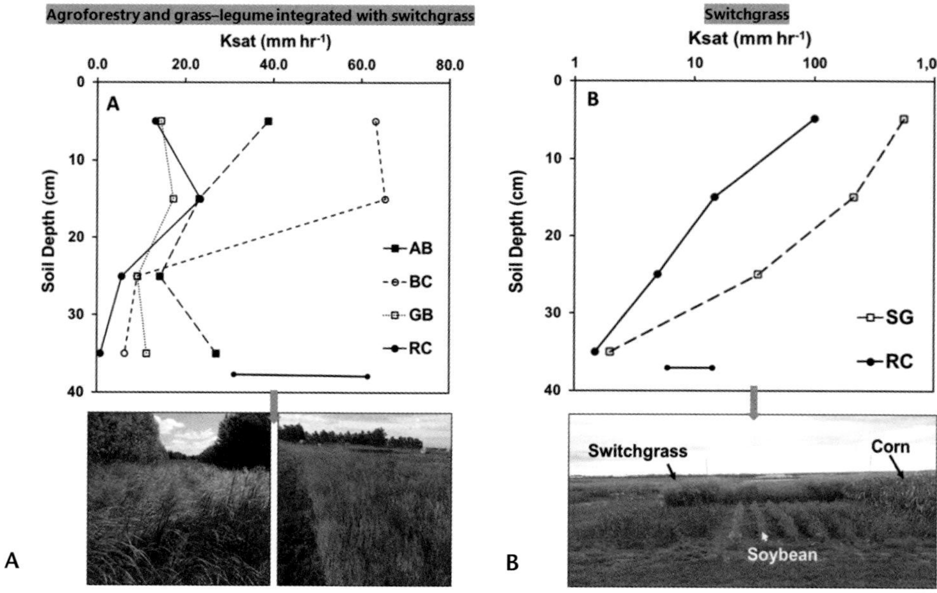

Fig. 7.2. Effects of conservation buffers and soil depth on saturated hydraulic conductivity (K_{sat}) for agroforestry buffers (AB), biomass crops (BC), grass buffers (GB), switchgrass (SG) and row crops (RC). The horizontal bar indicates the least significant difference ($P \leq 0.05$) for K_{sat}. (a) Data from Alagele *et al.* (2019b) for a study conducted on Putnam silt loam. (b) Data from Zaibon *et al.* (2016) for a study conducted on Mexico silt loam.

2017). Perennial CBs enrich below-ground SOC through the decomposition of older roots (Bonin *et al.* 2012). Soils under long-term CBs have had higher SOC inputs, which subsequently enhance soil macroporosity, infiltration and water storage compared with short-term annual management (Sahin *et al.* 2016; Alagele *et al.* 2019a). Increased SOC inputs stimulate the activity of soil organisms that form water-conducting networks of biopores (Kladivko and Clapperton 2011). Biopores are a special type of macropore, and their presence has a significant effect on the K_{sat} in different soil horizons (Reck *et al.* 2018).

Soil water retention and available water capacity

Capillarity in soils is the phenomenon responsible for soil water retention (SWR). SWR can be defined as the capacity of a soil to retain water at selected pressures (Hillel 2004). This soil property is important for predicting the required amount of irrigation to apply on farms and other important applications, such as plant water use efficiency. Plant-available water is the portion of water in the soil that can be easily used by plants and is defined as the amount of water between field capacity and the permanent wilting point (Brady and Weil 2008). Because there is a relationship between soil carbon and soil hydraulic properties, the increase in SOC within established long-term CBs can increase SWR. Several authors have emphasised that SWR in the topsoil can be affected by CBs (Rachman *et al.* 2004a; Seobi *et al.* 2005; Kumar *et al.* 2008; Mudgal *et al.* 2010; Zaibon *et al.* 2016; Alagele *et al.* 2019b). These studies showed that CBs can produce more diverse plant residues, increasing SOC, improving SWR and enhancing available water capacity.

The benefits of ABs and GBs on SWR were assessed by Seobi *et al.* (2005), who found that soil water storage in the upper 30-cm layer was of 9 and 11 mm higher under GB and AB respectively than under RC. Alagele *et al.* (2019b) reported that ABs, GBs and biomass crop had higher water content at soil water pressures greater than –2.5 kPa compared with RC. Soil water at saturation and –20 kPa was significantly greater (8%) for the biomass crop than the other practices (Alagele *et al.* 2019b). Soil water as a function of water pressure was greater for the 0–10 cm depth at 0.0, –0.4 and –1.0 kPa than 10–20 and 20–30 cm depths (Alagele *et al.* 2019b). The improvements in SWR are likely due to the greater rooting that occurred under the CBs, which resulted in greater macroporosity in the topsoil.

Kumar *et al.* (2008) evaluated the effects of ABs and GBs on SWR compared with rotationally and continuously grazed pastures in silt loam soil in central Missouri, USA. The SWR evaluated at 0.0 and −0.4 kPa was higher for ABs and GBs than for rotationally and continuously grazed pasture treatments for the 0–10 cm soil depth. Evaluating SWR in a field in Iowa managed with switchgrass hedges 10 years after establishment, Rachman *et al.* (2004a) found that the saturated water content for the grass hedge position was 11% and 21% higher than in the RC and deposition positions respectively. Six years after the establishment of multispecies riparian buffer vegetation (trees, shrubs and grass), Bharati *et al.* (2002) reported fivefold higher soil water storage in the buffer than in cultivated field and pasture. These benefits can be attributed to the increase in SOC inputs under these perennial systems (Udawatta *et al.* 2008a; Weerasekara *et al.* 2016; Alagele *et al.* 2019a). Because the surface of organic particles has a higher specific area than that of inorganic particles, the soils under CBs can hold greater amounts of water, concomitantly increasing the proportion of plant-available water (Brady and Weil 2008). Hudson (1994) reported that plant-available water in soil containing 4% SOC was double that in soil containing 1% SOC. Thus, CBs contribute to increasing a soil's ability to retain water.

Soil Db and porosity

Changes in Db and porosity are highly correlated with soil hydraulic properties and soil quality (Brady and Weil 2008). Increased Db and decreased porosity lead to reduced soil hydrological function, which includes reduced infiltration and water storage. High Db is related to low porosity and high soil compaction (Hillel 2004). High Db reduces available water capacity, root growth and the movement of air and water through the soil layers. The adverse effects of increased Db on soil properties reduce crop yields and vegetative cover available to protect soil against erosion (Durán Zuazo and Rodríguez Pleguezuelo 2008).

Establishing CBs can increase SOC inputs and soil microbes and may reduce Db and increase total porosity. Several studies have reported that establishing a series of tree, grass or shrub strips, especially within cropping systems, can improve SOC inputs, reduce Db and increase total porosity (Rachman *et al.* 2004a; Seobi *et al.* 2005; Kumar *et al.* 2008; Mudgal *et al.* 2010; Blanco-Canqui *et al.* 2014; Zaibon *et al.* 2016; Alagele *et al.* 2019b; Table 7.1). Most of these studies show that the use of perennial CBs for a longer period in buffers reduces soil degradation by intercepting raindrops and improving soil physical and hydraulic properties, especially Db and total porosity.

The effects of AB, GB and switchgrass on Db versus no-till RC (corn (*Zea mays* L.)–soybean (*Glycine max* L.) rotation) were assessed by Seobi *et al.* (2005) and Alagele *et al.* (2019b). These experiments were implemented on claypan soil in Missouri, USA. Seobi *et al.* (2005) showed that AB and GB reduced Db by 2.3% and increased porosity by 3% compared with RC. For AB, GB and biomass crop treatments, Db was 6%, 12% and 31% lower and porosity was 6%, 12% and 25% higher respectively than that of the RC topsoil (0–10 cm; Alagele *et al.* 2019b). Rachman *et al.* (2004a), Blanco-Canqui *et al.* (2014) and Zaibon *et al.* (2016) reported that switchgrass hedges reduced Db by 4%, 8% and 19%

Table 7.1. Effects of selected conservation buffer (CB) systems on bulk density (Db), total porosity (TP), macropores (Mac), coarse mesopores (CMes) and micropores (Mic) at a depth of 0–10 cm.

Within columns, different letters indicate significant differences between CBs ($P \leq 0.05$, Duncan's test). AB, agroforestry buffer; GB, grass buffer; RC, row crop; SG, switchgrass

CB	Db (g cm^{-3})	TP (%)	Mac (m^3 m^{-3})	CMes (m^3 m^{-3})	Mic (m^3 m^{-3})	References
AB	1.27c	52a	0.036a	0.14a	0.31a	Seobi *et al.* (2005)
GB	1.21b	55a	0.028ab	0.18a	0.30a	
RC	1.32a	50b	0.026a	0.076b	0.32a	
AB	1.28b	52ab	0.032ab	0.097b	0.300a	Alagele *et al.* (2019b)
GB	1.21b	54ab	0.034ab	0.145a	0.286a	
SG	1.04c	61a	0.053a	0.072bc	0.313a	
RC	1.36a	49b	0.027b	0.067c	0.327a	
SG	1.02b	62a	0.100a	0.030a	0.35a	Zaibon *et al.* (2016)
RC	1.2a	55b	0.020b	0.020b	0.35a	
AB	1.03c	61a	0.080a	0.100a	0.380a	Kumar *et al.* (2008)
GB	1.15b	57a	0.060b	0.110a	0.400a	
Grazed pasture						
Rotational	1.35a	49ab	0.050b	0.050b	0.430a	
Continuous	1.45a	45b	0.010c	0.030c	0.420a	
Native grass	1.13c	57a	0.048a	0.170a	0.340a	Chandrasoma *et al.* (2016)
Restored prairie	1.23b	54a	0.022b	0.090b	0.360a	
Timothy grass	1.29a	51ab	0.038ab	0.010c	0.340a	
Row crop	1.35a	49b	0.020b	0.050c	0.340a	

respectively relative to the RC management, and increased porosity. The authors of these studies reported that active and decaying root systems of the stiff-stemmed switchgrass may enhance porosity and contribute to increased K_{sat} within the grass hedge.

Kumar *et al.* (2008) found that Db was 13% greater for rotationally and continuously grazed pasture treatments (1.41 and 1.45 Mg m^{-3} respectively) compared with that under GB and AB (1.25 and 1.29 Mg m^{-3} respectively). Mudgal *et al.* (2010), comparing 100 years of RC with remnant uncultivated prairie, showed that the Db of the surface layer (0–10 cm) in the remnant prairie was two-thirds of that in the continuous RC area (0.81 vs 1.44 Mg m^{-3} respectively). The total porosity was 69% at the remnant prairie, compared with 46% for the RC management. Establishing CBs results in a higher root density and a greater addition of plant residues, particularly in the topsoil, which leads to a reduction in Db (Udawatta *et al.* 2008a; Bonin *et al.* 2012; Paudel *et al.* 2012; Alagele *et al.* 2019a). Thus, decomposing plant residues in CBs produce greater SOC inputs. In support of this, in a study of soil quality in Missouri, USA, Alagele *et al.* (2019a) found 19%, 30%, 96% and 40% greater SOC in GB, biomass crops, GWW and AB respectively than in RC. These increases in SOC show that various perennial CBs effectively contribute to the addition of SOC, which leads to a decrease in Db and an increase in porosity.

Pore size distribution

PSD is an important soil property that affects soil water retention and K_{sat}. Generally, pores are classified into three major size groups (Anderson *et al.* 1990; Zaibon *et al.* 2016):

macropores (effective diameter >1000 μm), mesopores (effective diameter 10–1000 μm) and micropores (effective diameter <10 μm).

Macropores are important for infiltration and are responsible for the rapid channel flow of water through the soil profile. Macropores that are created by plant roots, earthworms and other soil biota are termed biopores. Biopores are usually continuous along their length and include branched channels and some tubular shapes (Brady and Weil 2008). Mesopores contribute to rapid infiltration within the soil profile without macropores being filled up (Luxmoore 1981). The primary phenomena that involve mesopores are drainage, hysteresis and gravitational driving forces for water dynamics (Luxmoore 1981). Micropores are responsible for the part of the soil matrix flow that is driven by changes in matric pressure. The primary phenomena that involve micropores are ET and matric pressure for water redistribution (Hillel 2004). The size and shape of soil pores vary from micropores, which are the intra-aggregate capillaries and responsible for the retention of water and solutes, to macropores, which are mostly the interaggregate spaces and responsible for the infiltration and drainage of water and for aeration (Hillel 2004).

Perennial CBs alter PSD by increasing the proportion of macropores and mesopores and reducing the proportion of micropores. Increased macroporosity has significant benefits for enhanced infiltration into and within the soil and may reduce surface run-off through soil managed with various CBs compared with soil managed with only monocropping systems (Udawatta *et al.* 2006). In their study, Udawatta *et al.* (2006) reported that macropores can conduct water more rapidly than micropores. The effects of ABs and GBs on PSD determined from SWR data were assessed by Seobi *et al.* (2005), who found that mesoporosity was 33% and 3% higher for the GBs and ABs respectively compared with RC 6 years after their establishment in Missouri, USA. In a study conducted 14 years after the study of Seobi *et al.* (2005), Alagele *et al.* (2019b) evaluated the effects of three CBs (ABs, GBs and switchgrass) on PSD relative to RC. In that study, ABs, GBs and switchgrass increased the proportion of macropores (19%, 26% and 96% respectively) and mesopores (45%, 7% and 116% respectively) compared with the RC. Similarly, Rachman *et al.* (2004a) and Zaibon *et al.* (2016) found that grass hedges contributed to an increase in macropores (55–125%) and mesopores (25–35%) relative to RC. In both studies, switchgrass hedges were found to have a deeper root system, which increased PSD compared with the RC (Table 7.1). In a study investigating the effects of AB and GB on PSD compared with rotationally and continuously grazed pasture treatments in deep loess soils, Kumar *et al.* (2008) found that macroporosity was six-, four- and fourfold higher under AB, GB and rotationally grazed pastures respectively than under continuously grazed pasture. The rotationally and continuously grazed pasture had higher microporosity than the AB and GB, but lower macroporosity and mesoporosity (Kumar *et al.* 2008). Jiang *et al.* (2007) and Mudgal *et al.* (2010) reported an increase in macropores and mesopores with long-term perennial CBs, with a concomitant improvement in infiltration and a reduction in NPSP.

Soil structural stability

Climate factors (e.g. precipitation, freeze–thaw and wet–dry cycles) and management (e.g. tillage and cropping) are major factors affecting soil structure and aggregate stability (Chan 2011). Intensive crop production systems, which return fewer plant residues to the soil, lead to a reduction in soil aggregate stability and compact the soil under long-term RC (Castro Filho *et al.* 2002). This leads to an increased possibility of soil erosion and run-off, which negatively affect soil quality. Continuous cropping with any single crop and leaving the soil bare and more susceptible to erosion effects of rainfall will lead to a loss of natural soil aggregates, which reduces the accumulation and decomposition of SOC (Magdoff and Van Es 2009). These consequences decrease SOC inputs, particularly in the topsoil, and diminish soil organisms, which play vital roles in forming and stabilising aggregates (Alagele *et al.* 2020a). Fungal hyphae form networks and bind soil particles together with polysaccharides and glycoproteins secretions to form larger aggregates (Lupwayi *et al.* 2001).

Soil structural stability is one significant physical characteristic of the soil affecting soil sustainability and crop production (Amézketa 1999). Soil structural stability affects PSD and soil water flow (Hillel 2004). SOC is considered a major factor in the formation and stabilisation of secondary particles formed through the combination of mineral particles (Tisdall and Oades 1982). Thus, the amount of SOC affects aggregate stability and can enhance the formation of soil aggregates (Brady and Weil 2008). Organic matter, which may decompose from residues of trees (e.g. twigs, leaves and branches), grasses and shrubs, acts as a binding agents for aggregates near the surface and plays a vital role in reducing soil erosion (Alagele *et al.* 2019a). Because SOC levels are directly affected by perennial CBs (Anderson and Udawatta 2019), the establishment of long-term CBs should be part of most suitable management strategies to enrich the soil with organic matter and concomitantly improve the formation of soil aggregates (Udawatta *et al.* 2008a). Udawatta *et al.* (2008a) confirmed that there are mutual positive interrelationships between soil aggregates and SOC. Tree and grass roots of CBs are believed to contribute more to increased SOC than a crop's shoot and leaf systems.

Udawatta *et al.* (2008a) investigated the effects of various CBs (AB, GWW and GB) versus the RC management on water-stable aggregates (WSA) in claypan soil in Missouri, USA, finding that WSA were highest in GWW (20%), followed by the AB and GB (15%), with the lowest in the RC (8%). Alagele *et al.* (2019a) conducted a study that evaluated the effects of management on WSA 21 years after establishing AB, GB, and GWW treatments and 6 years after establishing biomass crops. Alagele *et al.* (2019a) found that perennial CBs had 2.6- to 3.4-fold more WSA than the RC. Establishing long-term CBs was found to contribute to continuous SOC inputs from fibrous root, as well as from the leaves, twigs and branches of these systems, and may increase WSA relative to no-till RC (Alagele *et al.* 2019a). These studies explain why soil health is enhanced, with increased WSA and SOC, after the establishment of these CBs (Fig. 7.3).

In Missouri, USA, soils, Kremer and Li (2003) found that WSA was 18% and 69% in soils managed with cool-season grasses and a remnant prairie of native warm-season grasses respectively, compared with 10% under conventional tillage RC. Another study in Missouri reported 56–71% greater WSA under grazed pasture, GB and AB than under RC management (18%; Paudel *et al.* 2011, 2012). In a study evaluating the effects of long-term (>10 years) switchgrass, RC, cool-season grass pasture and forest management on soil aggregates in the south-eastern US, Blanco-Canqui *et al.* (2005) found that the establishment of long-term switchgrass enhanced soil aggregates relative to RC and cool-season grass pasture systems. In Nebraska, USA, Blanco-Canqui *et al.* (2014) found that after 15 years of management, switchgrass hedges promoted soil structural stability. Further, WSA was increased in switchgrass hedges by 70%, 50% and 20% at depths of 0–15, 15–45 and 45–60 cm respectively compared with the RC. Blanco-Canqui *et al.* (2014) confirmed that enhanced aggregate stability under perennial CBs can be attributed to the higher root biomass, root exudates, microbial activity and SOC of these systems. Organic chemical compounds produced during the decomposition of fresh residues in CBs contribute to binding soil particles together, resulting in better aggregate stability in the soil (Zhang *et al.* 2008; Veum *et al.* 2012; Udawatta *et al.* 2014a).

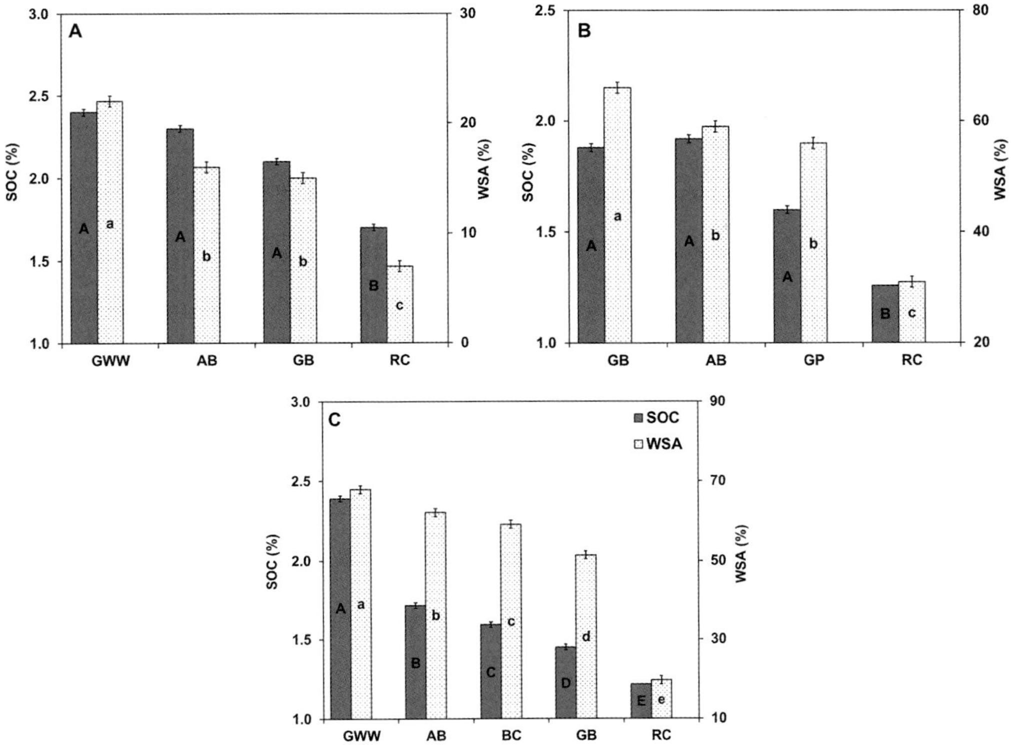

Fig. 7.3. Results from selected studies evaluating the effects of different conservation buffer systems on soil organic carbon (SOC) and water stable aggregates (WSA): (a) Udawatta *et al.* (2008a), (b) Paudel *et al.* (2012) and (c) Alagele *et al.* (2019a). Data are the mean ± s.d. Different letters represent significant differences in SOC (uppercase letters) and WSA (lowercase letters) among treatments ($P \leq 0.05$, Duncan's least significant difference test). AB, agroforestry buffers; BC, biomass crops; GB, grass buffers; GP, grazed pasture; GWW, grass waterways; RC, row crop.

Effects of CBs on soil hydrology

Perennial CBs play a major role in buffering the hydrology of soil. Increasing SOC inputs by establishing various long-term CBs leads to improved water infiltration and storage, thereby increasing plant water availability and reducing erosion (Anderson *et al.* 2009). The CBs transpire more water, and this process can have a positive or negative effect on water relations depending on the climatic region (e.g. arid, temperate and wet regions). Several studies have documented that hydraulically lifted water contributes to solving many challenges, such as supplying water to irrigate neighbouring plants, mitigating climate change and enhancing crop production (Hirota *et al.* 2004). Understanding the relationship between hydraulically lifted water and water-lifting plant species helps explain the movement of soil water from the moist deep to dry shallow soil zones and to develop crop–grass–tree designs for optimum land production. The effects of CBs on soil hydrology are discussed in the following sections.

Soil water storage and dynamics

Soil water influences the engineering, agronomic, geological, ecological and hydrological properties of the soil mass (Susha Lekshmi *et al.* 2014). Soil water is significant in many plant–soil–water investigations and larger-scale hydrological studies (Kelleners *et al.* 2005). Soil water flow properties, which include infiltration and storage in the root zone, are closely related to water use efficiency (Mulebeke *et al.* 2010). Soil water affects many fundamental biophysical processes that influence the germination of seeds and microbial decomposition of SOC (Bittelli 2011). Using appropriate soil management systems can enhance infiltration and storage and may improve soil and water quality (Blanco-Canqui 2018).

Improved soil water storage and dynamics are important in various climatic regions, especially in dry regions for the irrigation of agricultural crops and under temperate and wet climatic conditions to reduce run-off. In dryland cropping, soil water availability is often the most limiting resource. Producers can manage and adapt to this limitation by choosing suitable crops depending on soil water availability in a specific season (e.g. spring; Shiferaw *et al.* 2009). For example, most agricultural lands in the western USA (e.g. California) and other dry climatic areas around the world would not be productive without the application of irrigation (Howell 2001). In addition, in humid regions, farms need to use supplemental irrigation because their soils have become less drought resistant as a result of intensive RC agricultural management and have depleted SOC (Magdoff and Van Es 2009). Investigations of soil water use by crop species in dryland areas reveal that longer-season and more deeply rooted crops (e.g. alfalfa) consume more soil water than shorter-season crops (e.g. dry pea), which depletes less soil water. Cropping systems that store more water in the rooting zone should be explored to improve crop productivity in regions where water resources are scarce (Basche and Edelson 2017). This challenge is of much more widespread consequence in developing countries (e.g. regions in Africa and the Middle East), where water resources are scarce. The selection of suitable tree and grass species and integrating them with alternative drought-tolerant crops is considered the best strategy and an applicable agriculture practice to improve soil water storage and use (Montgomery 2007; Evans and Sadler 2008).

The incorporation of CBs (tree and/or grasses) with crops contributes to the increased ability of the soil to store more water and decreases surface run-off (Udawatta *et al.* 2002; Lovell and Sullivan 2006). A soil managed under perennial CBs has a favourable soil structure that promotes rainfall infiltration and water storage for plants to use later. The practical benefit of using CBs is that these practices decrease soil water content through improved water consumption. This decreased soil water promotes infiltration during recharge periods and may decrease NPSP in temperate or wet climatic regions, particularly on claypan landscapes, which experience major run-off and erosion issues (Conway *et al.* 2017).

Anderson *et al.* (2009), Sahin *et al.* (2016) and Alagele *et al.* (2020b) evaluated soil water content, water use and recharge as influenced by AB, GB and biomass crops compared with RC management. Water content sensors were installed at multiple depths (5, 10, 20 and 40 cm) to continuously estimate soil water during the growing season from April to November for the selected systems throughout several years. The findings of these studies indicated that soil water decreased more rapidly in the pre-recharge period under CBs compared with RC. This decrease in soil water was probably due to higher transpiration and more water depletion by the tree, grass and biomass crop treatments compared with RC management (Anderson *et al.* 2009). In the post-recharge periods (during and after rainfall events), higher soil water noted in the perennial CB treatments than the RC could be attributed to better infiltration through the deep and extensive roots in these perennial systems (Sahin *et al.* 2016). These changes provide potential benefits that include the storage of more water for plants to use later, a reduction in run-off and an increase in SOC inputs compared with RC (Zaibon *et al.* 2016; Blanco-Canqui *et al.* 2017; Alagele *et al.* 2019a).

Infiltration and surface run-off

Infiltration is an important process for soils to store sufficient water for plant growth and nutrient cycling (Kumar 2009). A decrease in infiltration may lead to reduced water storage in the soil profile for later use by crops, and may often reduce agricultural crop production (Connolly *et al.* 1997). The movement of water through the soil profile as infiltration rather than as run-off provides many environmental benefits, including increasing groundwater recharge and decreasing NPSP (Arshad *et al.* 1999). Soil management, soil texture, soil structure, landscape position, SOC, antecedent soil moisture, canopy cover, rainfall intensity and pore parameters are predominant factors that determine infiltration (Kumar *et al.* 2012).

Antecedent soil water in the soil profile, particularly in the upper soil layers before a rainfall event, can affect the relationship between rainfall and run-off (Tramblay *et al.* 2010). The intensity and frequency of rainfall events are considered important dynamic factors for run-off and chemical losses under continuous cropping management (Udawatta *et al.* 2011; Han *et al.* 2018). When a rainfall event exceeds a soil's infiltration capacity, run-off is generated (Magdoff and Van Es 2009). Intensive long-term RC management, which maintains less surface residue, may cause the sealing of water-conducting surface pores (Blanco-Canqui 2010). These agricultural issues generate and accelerate water movement as run-off, which causes the soil to be more prone to soil erosion. As a result of run-off, sediments may be carried into off-site waterbodies (e.g. streams, rivers and lakes) and may consequently degrade water quality (Udawatta *et al.* 2002). Levels and inputs of SOC may be depleted due to the accelerated run-off, which removes surface soils and organic matter (Udawatta *et al.* 2004). Generally, infiltration capacity, drainage and K_{sat} are important factors that determine the run-off potential (Brady and Weil 2008). Thus, NPSP produced from agricultural lands needs to be managed to improve soil, water and environmental quality (Blanco-Canqui 2018). The integration of CBs with RC is an appropriate approach to control run-off and erosion (Udawatta *et al.* 2011). Studies have noted that 100% of soils covered with living CBs (e.g. trees and grass) for a longer period substantially enhance infiltration and reduce run-off on most agricultural crop lands.

The effects of CBs such as AB, GB, switchgrass hedges, restored prairies, riparian buffers and GWW on infiltration have been widely evaluated (Bharati *et al.* 2002; Rachman *et al.* 2004b; Anderson *et al.* 2009; Kumar *et al.* 2012; Sahin *et al.* 2016; Zaibon *et al.* 2017). These systems enhance the proportion of soil macropores and improve infiltration, especially because macropores are primarily responsible for higher infiltration rates (Kumar 2009). Kumar *et al.* (2012) found that GB and AB had the highest steady infiltration flux (221 mm h^{-1}), whereas the continuously and rotationally grazed pasture systems had the lowest (3.7 mm h^{-1}). Kumar *et al.* (2012) demonstrated that infiltration was enhanced because these buffers had more macropores, and this, in turn, led to increased infiltration in the buffer areas compared with grazed pasture. Bharati *et al.* (2002) compared infiltration in the Midwestern USA under multispecies riparian buffers to that of cultivated fields and a grazed pasture. In that study, the mean 60-min cumulative infiltration was fivefold higher under the buffer treatments (silver maple (*Acer saccharinum*), grass filter, switchgrass) than under the RC and pasture (Bharati *et al.* 2002; see also Table 7.2).

Rachman *et al.* (2004b) observed that, on Monona silt loam soil, the infiltration rate under switchgrass (grass hedge) was significantly greater than for RC. In addition,

Table 7.2. Effects of various conservation buffers (CB) on mean quasi-steady state infiltration rates (qs), field-saturated hydraulic conductivity values (Kfs), run-off (RO) depth and 60-min cumulative infiltration (CI)

Different letters represent significant differences among treatments ($P \leq 0.05$. Duncan's least significant difference test)

CB	q_s (mm h^{-1})	K_{fs} (mm h^{-1})	RO depth (cm)	CI (cm)	Duration (years)	Soil	Location	Reference
Grass hedge	230a	154a			10	Monona silt loam	Treynor, Iowa, USA	Rachman *et al.* (2004b)
Row crop	40b	23b						
Switchgrass	25a	18a	5b		> 30	Mexico silt loam	New Haven, Missouri, USA	Zaibon *et al.* (2017)
Row crop	12b	7b	9a					
Agroforestry buffer	106b	78b			8	Menfro silt loam	New Franklin, Missouri, USA	Kumar *et al.* (2012)
Grass buffer	245a	181a			8			
Grazed pasture								
Rotational	15c	9c			7			
Continuous	10c	4c			7			
Agroforestry buffer	17a				6	Putnam	Knox County, Missouri, USA	Anderson *et al.* (2009)
Grass buffer	14a				6			
Row crop	10ab				12			
Silver Maple				40a	5	Coland	Story County, Iowa, USA	Bharati *et al.* (2002)
Grass Filter				30b				
Switchgrass				29b				
Corn				4c				

Rachman *et al.* (2004b) found that the field K_{sat} for the switchgrass buffer was sevenfold higher than that of RC. Zaibon *et al.* (2017) assessed the effects of switchgrass on infiltration after 24-h mean duration storms (rainfall 11.8, 14.2 and 16.2 cm) in the eroded Mexico silt loam soil, and found that switchgrass improved estimated infiltration and reduced estimated run-off compared with RC. Udawatta *et al.* (2002, 2004) evaluated the effect of establishing GB and AB watersheds on run-off, sediment and nutrient losses compared with RC in Missouri, USA and found that both types of watersheds efficiently decreased NPSP from the cropping system in the claypan landscapes.

To provide support of the idea that CBs improve infiltration and reduce run-off, Udawatta *et al.* (2006, 2008b) and Udawatta and Anderson (2008) examined the effects of GB and AB on PSD using computed tomography in claypan soils. These studies confirmed that GB and AB have 26 and 36 macropores per 2500-mm^2 area respectively, and these numbers were 2- and 2.6-fold higher than the number of macropores under RC (14 macropores per 2500-mm^2 area). Because water can easily move through macropores, increased infiltration and reduced run-off occurred in soils managed with CBs, which have a greater number of macropores than soils managed with RC, particularly in eroded claypan soils (Rachman *et al.* 2005).

Soil water buffering by ET and hydraulic redistributions

Soil water use and storage vary according to vegetation management under different climates, and these systems influence land productivity and hydrological processes (Magdoff and Van Es 2009). The effect of climate change on water quantity may critically affect the amount of water stored in the soil profile (Delgado *et al.* 2011). Increasing air temperatures and net radiation caused by climate change are expected to accelerate ET processes and water use by plants. ET reduces water storage (Hillel 2004). Plants may experience water stress because of increased water loss by ET processes and transpiration may consume stored soil water. Some stored water moves to the soil surface upward by capillarity forces and is lost by evaporation in the soil surface, and most water used by plants comes through the roots and stems to the leaves, where it is lost by transpiration (Priyadharshini and Canessane 2015).

Although establishing CBs in combination with crops in an alley cropping system improves soil properties and crop production, grain yield can be affected in different climatic areas (e.g. arid, semi-arid, humid and temperate regions) as a result of competition for water between the CB species and crops (Udawatta *et al.* 2014b, 2016). For example, the main objectives for landowners, soil scientists and policy makers in temperate and humid regions in the USA (e.g. Nebraska, Iowa and Missouri) are to reduce erosion and run-off, to keep sediments and other chemicals away from water sources and to improve soil and water quality. Thus, establishing CBs in semihumid and humid regions of the USA, where average annual precipitation is ~900–1000 mm $year^{-1}$, is an important approach. The CBs can consume more water but also create more macropores by these systems. This decrease in soil water allows much higher infiltration and more stored water through the root system in the soil profile during heavy rains and may reduce NPSP (Anderson *et al.* 2009).

Establishing CBs provides differential soil water extraction to reduce NPSP and may enhance soil quality (Sahin *et al.* 2016). Several management practices require consideration of a water budget, particularly the ET component (Svoma *et al.* 2016; Mohammed and Trauth 2019). Campi *et al.* (2009) indicated that energy levels are reduced under buffer systems and adjacent areas, and these reductions promote less ET and probably higher soil water storage. In Missouri, USA, Svoma *et al.* (2016) found significant changes in microclimate in the CBs, and these changes may influence ET and soil water dynamics. In addition, Svoma *et al.* (2016) found higher temperatures and net radiation, along with lower wind speed, in ABs than GBs. Trees in the AB as part of an alley cropping system can act as a barrier to reduce wind speed and affect ET and other energy fluxes (Campi *et al.* 2009). Perennial CBs can buffer and reduce heat flow by increasing the thermal heat capacity and thereby help crops to adapt to the impacts of climate change (Adhikari *et al.* 2014). Brandle *et al.* (2004) stated that the radiation budget (both scattered short-wave and emitted long-wave radiation) in RC areas may also be affected by trees in ABs. Zaibon *et al.* (2019) found that switchgrass reduced thermal conductivity and diffusivity by 5–7% and 8–9% respectively and increased volumetric heat capacity by 2–3% compared with RC. Zaibon *et al.* (2019) reported that CBs significantly affected soil thermal properties, which can buffer heat transport within the soil profile. More studies are required to understand how different CBs affect patterns of dewatering rates (decay; ET) to decrease run-off and increase infiltration compared with RC.

Another important benefit of integrating CBs (trees, grass) into crop areas is that more hydraulically lifted water could be transferred by deep roots to drier topsoil layers (Espeleta *et al.* 2004). These systems may provide additional water resources for plants, especially during drought periods, to irrigate adjacent shallow-rooted plants (Liste and White 2008). Soil scientists have documented that root systems move significant amounts of water from moist to dry soil layers, a process termed hydraulic redistribution or hydraulic lift (HL; Brady and Weil 2008). Scientists have confirmed that deep tree, grass and plant roots can serve as channels for transferring water upward through the root system from a wet subsurface to dry soil surface layers; this water can irrigate neighbouring shallower-rooted crops during long dry periods, ensuring their survival (Hirota *et al.* 2004).

Many reviews have evaluated the effects of various plant species on the HL process (Richards and Caldwell 1987; Dawson 1993, 1996). For example, sage brush, a deep-rooted woody plant found in deserts of North America, conducts enough hydraulic redistribution and provides one-third to one-half of the water used by relatively shallow-rooted crested wheatgrass growing nearby (Brady and Weil 2008). In field plots of interspersed sagebrush (*Artemisia tridentata* ssp. *vaseyana* (Rydb.) Beetle) and tussock grasses (*Agropyron desertorum* and *Agropyron spicatum*), Richards and Caldwell (1987) observed that water was absorbed and transported at night from moist to drier upper soil layers by deeper roots. Deep-rooted tree species, such as casuarina (*Casuarina equisetifolia*), which is a drought-tolerant species, can be used in alley cropping systems to reduce evaporation and supplement irrigation by HL, and may improve soil quality and productivity.

Potential challenges following the integration of CBs with crops, especially in arid and semi-arid regions, include a reduction in crop yields due to water demands for irrigation and increased competition for water resources (Magdoff and Van Es 2009). However, Senaviratne *et al.* (2012) suggested that as trees mature, root pruning, removing branches and thinning may be required to decrease competition for water resources and to avoid potential crop production losses in alley cropping systems.

To summarise, selecting appropriate plants, namely those with deeper root systems and drought-tolerant species, under fluctuating climate conditions can be an efficient strategy to increase crop yields and improve soil, water and environmental quality. The plants selected should be able to redistribute water, via their roots, from moist to drier zones and thereby provide enough water to shallow-rooted plants, possibly enhancing crop production.

Summary and conclusions

The specific objective of this chapter was to describe the effects of various CBs on selected soil hydraulic properties and soil hydrology under fluctuating climates. A strategy of establishing long-term CBs should result in the reduction of many unfavourable environmental and agricultural issues. For example, run-off and soil erosion are closely linked with soil hydraulic and physical properties, soil hydrology and SOC. These challenges (soil degradation, impaired water quality, scarce water resources with concomitant increases in population growth under a changing climate) lead to loss of soil, especially under RC management. Narrow strips of permanent vegetation (trees, grasses or shrubs) established as CBs around or among crops can decrease NPSP and improve SOC and soil productivity. Therefore, the establishment of CBs must be promoted because they provide many agricultural and environmental benefits.

Integrating perennial CBs (e.g. trees, grasses and shrubs) with annual crops can add plant residues, such as roots, twigs, leaves and branches, to the soil. The decomposition of these residues can contribute to greater SOC and promote a greater quantity and diversity of soil microorganisms. Increasing SOC levels contribute to improved performance and functionality of the soil. Greater SOC leads to better soil aggregation and higher water storage, infiltration and redistribution, as well as enhanced porosity, macropores, K_{sat} and favourable Db, leading to greater soil and crop productivity. Improvements in the soil system can also provide benefits such as increasing farm income, promoting landscape aesthetics and enhancing wildlife habitat.

Two principal strategies are required to improve soil and water quality: (1) to permit more time for infiltration to occur by increasing soil water storage; and (2) to retain greater vegetation cover on the surface for the entire year, especially during high-frequency and intense rainfall events, to protect the soil against water erosion. For example, establishing CBs enhances infiltration by creating open root channels, stimulating earthworm activity, preserving the soil structure, increasing water storage and reducing run-off and erosion. Increasing soil water use by plants benefits watershed restoration, especially in humid, semihumid and temperate climatic regions, because of improved infiltration capacity and reduced run-off. Another significant benefit is that tree roots absorb and transport a considerable amount of water upward through the root system from deep roots, and later pass it into the shallow soil layer at night, irrigating neighbouring shallow-rooted plants. This has important agricultural and environmental implications for improved crop yields, especially in dry climates where water resources are scarce or during short dry periods in humid areas. Thus, the use of CBs can have many potential benefits to soil health, agricultural productivity and sustainability.

The widespread adoption of perennial CBs under an unstable climate needs to be assessed carefully to better understand the role and efficiency of these systems for managing agricultural issues of run-off, erosion and sediment and chemical losses from lands at broad watershed scales, and their effects on soil, water and environmental quality. Finding useful management systems that can contribute to climate change mitigation and adaptation is important for conservation, the sustainability of cropping systems, soil quality and food security. Various CB systems have been successfully and effectively established in different climatic regions (e.g. humid and semi-arid regions) over the long term and numerous benefits of these systems have been quantified. Managing soils with various long-term CBs that reduce soil degradation and improve carbon sequestration will help crops better mitigate and adapt to climate change. This chapter has reviewed studies available in the literature, with the results of those studies indicating that the long-term strategy of using CBs compared with RC management can provide many soil ecosystem services, including enhanced soil properties and overall agricultural productivity. Overall, the successful establishment of CBs can be a management strategy to promote sustainable crop production and may mitigate global climate change.

References

Adhikari P, Udawatta RP, Anderson SH (2014) Soil thermal properties under prairies, conservation buffers, and corn–soybean land use systems. *Soil Science Society of America Journal* **78**, 1977–1986. doi:10.2136/sssaj2014.02.0074

Alagele SM, Anderson SH, Udawatta RP, Veum KS, Rankoth LM (2019a) Effects of conservation practices on soil quality compared with a corn–soybean rotation on a claypan soil. *Journal of Environmental Quality* **48**, 1694–1702. doi:10.2134/jeq2019.03.0121

Alagele SM, Anderson SH, Udawatta RP (2019b) Biomass and buffer management practice effects on soil hydraulic properties compared to grain crops for claypan landscapes. *Agroforestry Systems* **93**, 1609–1625. doi:10.1007/s10457-018-0255-1

Alagele SM, Anderson SH, Udawatta RP, Veum KS, Rankoth LM (2020a) Long-term perennial management and cropping effects on soil microbial biomass for claypan watersheds. *Agronomy Journal* **112**, 815–827. doi:10.1002/agj2.20116

Alagele SM, Anderson SH, Udawatta RP (2020b) Agroforestry, grass, biofuel crop, and row-crop management effects on soil water dynamics for claypan landscapes. *Soil Science Society of America Journal* **84**, 203–219. doi:10.1002/saj2.20026

Amézketa E (1999) Soil aggregate stability: a review. *Journal of Sustainable Agriculture* **14**, 83–151. doi:10.1300/J064v14n02_08

Anderson SH, Udawatta RP (2019) Agroforestry: a system for improving soil health. In *Agroforestry for Sustainable Agriculture.* (Eds MR Mosquera-Losada and R Prabhu) pp. 1–18. Burleigh Dodds Science Publishing, Cambridge.

Anderson SH, Gantzer CJ, Brown JR (1990) Soil physical properties after 100 years of continuous cultivation. *Journal of Soil and Water Conservation* **45**, 117–121.

Anderson SH, Udawatta RP, Seobi T, Garrett HE (2009) Soil water content and infiltration in agroforestry buffer strips. *Agroforestry Systems* **75**, 5–16. doi:10.1007/s10457-008-9128-3

Arshad MA, Franzluebbers AJ, Azooz RH (1999) Components of surface soil structure under conventional and no-tillage in northwestern Canada. *Soil & Tillage Research* **53**, 41–47. doi:10.1016/S0167-1987(99)00075-6

Basche AD, Edelson OF (2017) Improving water resilience with more perennially based agriculture. *Agroecology and Sustainable Food Systems* **41**, 799–824. doi:10.1080/21683565.2017.1330795

Becker R, Gebremichael M, Märker M (2018) Impact of soil surface and subsurface properties on soil saturated hydraulic conductivity in the semi-arid Walnut Gulch Experimental Watershed, Arizona, USA. *Geoderma* **322**, 112–120. doi:10.1016/j.geoderma.2018.02.023

Bharati L, Lee KH, Isenhart TM, Schultz RC (2002) Soil–water infiltration under crops, pasture, and established riparian buffer in Midwestern USA. *Agroforestry Systems* **56**, 249–257. doi:10.1023/A:1021344807285

Bittelli M (2011) Measuring soil water content: a review. *HortTechnology* **21**, 293–300. doi:10.21273/HORTTECH.21.3.293

Blanco-Canqui H, Lal R (2008) *Principles of Soil Conservation and Management.* Springer, New York.

Blanco-Canqui H (2010) Energy crops and their implications on soil and environment. *Agronomy Journal* **102**, 403–419. doi:10.2134/agronj2009.0333

Blanco-Canqui H (2016) Growing dedicated energy crops on marginal lands and ecosystem services. *Soil Science Society of America Journal* **80**, 845–858. doi:10.2136/sssaj2016.03.0080

Blanco-Canqui H (2018) Cover crops and water quality. *Agronomy Journal* **110**, 1633–1647. doi:10.2134/agronj2018.02.0077

Blanco-Canqui H, Lal R, Lemus R (2005) Soil aggregate properties and organic carbon for switchgrass and traditional agricultural systems in the southeastern United States. *Soil Science* **170**, 998–1012. doi:10.1097/01.ss.0000187342.07331.a6

Blanco-Canqui H, Gilley JE, Eisenhauer DE, Jasa PJ, Boldt A (2014) Soil carbon accumulation under switchgrass barriers. *Agronomy Journal* **106**, 2185–2192. doi:10.2134/agronj14.0227

Blanco-Canqui H, Mitchell RB, Jin VL, Schmer MR, Eskridge KM (2017) Perennial warm-season grasses for producing biofuel and enhancing soil properties: an alternative to corn residue removal. *Global Change Biology. Bioenergy* **9**, 1510–1521. doi:10.1111/gcbb.12436

Bonin C, Lal R, Schmitz M, Wullschleger S (2012) Soil physical and hydrological properties under three biofuel crops in Ohio. *Acta Agriculturæ Scandinavica. Section B, Soil and Plant Science* **62**, 595–603. doi:10.1080/09064710.2012.679309

Bossio D, Geheb K, Critchley W (2010) Managing water by managing land: addressing land degradation to improve water productivity and rural livelihoods. *Agricultural Water Management* **97**, 536–542. doi:10.1016/j.agwat.2008.12.001

Brady NC, Weil RR (2008) *The Nature and Properties of Soils*. Prentice Hall, Upper Saddle River.

Brandle JR, Hodges L, Zhou XH (2004) Windbreaks in North American agricultural systems. *Agroforestry Systems* **61**, 65–78. doi:10.1023/B:AGFO.0000028990.31801.62

Campi P, Palumbo AD, Mastrorilli M (2009) Effects of tree windbreak on microclimate and wheat productivity in a Mediterranean environment. *European Journal of Agronomy* **30**, 220–227. doi:10.1016/j.eja.2008.10.004

Castro Filho D, Lourenço A, Guimarães M, Fonseca IC (2002) Aggregate stability under different soil management systems in a red latosol in the state of Parana, Brazil. *Soil & Tillage Research* **65**, 45–51. doi:10.1016/S0167-1987(01)00275-6

Chan KY (2011) Climate change on soil structure and soil health: impacts and adaptation. In *Soil Health and Climate Change*. (Eds BP Singh, AL Cowie and KY Chan) pp. 49–67. Springer Science and Business Media, Berlin.

Chandrasoma JM, Udawatta RP, Anderson SH, Thompson AL, Abney MA (2016) Soil hydraulic properties as influenced by prairie restoration. *Geoderma* **283**, 48–56. doi:10.1016/j.geoderma.2016.08.001

Connolly RD, Freebairn DM, Bridge BJ (1997) Change in infiltration characteristics associated with cultivation history of soils in south-eastern Queensland. *Soil Research* **35**, 1341–1358. doi:10.1071/S97032

Conway LS, Yost M, Kitchen N, Sudduth K, Thompson A, Massey RE (2017) Topsoil thickness effects on corn, soybean, and switchgrass production on claypan soils. *Agronomy Journal* **109**, 782–794. doi:10.2134/agronj2016.06.0365

Dawson TE (1993) Hydraulic lift and water use by plants: implications for water balance, performance, and plant–plant interactions. *Oecologia* **95**, 565–574. doi:10.1007/BF00317442

Dawson TE (1996) Determining water use by trees and forests from isotopic, energy balance, and transpiration analyses: the roles of tree size and hydraulic lift. *Tree Physiology* **16**, 263–272. doi:10.1093/treephys/16.1-2.263

Delgado JA, Groffman PM, Nearing MA, Goddard T, Reicosky D, Lal R, Kitchen NR, Rice CW, Towery D, Salon P (2011) Conservation practices to mitigate and adapt to climate change. *Journal of Soil and Water Conservation* **66**, 118A–129A. doi:10.2489/jswc.66.4.118A

Derpsch R (2008) No-tillage and conservation agriculture: a progress report. In *No-till Farming Systems*. (Eds YT Gan, W Ellis and A Watson) pp. 7–39. World Association of Soil and Water Conservation, Bangkok.

Dhyani SK, Newaj R, Sharma AR (2009) Agroforestry: its relation with agronomy, challenges and opportunities. *Indian Journal of Agronomy* **54**, 249–266.

Dollinger J, Jose S (2018) Agroforestry for soil health. *Agroforestry Systems* **92**, 213–219. doi:10.1007/s10457-018-0223-9

Durán Zuazo VH, Rodríguez Pleguezuelo CR (2008) Soil-erosion and runoff prevention by plant covers: a review. *Agronomy for Sustainable Agriculture* **28**, 785–811. doi:10.1051/agro:2007062

Espeleta JF, West JB, Donovan LA (2004) Species-specific patterns of hydraulic lift in co-occurring adult trees and grasses in a sandhill community. *Oecologia* **138**, 341–349. doi:10.1007/s00442-003-1460-8

Evans RG, Sadler EJ (2008) Methods and technologies to improve efficiency of water use. *Water Resources Research* **44**, 1–15. doi:10.1029/2007WR006200

Fuentes JP, Flury M, Bezdicek DF (2004) Hydraulic properties in a silt loam soil under natural prairie, conventional till, and no-till. *Soil Science Society of America Journal* **68**, 1679–1688. doi:10.2136/sssaj2004.1679

Garrity DP (2004) Agroforestry and the achievement of the Millennium Development Goals. *Agroforestry Systems* **61**, 5–17. doi:10.1023/B:AGFO.0000028986.37502.7c

Gold MA, Garrett HE (2009) Agroforestry nomenclature, concepts, and practices. In *North American Agroforestry: An Integrated Science and Practice*. (Ed. HE Garrett) pp. 45–55. America Society Agronomy, Madison.

Han Y, Feng G, Ouyang Y (2018) Effects of soil and water conservation practices on runoff, sediment and nutrient losses. *Water (Basel)* **10**, 1333. doi:10.3390/w10101333

Hickey MB, Doran B (2004) A review of the efficiency of buffer strips for the maintenance and enhancement of riparian ecosystems. *Water Quality Research Journal* **39**, 311–317. doi:10.2166/wqrj.2004.042

Hillel D (2004) *Introduction to Environmental Soil Physics*. Elsevier Science, San Diego.

Hirota I, Sakuratani T, Sato T, Higuchi H, Nawata E (2004) A split-root apparatus for examining the effects of hydraulic lift by trees on the water status of neighboring crops. *Agroforestry Systems* **60**, 181–187. doi:10.1023/B:AGFO.0000013293.77907.64

Hobbs PR (2007) Paper presented at international workshop on increasing wheat yield potential, CIMMYT, Obregon, Mexico, 20–24 March 2006 Conservation agriculture: what is it and why is it important for future sustainable food production? *Journal of Agricultural Science* **145**, 127–137. doi:10.1017/S0021859607006892

Howell TA (2001) Enhancing water use efficiency in irrigated agriculture. *Agronomy Journal* **93**, 281–289. doi:10.2134/agronj2001.932281x

Hudson BE (1994) Soil organic matter and available water capacity. *Journal of Soil and Water Conservation* **49**, 189–194.

Jiang P, Anderson SH, Kitchen NR, Sadler EJ, Sudduth KA (2007) Landscape and conservation management effects on hydraulic properties of a claypan-soil toposequence. *Soil Science Society of America Journal* **71**, 803–811. doi:10.2136/sssaj2006.0236

Jose S (2009) Agroforestry for ecosystem services and environmental benefits: an overview. *Agroforestry Systems* **76**, 1–10. doi:10.1007/s10457-009-9229-7

Jose S (2012) Agroforestry for conserving and enhancing biodiversity. *Agroforestry Systems* **85**, 1–8. doi:10.1007/s10457-012-9517-5

Kelleners TJ, Seyfried M, Blonquist JM, Bilskie J, Chandler D (2005) Improved interpretation of water content reflectometer measurements in soils. *Soil Science Society of America Journal* **69**, 1684–1690. doi:10.2136/sssaj2005.0023

Kladivko EJ, Clapperton MJ (2011) Soil biology. In *Soil Management: Building A Stable Base for Agriculture*. (Eds JL Hatfield and TJ Sauer) pp. 145–160. American Society of Agronomy and Soil Science Society of America, Madison.

Kremen C, Miles A (2012) Ecosystem services in biologically diversified versus conventional farming systems: benefits, externalities, and tradeoffs. *Ecology and Society* **17**, art40. doi:10.5751/ES-05035-170440

Kremer RJ, Li J (2003) Developing weed-suppressive soils through improved soil quality management. *Soil & Tillage Research* **72**, 193–202. doi:10.1016/S0167-1987(03)00088-6

Kumar S (2009) Agroforestry and grass buffers for improving soil hydraulic properties and reducing runoff and sediment losses from grazed pastures. PhD thesis. The University of Missouri–Columbia, USA.

Kumar S, Anderson SH, Bricknell LG, Udawatta RP, Gantzer CJ (2008) Soil hydraulic properties influenced by agroforestry and grass buffers for grazed pasture systems. *Journal of Soil and Water Conservation* **63**, 224–232. doi:10.2489/jswc.63.4.224

Kumar S, Anderson SH, Udawatta RP, Kallenbach RL (2012) Water infiltration influenced by agroforestry and grass buffers for a grazed pasture system. *Agroforestry Systems* **84**, 325–335. doi:10.1007/s10457-011-9474-4

Lal R, Delgado JA, Groffman PM, Millar N, Dell C, Rotz A (2011) Management to mitigate and adapt to climate change. *Journal of Soil and Water Conservation* **66**, 276–285. doi:10.2489/jswc.66.4.276

Liste HH, White JC (2008) Plant hydraulic lift of soil water – implications for crop production and land restoration. *Plant and Soil* **313**, 1–7. doi:10.1007/s11104-008-9696-z

Liu X, Zhang X, Zhang M (2008) Major factors influencing the efficacy of vegetated buffers on sediment trapping: a review and analysis. *Journal of Environmental Quality* **37**, 1667–1674. doi:10.2134/jeq2007.0437

Lorenz K, Lal R (2014) Soil organic carbon sequestration in agroforestry systems. a review. *Agronomy for Sustainable Development* **34**, 443–454. doi:10.1007/s13593-014-0212-y

Lovell ST, Sullivan WC (2006) Environmental benefits of conservation buffers in the United States: evidence, promise, and open questions. *Agriculture, Ecosystems & Environment* **112**, 249–260. doi:10.1016/j.agee.2005.08.002

Lupwayi NZ, Rice WA, Clayton GW (1998) Soil microbial diversity and community structure under wheat as influenced by tillage and crop rotation. *Soil Biology & Biochemistry* **30**, 1733–1741. doi:10.1016/S0038-0717(98)00025-X

Lupwayi NZ, Arshad MA, Rice WA, Clayton GW (2001) Bacterial diversity in water-stable aggregates of soils under conventional and zero tillage management. *Applied Soil Ecology* **16**, 251–261. doi:10.1016/S0929-1393(00)00123-2

Luxmoore RJ (1981) Micro-, meso-, and macroporosity of soil. *Soil Science Society of America Journal* **45**, 671–672. doi:10.2136/sssaj1981.03615995004500030051x

Ma Z, Wood CW, Bransby DI (2000) Impacts of soil management on root characteristics of switchgrass. *Biomass and Bioenergy* **18**, 105–112. doi:10.1016/S0961-9534(99)00076-8

Magdoff F, Van Es H (2009) *Building Soils for Better Crops*. 3rd edn. Handbook Series Book 10. Sustainable Agriculture Research and Education (SARE) Program, Beltsville.

Matocha J, Schroth G, Hills T, Hole D (2012) Integrating climate change adaptation and mitigation through agroforestry and ecosystem conservation. In *Agroforestry the Future of Global Land Use*. (Ed. D Garrity) pp. 105–126. Springer, Dordrecht.

Mbow C, Smith P, Skole D, Duguma L, Bustamante M (2014) Achieving mitigation and adaptation to climate change through sustainable agroforestry practices in Africa. *Current Opinion in Environmental Sustainability* **6**, 8–14. doi:10.1016/j.cosust.2013.09.002

McKergow LA, Weaver DM, Prosser IP, Grayson RB, Reed AE (2003) Before and after riparian management: sediment and nutrient exports from a small agricultural catchment, Western Australia. *Journal of Hydrology* **270**, 253–272. doi:10.1016/S0022-1694(02)00286-X

Mohammed MG, Trauth KM (2019) A python-based GIS simulation of the spatial and temporal variation in evapotranspiration. *Applied Engineering in Agriculture* **35**, 759–765. doi:10.13031/aea.13139

Montgomery DR (2007) Soil erosion and agricultural sustainability. *Proceedings of the National Academy of Sciences of the United States of America* **104**, 13268–13272. doi:10.1073/pnas.0611508104

Monti A (2012) *Switchgrass: A Valuable Biomass Crop for Energy*. Springer Science and Business Media, London.

Mudgal A, Anderson SH, Baffaut C, Kitchen NR, Sadler EJ (2010) Effects of long-term soil and crop management on soil hydraulic properties for claypan soils. *Journal of Soil and Water Conservation* **65**, 393–403. doi:10.2489/jswc.65.6.393

Mulebeke R, Kironchi G, Tenywa MM (2010) Enhancing water use efficiency of cassava and sorghum based cropping systems in drylands. In *Second RUFORUM Biennial Meeting*, 20–24 September 2010, Entebbe. <http://news2.mak.ac.ug/documents/RUFORUM/Mulebeke.pdf>.

Nema P, Nema S, Roy P (2012) An overview of global climate changing in current scenario and mitigation action. *Renewable & Sustainable Energy Reviews* **16**, 2329–2336. doi:10.1016/j.rser.2012.01.044

Novák V, Hlaváčiková H (2019) *Applied Soil Hydrology: Theory and Applications of Transport in Porous Media*. Springer Nature Switzerland, Cham.

Paudel BR, Udawatta RP, Anderson SH (2011) Agroforestry and grass buffer effects on soil quality parameters for grazed pasture and row-crop systems. *Applied Soil Ecology* **48**, 125–132. doi:10.1016/j.apsoil.2011.04.004

Paudel BR, Udawatta RP, Kremer RJ, Anderson SH (2012) Soil quality indicator responses to row crop, grazed pasture, and agroforestry buffer management. *Agroforestry Systems* **84**, 311–323. doi:10.1007/s10457-011-9454-8

Pramanik P, Bhattacharya P, Chakrabarti B, Ghosh T (2019) Improved soil environment under conservation agriculture. In *Sustainable Management of Soil and Environment*. (Eds RS Meena, S Kumar, JS Bohra and ML Jat) pp. 169–192. Springer, Singapore.

Priyadharshini S, Canessane RA (2015) Evaluation of evapotranspiration for agricultural crops. In *2015 IEEE International Conference on Computational Intelligence and Computing Research*, 10–12 December 2015, Madurai. IEEE, <https://ieeexplore.ieee.org/document/7435702>.

Rachman A, Anderson SH, Gantzer CJ, Alberts EE (2004a) Soil hydraulic properties influenced by stiff-stemmed grass hedge systems. *Soil Science Society of America Journal* **68**, 1386–1393. doi:10.2136/sssaj2004.1386

Rachman A, Anderson SH, Gantzer CJ, Thompson AL (2004b) Influence of stiff-stemmed grass hedge systems on infiltration. *Soil Science Society of America Journal* **68**, 2000–2006. doi:10.2136/sssaj2004.2000

Rachman A, Anderson SH, Gantzer CJ (2005) Computed-tomographic measurement of soil macroporosity parameters as affected by stiff-stemmed grass hedges. *Soil Science Society of America Journal* **69**, 1609–1616. doi:10.2136/sssaj2004.0312

Ramachandran Nair P (1993) *An Introduction to Agroforestry*. Kluwer Academic Publishers, Dordrecht.

Ramachandran Nair P (2007) The coming of age of agroforestry. *Journal of the Science of Food and Agriculture* **87**, 1613–1619. doi:10.1002/jsfa.2897

Ramachandran Nair P, Mohan Kumar B, Nair VD (2009) Agroforestry as a strategy for carbon sequestration. *Journal of Plant Nutrition and Soil Science* **172**, 10–23. doi:10.1002/jpln.200800030

Reck A, Jackisch C, Hohenbrink TL, Schröder B, Zangerlé A, Van Schaik L (2018) Impact of temporal macropore dynamics on infiltration: field experiments and model simulations. *Vadose Zone Journal* **17**, 170147. doi:10.2136/vzj2017.08.0147

Richards JH, Caldwell MM (1987) Hydraulic lift: substantial nocturnal water transport between soil layers by *Artemisia tridentata* roots. *Oecologia* **73**, 486–489. doi:10.1007/BF00379405

Sahin H, Anderson SH, Udawatta RP (2016) Water infiltration and soil water content in claypan soils influenced by agroforestry and grass buffers compared to row crop management. *Agroforestry Systems* **90**, 839–860. doi:10.1007/s10457-016-9899-x

Sauer TJ, Hernandez–Ramirez G (2011) Agroforestry. In *Soil Management: Building a Stable Base for Agriculture*. (Eds JL Hatfield and TJ Sauer) pp. 351–370. American Society of Agronomy and Soil Science Society of America, Madison.

Schoonover JE, Crim JF (2015) An introduction to soil concepts and the role of soils in watershed management. *Journal of Contemporary Water Research & Education* **154**, 21–47. doi:10.1111/j.1936-704X.2015.03186.x

Senaviratne GA, Udawatta RP, Nelson KA, Shannon K, Jose S (2012) Temporal and spatial influence of perennial upland buffers on corn and soybean yields. *Agronomy Journal* **104**, 1356–1362. doi:10.2134/agronj2012.0081

Seobi T, Anderson SH, Udawatta RP, Gantzer C (2005) Influence of grass and agroforestry buffer strips on soil hydraulic properties for an Albaqualf. *Soil Science Society of America Journal* **69**, 893–901. doi:10.2136/sssaj2004.0280

Shiferaw BA, Okello J, Reddy RV (2009) Adoption and adaptation of natural resource management innovations in smallholder agriculture: reflections on key lessons and best practices. *Environment, Development and Sustainability* **11**, 601–619. doi:10.1007/s10668-007-9132-1

Spiertz JH, Ewert F (2009) Crop production and resource use to meet the growing demand for food, feed and fuel: opportunities and constraints. *NJAS Wageningen Journal of Life Sciences* **56**, 281–300. doi:10.1016/S1573-5214(09)80001-8

Stutter MI, Chardon WJ, Kronvang B (2012) Riparian buffer strips as a multifunctional management tool in agricultural landscapes: introduction. *Journal of Environmental Quality* **41**, 297–303. doi:10.2134/jeq2011.0439

Susha Lekshmi SU, Singh DN, Baghini MS (2014) A critical review of soil moisture measurement. *Measurement* **54**, 92–105. doi:10.1016/j.measurement.2014.04.007

Svoma BM, Fox NI, Pallardy Q, Udawatta RP (2016) Evapotranspiration differences between agroforestry and grass buffer systems. *Agricultural Water Management* **176**, 214–221. doi:10.1016/j.agwat.2016.06.018

Tisdall JM, Oades J (1982) Organic matter and water-stable aggregates in soils. *Journal of Soil Science* **33**, 141–163. doi:10.1111/j.1365-2389.1982.tb01755.x

Tramblay Y, Bouvier C, Martin C, Didon–Lescot JF, Todorovik D, Domergue JM (2010) Assessment of initial soil moisture conditions for event-based rainfall–runoff modelling. *Journal of Hydrology* **387**, 176–187. doi:10.1016/j.jhydrol.2010.04.006

Udawatta RP, Anderson SH (2008) CT-measured pore characteristics of surface and subsurface soils influenced by agroforestry and grass buffers. *Geoderma* **145**, 381–389. doi:10.1016/j.geoderma.2008.04.004

Udawatta RP, Krstansky JJ, Henderson GS, Garrett HE (2002) Agroforestry practices, runoff, and nutrient loss: a paired watershed comparison. *Journal of Environmental Quality* **31**, 1214–1225. doi:10.2134/jeq2002.1214

Udawatta RP, Motavalli PP, Garrett HE (2004) Phosphorus loss and runoff characteristics in three adjacent agricultural watersheds with claypan soils. *Journal of Environmental Quality* **33**, 1709–1719. doi:10.2134/jeq2004.1709

Udawatta RP, Anderson SH, Gantzer CJ, Garrett HE (2006) Agroforestry and grass buffer influence on macropore characteristics. *Soil Science Society of America Journal* **70**, 1763–1773. doi:10.2136/sssaj2006.0307

Udawatta RP, Kremer RJ, Adamson BW, Anderson SH (2008a) Variations in soil aggregate stability and enzyme activities in a temperate agroforestry practice. *Applied Soil Ecology* **39**, 153–160. doi:10.1016/j.apsoil.2007.12.002

Udawatta RP, Anderson SH, Gantzer CJ, Garrett HE (2008b) Influence of prairie restoration on CT-measured soil pore characteristics. *Journal of Environmental Quality* **37**, 219–228. doi:10.2134/jeq2007.0227

Udawatta RP, Garrett HE, Kallenbach R (2011) Agroforestry buffers for nonpoint source pollution reductions from agricultural watersheds. *Journal of Environmental Quality* **40**, 800–806. doi:10.2134/jeq2010.0168

Udawatta RP, Kremer RJ, Nelson KA, Jose S, Bardhan S (2014a) Soil quality of a mature alley cropping agroforestry system in temperate north America. *Communications in Soil Science and Plant Analysis* **45**, 2539–2551. doi:10.1080/00103624.2014.932376

Udawatta RP, Motavalli PP, Jose S, Nelson KA (2014b) Temporal and spatial differences in crop yields of a mature silver maple alley cropping system. *Agronomy Journal* **106**, 407–415. doi:10.2134/agronj2013.0429

Udawatta RP, Gantzer CJ, Reinbott TM, Wright RL, Pierce RA (2016) Yield differences influenced by distance from riparian buffers and conservation reserve program. *Agronomy Journal* **108**, 647–655. doi:10.2134/agronj2015.0273

United Nations Department of Economic and Social Affairs (UN DESA) (2009) 'World population prospects: the 2008 revision.' Working Paper No. ESA/P/WP 210, UN DESA, New York.

Veum KS, Goyne KW, Kremer R, Motavalli PP (2012) Relationships among water stable aggregates and organic matter fractions under conservation management. *Soil Science Society of America Journal* **76**, 2143–2153. doi:10.2136/sssaj2012.0089

Weerasekara C, Udawatta RP, Jose S, Kremer RJ, Weerasekara C (2016) Soil quality differences in a row-crop watershed with agroforestry and grass buffers. *Agroforestry Systems* **90**, 829–838. doi:10.1007/s10457-016-9903-5

Zaibon S, Anderson SH, Kitchen NR, Haruna SI (2016) Hydraulic properties affected by topsoil thickness in switchgrass and corn–soybean cropping systems. *Soil Science Society of America Journal* **80**, 1365–1376. doi:10.2136/sssaj2016.04.0111

Zaibon S, Anderson SH, Thompson AL, Kitchen NR, Gantzer CJ, Haruna SI (2017) Soil water infiltration affected by topsoil thickness in row crop and switchgrass production systems. *Geoderma* **286**, 46–53. doi:10.1016/j.geoderma.2016.10.016

Zaibon S, Anderson SH, Veum KS, Haruna SI (2019) Soil thermal properties affected by topsoil thickness in switchgrass and row crop management systems. *Geoderma* **350**, 93–100. doi:10.1016/j.geoderma.2019.05.005

Zeri M, Hussain MZ, Anderson-Teixeira KJ, DeLucia E, Bernacchi CJ (2013) Water use efficiency of perennial and annual bioenergy crops in central Illinois. *Journal of Geophysical Research. Biogeosciences* **118**, 581–589. doi:10.1002/jgrg.20052

Zhang Z, Wei C, Xie D, Gao M, Zeng X (2008) Effects of land use patterns on soil aggregate stability in Sichuan Basin, China. *Particuology* **6**, 157–166. doi:10.1016/j.partic.2008.03.001

8

Forest land use and soil hydrology

Rebecca Abney, Nandita Gaur, Matthew R. Levi, Gary L. Hawkins

Forest hydrological cycle

> *Forest hydrology…deals with the effects of forests and associated wildland vegetation on the water cycle, including the effect on erosion, water quality, and micro-climate. (Hewlett 1982)*

Forest hydrology is often studied by quantifying the water movement, storage, distribution and quality of water through a forest ecosystem (Amatya *et al.* 2016). Although hydrological forcings (precipitation, temperature, hydroclimate) and geophysical attributes (soils, geology, landscape position) drive forest ecosystems (National Research Council (NRC) 2008; Neary *et al.* 2009; Meng *et al.* 2019), strong feedbacks also exist between forest types and management practices on forest hydrology that have the potential to alter the local and regional hydrological cycle (Hahmann and Dickinson 1997; Costa and Pires 2010; Wang *et al.* 2010; Sun *et al.* 2016). Forest hydrology also encompasses what happens at multiple scales within and across watersheds, with different processes dominating at different scales. In this chapter we describe forest hydrology in two ways: (1) how hydrological drivers affect forest hydrology, soil quality and forest productivity; and (2) the effects of forest management activities on forest hydrology. We also present some relevant case studies that highlight the impact of climate change on forest hydrology.

Drivers of the forest hydrological cycle

Different processes control water availability at different scales (Meng *et al.* 2019; Roebroek *et al.* 2020), which consequently make the hydrological drivers of forest hydrology scale dependent. For example, at the continental scale, hydroclimate and water availability are the main drivers of forest productivity, with higher water availability supporting higher energy absorption leading to more growth (Roebroek *et al.* 2020). At more local scales, access to plant-available water and groundwater is controlled by topographic redistribution of precipitation and subsurface geology (Fan 2015). Some significant drivers of forest hydrology active at all scales are described in detail below. Note that the impact of these drivers will vary with scale, and readers should refer to the scales of the studies cited when drawing inferences.

Precipitation

Precipitation inputs to forests can occur in the form of rain, snow or occult precipitation. Occult precipitation includes dew, fog and drizzle and can be defined as precipitation that a rain gauge will not normally measure (Holwerda *et al.* 2010). Occult precipitation has been reported as an important source of water input to forests located at high elevations (Katata 2014) or in the presence of low clouds (Díaz-Granados *et al.* 2005), whereas elsewhere precipitation in the form of rain or snow has a large effect on the hydrology of forests. Our discussion may not be relevant for forests whose primary source of water is occult precipitation.

As regions go through droughts, floods and increased or decreased snowfall and rainfall, the amount of water in the region changes. These changes in the amount of available moisture can result in changes to the forest mass balance (Fan 2015; Roebroek *et al.* 2020). Decreasing precipitation rates that cause drought conditions over a prolonged period can result in increased leaf drop, increased near-ground solar radiation, increased evapotranspiration (ET) and decreased soil moisture. Increasing leaf drop can also encourage increased understorey growth, leading to increased ET and a changing microclimate (Royer *et al.* 2011). In contrast, increased precipitation can increase both surface and subsurface water movement. Under conditions of low soil moisture, increased surface flow and subsurface flow (mat flow) can result in increased soil moisture through the process of infiltration at downstream locations. However, once surface soil becomes saturated, the percentage of water infiltrating and percolating into the subsurface decreases and run-off volumes increase (Hortonian overland flow). For example, many regions of the US have experienced heavy downpours that have been larger than natural variations from 1958 to 2011 (National Climate Assessment 2014; Kunkel 2020), which can drastically change the proportions of water distributed to the soil profile and discharge.

Topography

Forest watersheds commonly have lower volumes of water leaving the watershed based on the amount of precipitation and the presence of litter and duff compared with other land uses. One driver of water movement in a forested watershed is topography. Topography is defined by Merriam-Webster as 'the configuration of a surface including its relief and the position of its natural and man-made features' (Merriam-Webster 2020). The surface topography of a forested landscape affects water availability and redistribution to forests by impacting snowmelt characteristics (higher in south-facing slopes in the Northern Hemisphere) and rainfall patterns (orographic precipitation). Conversely, bedrock topography determines depth to water table and hence controls run-off processes like saturation, overland flow or variable source area run-off. The variable source area (VSA) concept implies that as different portions of the watershed become saturated due to rising water tables, they produce run-off. The size of the region contributing to run-off increases with the size of the region becoming saturated. For example, for flat woodlands, VSA size can vary from none (during a dry period) to the entire watershed in wet periods (Hewlett and Hibbert 1967). VSA is the dominant mechanism for run-off in forests (Hewlett and Hibbert 1967). The VSA can be thought of as the riparian zone (Fig. 8.1) within a forested watershed. Across different watershed topographies, water movement is controlled by a variety of hydrological properties. On portions of the landscape with lower slopes, water will move slower and infiltration will dominate over run-off (Kirkby *et al.* 2002). In comparison, in a steeper-sloped watershed, run-off will happen quicker and water will have less opportunity to infiltrate. Topography also varies within a watershed, so the mechanisms of water movement across a hillslope

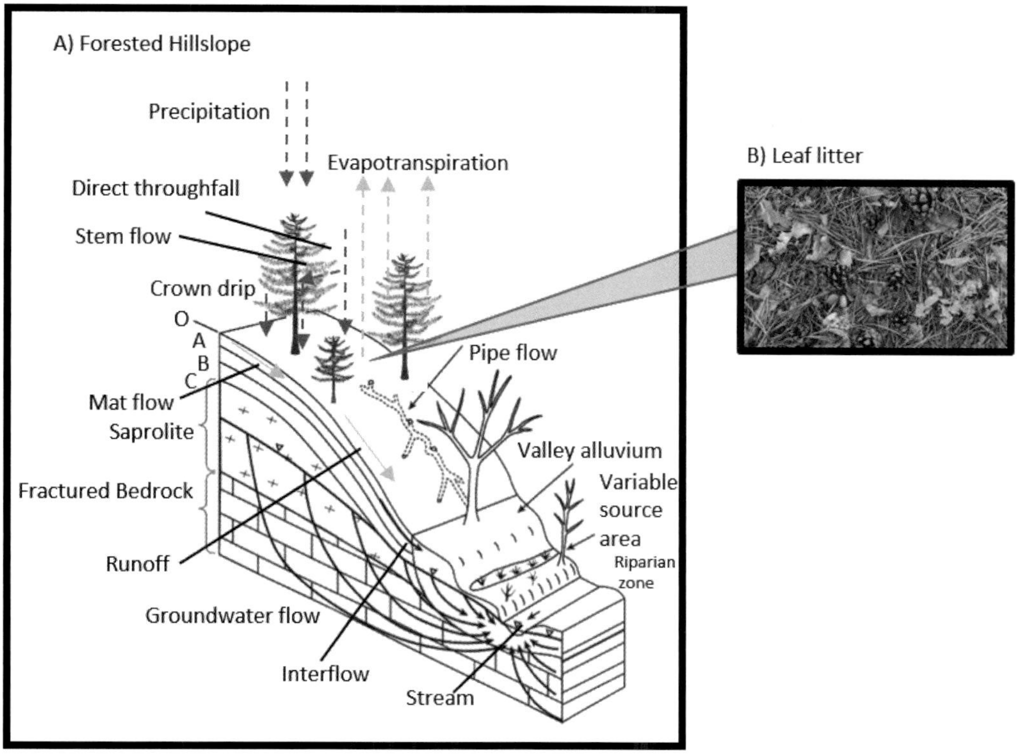

Fig. 8.1. (a) Forest water cycle at the hillslope scale. Dashed arrows pointing down indicate water input to the forest ecosystem, whereas the dashed arrows pointing up indicate the loss of water. Soil horizons are labelled (O, A, B, C). The small triangle in the Saprolite zone indicates top of the water table. (b) Image of leaf litter on the hillslope. Some forests also have a layer of duff underneath the top litter as a result of years of litter decomposition.

as discussed later will also change. Riparian zones may be flatter and have a slower velocity of water flow and higher rates of sedimentation and interaction with the subsurface water. However, this process is also dependent on the slope of the associated stream, with steeper slopes having more sediment transport and shallower slopes having more sedimentation. Thus, topography is an important driver for water availability in forests.

Vegetation type

Vegetation species compositions and demographics also have a large effect on water movement in a forested watershed. Vegetation species composition controls processes like canopy interception, ET, bark storage, stemflow, throughfall, crowndrip and accessibility to groundwater based on rooting depth, and thus exerts a strong control on water storage, the spatial distribution of precipitation within the forest and water losses.

The shape, layering patterns, density and characteristics of the leaves on the canopy, along with vegetation density itself, determine the impact of canopy interception (Pypker *et al.* 2011) and throughfall (the direct passage of rainfall to the ground through gaps in the canopy). Canopies intercept precipitation and either delay water transfer to the soil (through crowndrip or stemflow) or make the intercepted water unavailable for plant water use because of direct evaporation of intercepted water. Stemflow refers to the funnelling of intercepted water along the stem of the trees. The amount of stemflow is dependent on storm intensity, wind and atmospheric demand. Increased rainfall may first generate greater stemflow volumes (Van Stan *et al.* 2014), but eventually overloads stemflow pathways and reduces the efficiency of tree canopies (Carlyle-Moses and Price 2006; Staelens *et al.* 2008). Windier conditions tend to increase stemflow generation (Xiao *et al.* 2000; Koichiro *et al.* 2001), but increased atmospheric water demand (low relative humidity) reduces stem flow by increasing evaporation from the trunk (Llorens *et al.* 1997; Staelens *et al.* 2008). The amount of water intercepted varies based on canopy layering and density, as well as the different mechanisms by which leaves hold water (thin layer of water coating vs capillary pores between leaves). Thus, flat leaves, such as those found in many deciduous species, hold less water than trees with needles (Keim *et al.* 2006). Intercepted water that does not get evaporated can reach the ground as crowndrip or stemflow. Both these processes change the spatial distribution of precipitation by concentrating precipitation around trees, thus making more water available in concentrated areas within the forests.

Bark storage also forms an important water storage mechanism in forest canopies. Some of the water that is intercepted by the bark does not reach the ground. Levia and Herwitz (2005) reported that forests can store tens of thousands of litres of water as bark storage. It is important to note that water storage characteristics in the forest canopies and bark depend on the texture of the bark. Tree barks are texturally variable with height (Klamerus-Iwan *et al.* 2020). Bark near the bottom of the tree tends to be rough with numerous fissures (holds more water), whereas the top bark is smoother (Levia and Wubbena 2006) and consequently holds less water. This causes vertical variability in bark water storage. In addition, bark morphology, and consequently the amount of water stored in it, depends on tree age, tree size and site conditions (Legates *et al.* 2014).

Water losses in forests also occur through ET and can be significant. Schlesinger and Jasechko (2014) indicated that the percentage of transpiration to ET ranges from a mean (±s.d.) of 47 ± 10% to 70 ± 14%. However, under the tree canopy, the understorey vegetation has the potential of evapotranspiring water as well (Schlesinger and Jasechko 2014). The understorey consists of many different varieties of brush, shrubs and grasses, all of which have different transpiration rates. Based on the mechanisms of water movement, precipitation rates, topography and available water in the soil profile, the movement of water can be highly variable. The age of the vegetation can also affect water movement in the watershed, with transpiration patterns being determined by tree type and age. For example, Liu *et al.* (2017) found that younger eucalyptus trees transpired less than the older trees, whereas Delzon and Loustau (2005) reported lower transpiration from older pine trees. Therefore, vegetation has a direct effect on water movement in a watershed, along with the amount of precipitation and the topography of the landscape. These processes are illustrated on an example hillslope in Fig. 8.1a.

Mechanisms of water movement in forest soils

Water movement in forest soils follows the same basic principles as landscapes that support grasslands or other crops, with the primary difference being that forest soils support a 'crop' with a larger root system, greater vertical separation from the land surface and, when highly managed, with a typical cycle of 30 years (NRC 2008; Ouyang *et al.* 2019). This creates distinct conditions for the input and distribution of water within forests. As an example, the amount and intensity of water reaching the ground surface in forests depends on the type of forest (broadleaf vs needle leaf), vegetation density and canopy thickness and season (leaf off vs leaf on). During the typical lifecycle of a forest, many factors can influence forest hydrology by changing soil infiltration rates, percolation rates and ET of water into and out of the soil profile. These factors include, but are not limited to, soil properties, plant community composition, canopy layering and structure, leaf drop and accumulation, fire, erosion, insects and logging operations (Hesslerová *et al.* 2013; Ajaz Ahmed *et al.* 2017; Boisramé *et al.* 2018; Blount *et al.* 2020; Kopáček *et al.* 2020). In unmanaged forests, vegetation age structure may be mixed, resulting in different rates of ET.

A characteristic property of forest soils is that they commonly have an organic soil horizon that protects the soil surface from erosion and provides a steady supply of organic matter that promotes aggregate formation and subsequently increases infiltration at and near the surface. These same organic horizons can also cause a hydrophobic layer on the mineral soil due to the presence of organic compounds, particularly in burned soils. These conditions create subsurface flow and transport characteristics that are unique to forest soils. As the dynamics of the forest floor and canopy change, so do the different aspects of water movement (Jones 2000; Berg *et al.* 2009; Ouyang *et al.* 2019). The water cycle for a typical forest is shown in Fig. 8.1. The precipitating water either falls to the ground surface through the intercanopy spaces as throughfall and stemflow or is intercepted by the canopy. The water that reaches the ground surface can either enter the forest land surface or is transported to the nearest streams through the various processes described below.

Infiltration

Infiltration is the process of water entering the soil surface. The amount of infiltrated water depends on land cover, soil properties (texture, structure, porosity etc.) and landscape properties such as slope angle, aspect and shape (Neary *et al.* 2009; Ajaz Ahmed *et al.* 2017; Ouyang *et al.* 2019; Blount *et al.* 2020). The soil properties that determine infiltration include soil porosity, structure, mineralogy, compaction and texture (Plate 5c). Soils with coarser textures, good structure (big macropores), low bulk density and a lack of hydrophobic compounds enhance infiltration and reduce run-off. Soil infiltration in forests is enhanced because of an increase in organic matter in soils as a result of decomposing leaf litter, which increases soil porosity and decreases bulk density (Sayer 2006; Jourgholami *et al.* 2020), and because of the abundance of fauna that create burrows in the A and B soil horizons (Whipkey 1965; Neary *et al.* 2009), which enhance soil structure. Conversely, the deposition of hydrophobic substances on the soil surface as a result of microbial activity, plant exudates and subsurface waxes from plant leaves, all of which are typically high in forest soils, can cause reduced infiltration rates, particularly when soils are dry (Hallett and Young 1999; Doerr *et al.* 2000). Prescribed or natural burns may also result in hydrophobic conditions and reduced infiltration rates in soils.

Percolation

The percolation of water refers to water movement within and through the soil layers. Percolation helps recharge aquifers that form sources of water for many different uses, including irrigation and drinking water. Ouyang *et al.* (2019) modelled this recharge and indicated that forests have a slightly greater deep percolation rate than agricultural lands. This percolation is associated with forests having cover from trees, shrubs, litter and other organic materials year-round and lower run-off. The percolation of water is dependent on site-specific soil properties. Shallow confining layers tend to cause percolated water to move laterally to surface water sources, whereas deeper confining layers permit water to move deeper into the soil profile. In landscapes where high water tables are common, percolation rates are frequently controlled by precipitation events, antecedent soil moisture and microtopography (Liu *et al.* 2020).

Evapotranspiration

ET is a combination of evaporation and transpiration and, in forested watersheds, it can be a significant portion of the partitioning of the water balance (Schlesinger and Jasechko 2014; Blount *et al.* 2020). Evaporation of water occurs from a surface (soil, leaf litter and leaves of plants) and transpiration is the movement of water from the soil through the plant for purposes of biological activity of the plant (Hawkins and Thomas 2014; Romashchenko *et al.* 2020; Yu *et al.* 2020). This movement of water through plant transpiration accounts for 60–80% of ET and is a dominant force driving the global water cycle (Schlesinger and Jasechko 2014; Yu *et al.* 2020).

ET affects forest hydrology in a few different ways depending on the time scale at which it is being monitored. Some research indicates that in addition to moving water for life processes, ET contributes to the cooling of the environment around forested areas (Hesslerová *et al.* 2013). The removal of forest biomass, either via logging or fire, has a long-term effect on ET by decreasing infiltration and increasing run-off and streamflow (Andréassian 2004; Oudin *et al.* 2008; Stednick 2008; Schlesinger and Jasechko 2014), water yield (Jackson *et al.* 2005; Blount *et al.* 2020) and soil moisture (Boisramé *et al.* 2018) in small (<1000 km^2) watersheds. Such generalisations have not yet been made for large watersheds. Over a shorter time scale, such as at the interannual scale, there can be a trade-off between the impact of ET on forest hydrology and cooling. ET from forests may not directly mimic the interannual variability in precipitation (Wullschleger and Hanson 2006) and hence can sustain the cooling effect on the microclimate even during drought years. However, the lack of correlation between precipitation and ET losses at the interannual scale may lead to ET losses from a forest during a drought year and an excessively wet year being similar. This causes decreased replenishment of groundwater water supply to downstream ecosystems during a drought year, as was observed in a mature hickory oak forest over a 2-year period in North Carolina where the interannual variation in precipitation ranged from 934 to 1346 mm but the annual ET varied only between 610 and 668 mm (Oishi *et al.* 2010). The slightly higher ET during the wet year was attributed to increased canopy rainfall interception losses (Hubbart *et al.* 2007; Oishi *et al.* 2010).

Transport of water from soils to streams

Water moves from the land surface to local streams through both surface and subsurface flow paths (Fig. 8.1a). This is governed by run-off, infiltration, interflow and percolation of the water that falls on the surface through either natural rainfall or irrigation. Run-off moves along the surface to transport water to streams, and the speed at which it travels depends on the characteristics of the forest floor covering, such as the presence or absence of debris resulting from harvesting, erosion or fire. At the hillslope scale, interflow and variable source area run-off production due to bedrock topography also become significant for subsurface and surface water movement respectively (Tromp-van Meerveld and McDonnell 2006a, 2006b). Interflow, or lateral subsurface flow above an impeding or low-permeability soil layer, also often forms an important mechanism of subsurface water movement on forested hillslopes and depends on the depth of the impeding layer. In areas where bedrock topography is variable, temporary storage of water occurs in the microlows that fill up to intercept the surface and cause additional run-off. The movement of water through the processes of infiltration and percolation also contributes to the water entering streams as baseflow. Although all methods mentioned provide water to streams, the path they take and the time required to move from the forest floor are different, with run-off rates being higher than subsurface flowpaths. Although subsurface flow is generally slower than surface run-off, some subsurface flow paths that occur as preferential flow can also lead to the rapid movement of water to streams. Preferential flow paths form an important and often dominant mechanism of water movement through soils. Forest soils frequently support unique preferential flow paths because of large amounts of organic litter and duff, a hydrophobic or water-repellent layer on the mineral soil surface that develops as a result of a thick and persistent O horizon and pipe flow associated with decaying root channels. These preferential flow paths that are specific to forest environments are described below.

Mat flow

Mat flow, or biomat flow, is a lateral subsurface flow process that occurs as water movement within the loose litter layer, within the decomposed portion of the organic horizon or within and below pockets of colluvial organic and mineral materials deposited on hillslopes (Sidle *et al.* 2007). The large pore spaces associated with biomat litter layer enhance soil water storage, providing a high saturated hydraulic conductivity at the soil surface relative to the mineral soil and leading to rapid infiltration (Bauters *et al.* 2000). If a hydrophobic layer is present on the mineral soil surface beneath the litter layer, lateral distribution of infiltrating water is enhanced due to mat flow, which could lead to the preferential movement of water between the two layers. Water movement towards streams through this mechanism will be slower than surface run-off but faster than the subsurface flow path that water takes through groundwater. In case of a wet hydrophilic soil beneath the biomat, infiltration into the mineral soil increases.

Hydrophobicity-induced flow

Hydrophobic layers that may exist at the bottom of the litter layer require a threshold pressure head to allow infiltration of water into the mineral surface. In the case of a concave depression in the landscape, water tends to accumulate over a hydrophobic layer until the pressure threshold is reached or flows downhill into a depression where it will eventually locally percolate into the mineral soil creating spotty percolation zones. Thus, microtopography of the forested land surface can play a significant role in reducing run-off and promoting preferential infiltration. This will lead to reduced run-off to streams, thus mitigating the 'flashy' nature of streamflow after storms (Doerr *et al.* 2000; Shakesby *et al.* 2000).

Pipe flow

Pipe flow, as the name suggests, occurs through large, mostly lateral and continuous macropores in the forest subsurface. The pipes are typically >1 cm in diameter and are formed as a result of large, decayed tree roots, large animal burrows, chemical dissolutions or internal erosion (Uchida *et al.* 2002, 2005). Pipe flow channels mostly develop after large-scale deforestation or prescribed burns or wildfire, where the root channels are left behind (Leslie and Heinse 2013). These channels can act as conduits under saturated conditions and enable the rapid transport of water to streams.

Impact of forest hydrology on soil quality and productivity

Impact of forest hydrology on productivity, landscape factors and forest-specific composition

Landscape position plays a significant role in the distribution of soil properties, soil types and water table fluctuations (Bailey *et al.* 2014). As the slope of the land changes, run-off and associated erosion influence the soil profile, which can have positive or negative feedbacks on hydrological processes such as water distribution and soil wetness (Plate 5b). Collectively, soils, landscapes and climate affect the susceptibility of a particular site to erosion and potential loss of productivity. Soil loss in forested areas primarily occurs through three main types of erosion: sheet, rill and gully. Sheet erosion is the movement of water and soil downslope in a sheet with deposition of soil at the bottom of the slope. In the event that bare soil is exposed to raindrop impacts, splash erosion is likely to precede sheet erosion and dislodge soil particles. Rill erosion develops when sheets converge into small channels of water flow, and gully erosion is the formation of trenches that are too deep to till across. In the locations where any or all of these types of erosion occur in channelised conditions, the organic litter layer of the forest is removed and starts to cut into the various horizons of the soil. Slopes at the tops of rolling hills are often gentle and run-off occurs in the form of sheet flow, which is the shallow flow of water that carries soil and other particles downhill before depositing sediment in thin sheets as the flow decreases and stops. As the slope increases, the water in sheets starts to join and forms rills and gullies, which have higher velocities and energy to carry more soil and organic matter, thereby eroding the top soil layers. At the bottom of the hills, the slopes decrease and the sediment carried by the run-off is deposited. Another form of soil loss is mass wasting, where large volumes of soil are removed from a slope as the result of intense rainfall, snowmelt or saturated soil along steep slopes, to name but a few. Mass wasting typically occurs as a rapid movement of large volumes of soil or rock, whereas erosion is a slower process of soil removal and movement downslope.

Soils and hydrology work together to create favourable conditions for specific species. For example, in the south-eastern US, pines tend to grow in upland landscape positions and hardwoods tend to grow in bottomlands. Moving towards even wetter conditions would support species such as willows, box elder and other species that are adapted to wetter conditions.

Impact of forest hydrology on soil quality and function

Run-off potential and erosion

Run-off and erosion from well-established forests are typically lower than for most other land uses (Neary *et al.* 2009; Grace 2017). The potential for run-off varies with the condition of the forest itself. A well-established forest has understorey and organic matter on the forest floor to intercept precipitation and increase infiltration. As these factors work together, water has an increased potential to infiltrate, and thereby a reduced potential for run-off and erosion. However, based on the slope of the forest land, the intensity of rainfall and the locations of channelised flow, run-off from the landscape could be a large contributor to sediment transport and erosion. Like increased intensity of rainfall, other conditions can lead to the potential for increased run-off and erosion. These include, but are not limited to, dryer climate conditions, which can lead to decreased leaf production or the premature dropping of leaves, which reduces the potential to intercept rainfall and protect the soil. Drier conditions also reduce the microbial conversion of leaf matter into organic matter that, when incorporated into the forest floor, aids in the adsorption and infiltration of water into the soil profile. Therefore, as climate conditions change to decrease leaf litter or increase the intensity of rainfall and associated erosion, the soil quality has the potential to change and reduce the build-up of organic matter, which is vital in building soil quality, increasing infiltration and aiding in the improved water quality associated with forests.

Intensity, duration and frequency of rainfall

As the climate changes, the intensity of rainfall in different portions of the US, and globally, also changes (Adler *et al.* 2017; Easterling *et al.* 2017). Easterling *et al.* (2017) provide information on changes in the precipitation levels of the US for a 20-year period. As the rate of rainfall exceeds the infiltration rate of the soil, run-off and associated erosion begin. In forest systems, organic matter inputs, increased microbial action and increased soil biota actions improve soil quality and increase the infiltration rate, which can buffer the forest floor from some higher-intensity precipitation events. Overall, as water movement changes, erosion occurs and tree species change within a landscape, it should be remembered that land management can cause more significant changes to soil quality and function than individual soil types (Levi *et al.* 2010).

Carbon sequestration

Primary versus secondary forests

Stabilisation of carbon (C) in forested soils depends on the balance of vegetative, climatic and edaphic factors in a particular environment. Forested soils are responsible for more than half of total soil C storage (Lal 2005), but their management (i.e. maintenance of primary forests vs the establishment of secondary forests) can drive significant alterations in their soil properties, including C sequestration. Establishing pine plantations typically decreases soil C storage, particularly when the pine plantations replace primary forests, but as time since establishment increases, the overall soil C stock increases (Justine *et al.* 2017).

Land use change can also play a significant role in determining soil properties and C storage. However, these changes are often complex and depend on specific vegetation types, climatic factors, pre-existing land use and type of forest (Post and Kwon 2000; Guo and Gifford 2002). For example, the conversion of pasture and cropland that was previously forest to a secondary forest generally increases C sequestration in soil, but the conversion of primary forest to crop or pine plantation decreases soil C sequestration (Guo and Gifford 2002).

Soil organic matter and soil hydraulic properties

Soil organic matter is generally concentrated in the O and A soil horizons in forested soils. However, variability in the properties and thickness of these horizons depends significantly on landscape geomorphology, climate and vegetation. The accumulation of soil organic matter in depositional landform positions has been widely reported (Berhe *et al.* 2007, 2012; Doetterl *et al.* 2016). Erosion and the subsequent deposition of topsoil is also a mechanism for profile inversion in depositional landform positions (Berhe *et al.* 2007, 2012). Soil organic matter properties can dictate infiltration, percolation and water retention in soil.

Natural cycles of drought

Droughts and floods are natural parts of the water cycle. Droughts are characterised by a combination of lower-than-usual precipitation and warmer-than-usual temperatures. Han *et al.* (2019) discuss the effects of drought as it relates to the idea of groundwater droughts. As precipitation decreases in regions over short periods, vegetation can be affected, but in these agricultural droughts resumed precipitation can return the forest system back to normal. However, droughts over a long period are caused by extended periods of below-normal precipitation and hot weather, which increases ET from the trees and makes it hard for them to pull enough water from the soil into their crowns (Kirchen *et al.* 2017). These long-term droughts can start affecting the groundwater reserves and lead to 'groundwater droughts' (Han *et al.* 2019). Droughts are often self-sustaining, because with no moisture in the soil it is hard for convective rain to develop and spread moisture through the region, or even for clouds to form, which makes it hotter. Many droughts are ended by a flood event, which brings so much moisture to the area that it resets soil moisture conditions, allowing clouds and rain to develop again.

On the other side of the water cycle is flooding. Flooding can also damage trees through erosion or the suffocation of roots in flood zones if the water covers the ground for long periods of time. In the Southeastern US, these flood events are often tropical storms or hurricanes in summer or stationary fronts in winter, which concentrate a lot of rain in just a few days. In other areas, large-scale changes in atmospheric weather patterns, such as the El Niño Southern Oscillation may need to occur to swing from dry to wet conditions or vice versa (Yoon *et al.* 2015). Floods are usually more isolated in time and space and are easier to recover from because the water can be removed gradually by normal weather and drainage.

In the future, both droughts and floods are expected to occur more often and to be more extreme due to intensification of the water cycle because of rising temperatures, which can lead to either stronger droughts or heavier rainstorms. As the climate swings back and forth between wet and dry extremes, no one knows how or if forests will be able to survive or adapt to the new conditions. However, there is evidence that forest ecosystems are already shifting under the changing climate. In the future, it is likely that some species will not survive and that new forest types may develop as the climate continues to change.

Impact of forest land management on hydrology and water quality

Forest management activities like harvesting, re-afforestation, controlled burns, road building and site preparation for new plantings can affect flow processes by impacting the local and regional hydrology. This impact results from changes in the thickness and characteristics of the litter layer and surface soils, changes in the species composition of trees and changes in soil hydraulic properties. Forest management activities also affect the presence of the hydrophobic layer. An example of changes in the litter layer can be seen in the forested areas in Sweden, where 41 years of soil surveys showed an average 250 kg C ha^{-1} $year^{-1}$ accumulated in the forests (Berg *et al.* 2009); however, this number was higher in managed forests as opposed to old-growth undisturbed forests. Variations in the litter layer have the potential to affect several hydrological processes, as discussed earlier in this chapter.

Impact of land use and management activities on soil hydrology

Moore and Wondzell (2005) estimated that the time of recovery for forested watersheds to pre-disturbance hydrological conditions varies between 3 and 20 years in North America. Localised variability of management activities, like harvesting, can also be highly variable. For example, Fig. 8.2 shows contrasting levels of soil disturbance and soil cover from the same harvest event. Although management activities affect basic hydrological processes similarly, the actual impact of forest management activities on soil hydrology varies with the climate, dominant form of precipitation (rain/snow), hydrogeology, forest type, size of catchment, landscape position and spatial scale of the activity and the topography of the region (Lima *et al.* 2014; Zhang *et al.* 2017). For example, forest cover changes have a larger

A)

B)

Fig. 8.2. Contrasting levels of soil disturbance and vegetative cover for two adjacent areas harvested at the same time near Athens (GA, USA). More soil cover and vegetation prevented erosion (a), whereas exposed soil was subject to sheet erosion (b).

effect on changing annual run-off in water-limited watersheds than in energy-limited watersheds. Zhang *et al.* (2017) also showed that the type of forest is a more important factor when assessing annual run-off response to forest cover change in small watersheds, whereas the hydrological regime affects run-off more in larger watersheds. Coupling different management activities together can also alter the hydrological response of the forested land. As an example, Smerdon *et al.* (2009) showed how harvest activities reduce the amount of ET from a forested region. However, depending on the soil type, this can lead to either increased percolation and an elevated water table or enhanced run-off. Similarly, Smerdon *et al.* (2009) also showed that road building at the low points in a landscape may lead to the formation of a seepage face along the road cut and cause groundwater discharge, thus affecting regional hydrology by changing the timing and magnitude of peak flows, whereas building a road in a higher landscape position that is not intercepted by the water table would only have a local effect on the groundwater recharge and run-off generation process. However, if harvest activities are undertaken in a forest, the impact of existing or new roads would change as a function of the changing water table that results from harvest activities. Activities such as vegetation thinning can lead to an increase in underbrush development, which will decrease run-off, increase soil moisture and reduce soil loss due to erosion. This was observed in a tropical dry climate in the Caatinga Phytogeographical Domain, Brazil, which is characterised by 70% igneous-metamorphic and 30% sedimentary rock with shallow, low organic matter soils (Andrade *et al.* 2020).

Impact of management activities on streamflow quantity and quality

Forests have the potential to improve both the streamflow regime and the quality of water in the stream. As precipitation infiltrates into the soil profile, some of the water is transpired through the tree itself, but other portions have the potential to move laterally through the soil profile into surficial aquifers that are connected to local waterbodies (Fig. 8.1a). This connection provides a consistent and recharging baseflow to the creeks, streams and rivers adjacent to forested lands. Likewise, leaf litter and organic matter on the forest floor filter the water from precipitation and potentially run-off from sources outside the forest to improve the quality of the water entering streams, and thereby improving the water quality in creeks, streams and rivers (NRC 2008; Neary *et al.* 2009; Hawkins and Thomas 2014).

Fire, disease and run-off

Previous research demonstrated that fire alters soil physical, chemical and biological properties. The magnitude of the post-fire alterations in soil properties depends on the severity of the fire (i.e. charring intensity; Pyle *et al.* 2015) and ecosystem fire regime. Fire can have interactive effects with diseases and pests on forest health and productivity that drive shifts in soil hydrology.

Wildfire

Wildfires are characterised as uncontrolled burns in wildland areas and can range in severity from unburned patches to 100% tree mortality. Wildfires are associated with increases in the intensity and volume of run-off due to several factors, including the formation of a hydrophobic layer (see below), loss of stabilising vegetation and loss of surface cover (Cerdà and Doerr 2008; Larsen *et al.* 2009; Shakesby 2011). In forests with a high burn severity, the loss of soil stabilising and surface protection mechanisms exposes the soil to the full kinetic energy of precipitation, which can lead to significant erosion. When coupled with high-intensity precipitation, the baring of soil surfaces can lead to erosion associated with rill and gully formation. In addition, high-severity burns can leave hillslopes vulnerable to ash and debris flow or slope failure during initial post-fire precipitation events, as is frequently reported in the western US.

Prescribed fire

In contrast to wildfires, prescribed fires are controlled, lower-severity burns conducted to achieve a variety of management aims, including fuel reduction to prevent wildfire, pest management, game habitat management and site productivity (Ryan *et al.* 2013). Prescribed burns typically have much less loss of vegetation than wildfires, and considerably lower tree mortality. To moderate the severity of these burns, they are often conducted during seasons with higher precipitation and lower temperatures (Ryan *et al.* 2013); however, the higher precipitation and lower post-burn infiltration rates can leave these areas susceptible to higher run-off (Robichaud 2000). This higher run-off can lead to increased water use efficiency (Renninger *et al.* 2013) or increased water availability, where the thinning of forests reduces competition for water (Sala *et al.* 2005). An example of how a managed forest can be prescribed burned can be seen in Plate 5a.

Hydrophobicity

The formation of hydrophobic layers in the soil is widely reported after fires (prescribed and wildfires), and this shift can drive changes in infiltration and run-off in these soils. The formation of a hydrophobic layer generally occurs after low- and moderate-severity burns through the volatilisation of waxes and cutins, which are deposited onto the surfaces of soil particles deeper into the soil profile (Doerr *et al.* 2000; Shakesby *et al.* 2000). Hydrophobic layers can also form in some forest litters due to concentration of some organic compounds, including waxes, suberins and cutins, among others. These layers reduce the infiltration of precipitation into the soil profile, which results in increased run-off, erosion and transport of soil surface materials (Carroll *et al.* 2007) and decreased soil water storage. Hydrophobic layers can persist in the soil for decades (Abney *et al.* 2017) and can drive significant increases in local erosion rates (Abney *et al.* 2019).

Disease and pests

Diseases and pests of forests can have a range of effects on trees, including decreased growth, increased needle casting and tree mortality. The effects of diseases and pests on trees can alter the overall hydrology of a forested watershed, but the hydrological response depends on the interaction of several factors, such as the volume of needle cast, topography or the proportion of trees affected (Bladon *et al.* 2019). Hydrological responses to these disturbances may result in shifts in run-off, groundwater storage, infiltration, ET and soil moisture conditions (Mikkelson *et al.* 2013; Bearup *et al.* 2014; Bladon *et al.* 2019).

Case study: interaction of disease and wildfire as controls on soil hydrology

California, USA, experienced a historic drought from 2012 to 2015 that left many of the trees susceptible to a large invasion of bark beetles (McMahon *et al.* 2020). The combination of these two disturbances led to a massive die-off of trees in the Sierra Nevada during and after the drought, and this led to an increase in fuel load that was very dry (McMahon *et al.* 2020). From August to November 2013, the Rim Fire burned over 101 000 ha in Stanislaus National Forest and Yosemite National Park and, at the time, was the largest wildfire in the Sierra

Nevada. A bark beetle infestation within the perimeter of the Rim Fire contributed to conditions resulting in high burn severity and significant tree mortality (Hicke *et al.* 2016). The high-burn-severity areas of the Rim Fire produced significant sheet erosion (Abney *et al.* 2019), higher ET and discharge (Blasko 2020). The complex interactions between these disturbances can cause significant shifts in soil and watershed hydrology that persist for decades.

Interactions among climate change, silviculture and hydrology: case study

Increasing atmospheric CO_2 concentrations and the resulting climate changes have complex interacting effects on forest growth, competition among species and forest structure. With higher CO_2 levels, leaves can open their stoma for shorter periods of time to acquire the carbon needed for photosynthesis, making trees more water efficient, increasing productivity and altering the competitive balance among plants differing in photosynthetic pathways, mycorrhizal associations and xylem anatomy (Hwang *et al.* 2014, 2020). However, with longer growing seasons and warmer nights, total evaporation and respiration are increasing (Hobbins *et al.* 2004; Caldwell *et al.* 2016; Oishi *et al.* 2018) and accelerating hydrological and carbon cycles. Many places are experiencing more frequent intense rainfall events and more frequent severe droughts (Giorgi *et al.* 2011; Burt *et al.* 2018). These increasing hydrological extremes also affect forest growth and species composition.

Changes in precipitation intensity, amount and extremes

In the mixed hardwood forests of the southern Appalachian Mountains, USA, ecosystem responses to increasing CO_2 and hydroclimate variability have interacted with long-term fire suppression and successional responses to the loss of American chestnut in the 1930s to create complex changes in forest composition. In this region, rainfall is becoming more variable, with more frequent large storms and more frequent and severe dry periods (Laseter *et al.* 2012; Burt *et al.* 2018). Spring, summer and annual temperatures are increasing, with more hot summer days (Oishi *et al.* 2018). Contrary to expectations, less-drought-tolerant isohydric mesophytic species, predominantly tulip poplars and maples, have become more dominant over the past 40 years at the expense of more-drought-tolerant anisohydric xerophytic species, including oaks and hickories (Hwang *et al.* 2020). This unexpected change in forest composition appears to be mostly driven by long-term fire suppression (Nowacki and Abrams 2008, 2015), but other contributing factors include successional processes following the chestnut blight and enhanced biological nitrogen fixation following clearcutting and farming of the late 1800s. In this case, forest management and past disturbances seem to be pushing the forest ecosystem in directions counter to expectations informed by climate change alone. However, consistent with our expectations for how observed climate changes should affect forest processes, oaks and hickories in upslope positions are growing faster and are more resistant to drought effects (Hwang *et al.* 2020). In short, forest composition and structure result from interactions among climate, management, past disturbance, landscape position and hillslope water subsidies (Hawthorne and Miniat 2018; Lin *et al.* 2019) and will continue to respond to future climate regimes.

Conclusions

In the forest hydrological cycle, the same processes occur as in any other landscape; however, the presence of litter has the potential to change the infiltration and run-off rates. Typically, forest infiltration rates will be higher and run-off rates lower than for other land uses, which leads to an increased surface/subsurface connection through preferential flows paths. Several forest land management activities can have marked effects on soil hydrology. Shifts in soil water transport and storage can impact the productivity of a landscape, but understanding the potential for these hydrological shifts can lead to the implementation of more efficacious management strategies. In addition, appropriate management of forest resources can help protect existing forest health. For example, with an increased use of prescribed fire to manage the risks of larger wildfires and maintain stand productivity, careful planning needs to be taken to avoid erosion of essential soil nutrients, reductions in surface water quality and maintenance of soil water storage. Overall, forests have important places in the global water cycle, and they play a role in increasing infiltration, reducing run-off and erosion and providing resilience to a changing climate.

References

Abney RB, Sanderman J, Johnson D, Fogel ML, Berhe AA (2017) Post-wildfire erosion in mountainous terrain leads to rapid and major redistribution of soil organic carbon. *Frontiers of Earth Science* **5**, 99. doi:10.3389/feart.2017.00099

Abney RB, Kuhn TJ, Chow A, Hockaday W, Fogel ML, Berhe AA (2019) Pyrogenic carbon erosion after the Rim Fire, Yosemite National Park: the role of burn severity and slope. *Journal of Geophysical Research. Biogeosciences* **124**, 432–449. doi:10.1029/2018JG004787

Ajaz Ahmed MA, Abd-Elrahman A, Escobedo FJ, Cropper WP, Martin TA, Timilsina N (2017) Spatially-explicit modeling of multi-scale drivers of aboveground forest biomass and water yield in watersheds of the Southeastern United States. *Journal of Environmental Management* **199**, 158–171. doi:10.1016/j.jenvman.2017.05.013

Adler RF, Gu G, Sapiano M, Wang J, Huffman GJ (2017) Global precipitation: means, variations and trends during the satellite era (1979–2014). *Surveys in Geophysics* **38**, 679–699. doi:10.1007/s10712-017-9416-4

Amatya D, Williams T, Bren L, De Jong C (Eds) (2016) *Forest Hydrology: Processes, Management and Assessment.* CABI, Boston.

Andrade EM, Guerreiro MJS, Palácio HAQ, Campos DA (2020) Ecohydrology in a Brazilian tropical dry forest: thinned vegetation impact on hydrological functions and ecosystem services. *Journal of Hydrology. Regional Studies* **27**, 100649. doi:10.1016/j.ejrh.2019.100649

Andréassian V (2004) Waters and forests: from historical controversy to scientific debate. *Journal of Hydrology* **291**, 1–27. doi:10.1016/j.jhydrol.2003.12.015

Bailey SW, Brousseau PA, McGuire KJ, Ross DS (2014) Influence of landscape position and transient water table on soil development and carbon distribution in a steep, headwater catchment. *Geoderma* **226–227**, 279–289. doi:10.1016/j.geoderma.2014.02.017

Bauters TWJ, Steenhuis TS, DiCarlo DA, Nieber JL, Dekker LW, Ritsema CJ, Parlange J-Y, Haverkamp R (2000) Physics of water repellent soils. *Journal of Hydrology* **231**, 233–243. doi:10.1016/S0022-1694(00)00197-9

Bearup LA, Maxwell RM, Clow DW, McCray JE (2014) Hydrological effects of forest transpiration loss in bark beetle-impacted watersheds. *Nature Climate Change* **4**, 481–486. doi:10.1038/nclimate2198

Berg B, Johansson MB, Nilsson A, Gundersen P, Norell L (2009) Sequestration of carbon in the humus layer of Swedish forests – direct measurements. *Canadian Journal of Forest Research* **39**, 962–975. doi:10.1139/X09-022

Berhe AA, Harte J, Harden JW, Torn MS (2007) The significance of the erosion-induced terrestrial carbon sink. *Bioscience* **57**, 337–346. doi:10.1641/B570408

Berhe AA, Harden JW, Torn MS, Kleber M, Burton SD, Harte J (2012) Persistence of soil organic matter in eroding versus depositional landform positions. *Journal of Geophysical Research. Biogeosciences* **117**, G02019.

Bladon KD, Bywater-Reyes S, LeBoldus JM, Keriö S, Segura C, Ritóková G, Shaw DC (2019) Increased streamflow in catchments affected by a forest disease epidemic. *The Science of the Total Environment* **691**, 112–123. doi:10.1016/j.scitotenv.2019.07.127

Blasko C (2020) Assessing hydrologic impacts of the 2013 Rim Fire on the Tuolumne River watershed in Central Valley, California. PhD Thesis. Bowling Green State University, USA.

Blount K, Ruybal CJ, Franz KJ, Hogue TS (2020) Increased water yield and altered water partitioning follow wildfire in a forested catchment in the western United States. *Ecohydrology* **13**, 32170. doi:10.1002/eco.2170

Boisramé G, Thompson S, Stephens S (2018) Hydrologic responses to restored wildfire regimes revealed by soil moisture–vegetation relationships. *Advances in Water Resources* **112**, 124–146. doi:10.1016/j.advwatres.2017.12.009

Burt TP, Ford Miniat C, Laseter SH, Swank WT (2018) Changing patterns of daily precipitation totals at the Coweeta Hydrologic Laboratory, North Carolina, USA. *International Journal of Climatology* **38**, 94–104. doi:10.1002/joc.5163

Caldwell PV, Miniat CF, Elliott KJ, Swank WT, Brantley ST, Laseter SH (2016) Declining water yield from forested mountain watersheds in response to climate change and forest mesophication. *Global Change Biology* **22**, 2997–3012. doi:10.1111/gcb.13309

Carlyle-Moses DE, Price AG (2006) Growing-season stemflow production within a deciduous forest of southern Ontario. *Hydrological Processes* **20**, 3651–3663. doi:10.1002/hyp.6380

Carroll EM, Miller WW, Johnson DW, Saito L, Qualls RG, Walker RF (2007) Spatial analysis of a large magnitude erosion event following a Sierran Wildfire. *Journal of Environmental Quality* **36**, 1105–1111. doi:10.2134/jeq2006.0466

Cerdà A, Doerr SH (2008) The effect of ash and needle cover on surface runoff and erosion in the immediate post-fire period. *Catena* **74**, 256–263. doi:10.1016/j.catena.2008.03.010

Costa MH, Pires GF (2010) Effects of Amazon and Central Brazil deforestation scenarios on the duration of the dry season in the arc of deforestation. *International Journal of Climatology* **30**, 1970–1979. doi:10.1002/joc.2048

Delzon S, Loustau D (2005) Age-related decline in stand water use: sap flow and transpiration in a pine forest chronosequence. *Agricultural and Forest Meteorology* **129**, 105–119. doi:10.1016/j.agrformet.2005.01.002

Díaz-Granados MA, Navarrete González JD, Suárez López T (2005) Páramos: Hidrosistemas sensibles. *Revista de Ingenieria* **22**, 64–75.

Doerr SH, Shakesby RA, Walsh R (2000) Soil water repellency: its causes, characteristics and hydrogeomorphological significance. *Earth-Science Reviews* **51**, 33–65. doi:10.1016/S0012-8252(00)00011-8

Doetterl S, Berhe AA, Nadeu E, Wang Z, Sommer M, Fiener P (2016) Erosion, deposition and soil carbon: a review of process-level controls, experimental tools and models to address C cycling in dynamic landscapes. *Earth-Science Reviews* **154**, 102–122. doi:10.1016/j.earscirev.2015.12.005

Easterling DR, Kundel KE, Arnold JR, Knutson T, LeGrande AN, Leung LR, Vose RS, Waliser DE, Wehner MF (2017) Precipitation change in the United States. In *Climate Science Special Report: Fourth National Climate Assessment*. Vol. 1. (Eds DJ Wuebbles, DW Fahey, KA Hibbard, DJ Dokken, BC Stewart and TK Maycock) pp. 207–230. U.S. Global Change Research Program, Washington, DC.

Fan Y (2015) Groundwater in the Earth's critical zone: relevance to large-scale patterns and processes. *Water Resources Research* **51**, 3052–3069. doi:10.1002/2015WR017037

Giorgi F, Im ES, Coppola E, Diffenbaugh NS, Gao XJ, Mariotti L, Shi Y (2011) Higher hydroclimatic intensity with global warming. *Journal of Climate* **24**, 5309–5324. doi:10.1175/2011JCLI3979.1

Grace JM (2017) Predicting forest road surface erosion and storm runoff from high-elevation sites. *Transactions of the ASABE* **60**, 705–719. doi:10.13031/trans.11646

Guo LB, Gifford RM (2002) Soil carbon stocks and land use change: a meta analysis. *Global Change Biology* **8**, 345–360. doi:10.1046/j.1354-1013.2002.00486.x

Hahmann AN, Dickinson RE (1997) RCCM2–BATS model over tropical South America: applications to tropical deforestation. *Journal of Climate* **10**, 1944–1964. doi:10.1175/1520-0442(1997)010<1944:RBMOTS>2.0.CO;2

Hallett PD, Young IM (1999) Changes to water repellence of soil aggregates caused by substrate-induced microbial activity. *European Journal of Soil Science* **50**, 35–40. doi:10.1046/j.1365-2389.1999.00214.x

Han Z, Huang S, Huang Q, Leng G, Wang H, Bai Q, Zhao J, Ma L, Wang L, Du M (2019) Propagation dynamics from meteorological to groundwater drought and their possible influence factors. *Journal of Hydrology* **578**, 124102. doi:10.1016/j.jhydrol.2019.124102

Hawkins GL, Thomas D (2014) Protecting Georgia's surface water resources. University of Georgia (UGA) College of Agricultural and Environmental Sciences (CAES) Publication Bulletin number 1217, UGA CAES, Athens.

Hawthorne S, Miniat CF (2018) Topography may mitigate drought effects on vegetation along a hillslope gradient. *Ecohydrology* **11**, e1825. doi:10.1002/eco.1825

Hesslerová P, Pokorný J, Brom J, Rejškova-Procházková A (2013) Daily dynamic of radiation surface temperature of different land cover types in a temperate cultural landscape: consequences for the local climate. *Ecological Engineering* **54**, 145–154. doi:10.1016/j.ecoleng.2013.01.036

Hewlett JD (1982) *Principles of Forest Hydrology*. University of Georgia Press, Athens.

Hewlett JD, Hibbert AR (1967) Factors affecting the response of small watersheds to precipitation in humid areas. *Forest Hydrology* **1**, 275–290.

Hicke JA, Meddens AJ, Kolden CA (2016) Recent tree mortality in the western United States from bark beetles and forest fires. *Forest Science* **62**, 141–153. doi:10.5849/forsci.15-086

Hobbins MT, Ramírez JA, Brown TC (2004) Trends in pan evaporation and actual evapotranspiration across the conterminous US: paradoxical or complementary? *Geophysical Research Letters* **31**, L13503. doi:10.1029/2004GL019846

Holwerda F, Bruijnzeel LA, Muñoz-Villers LE, Equihua M, Asbjornsen H (2010) Rainfall and cloudwater interception in mature and secondary lower montane cloud forests, central Veracruz, Mexico. *Journal of Hydrology* **384**, 84–96. doi:10.1016/j.jhydrol.2010.01.012

Hubbart JA, Link TE, Gravelle JA, Elliot WJ (2007) Timber harvest impacts on water yield in the continental/maritime hydroclimatic region of the U.S. *Forest Science* **53**, 169–180.

Hwang T, Band LE, Miniat CF, Song C, Bolstad PV, Vose JM, Love JP (2014) Divergent phenological response to hydroclimate variability in forested mountain watersheds. *Global Change Biology* **20**, 2580–2595. doi:10.1111/gcb.12556

Hwang T, Band LE, Miniat CF, Vose JM, Knoepp JD, Song C, Bolstad PV (2020) Climate change may increase the drought stress of mesophytic trees downslope with ongoing forest mesophication under a history of fire suppression. *Frontiers in Forests and Global Change* **3**, 17. doi:10.3389/ffgc.2020.00017

Jackson RB, Jobbágy EG, Avissar R, Roy SB, Barrett DJ, Cook CW, Farley KA, Le Maitre DC, McCarl BA, Murray BC (2005) Trading water for carbon with biological carbon sequestration. *Science* **310**, 1944–1947. doi:10.1126/science.1119282

Jones JA (2000) Hydrologic processes and peak discharge response to forest removal, regrowth, and roads in 10 small experimental basins, western Cascades, Oregon. *Water Resources Research* **36**, 2621–2642. doi:10.1029/2000WR900105

Jourgholami M, Feghhi J, Picchio R, Tavankar F, Venanzi R (2020) Efficiency of leaf litter mulch in the restoration of soil physiochemical properties and enzyme activities in temporary skid roads in mixed high forests. *Catena* **198**, 105012.

Justine MF, Yang W, Wu F, Khan MN (2017) Dynamics of biomass and carbon sequestration across a chronosequence of masson pine plantations. *Journal of Geophysical Research: Biogeosciences* **122**, 578–591.

Katata G (2014) Fogwater deposition modeling for terrestrial ecosystems: a review of developments and measurements. *Journal of Geophysical Research, D, Atmospheres* **119**, 8137–8159. doi:10.1002/2014JD021669

Keim RF, Skaugset AE, Weiler M (2006) Storage of water on vegetation under simulated rainfall of varying intensity. *Advances in Water Resources* **29**, 974–986. doi:10.1016/j.advwatres.2005.07.017

Kirchen G, Calvaruso C, Granier A, Redon P-O, Van der Heijden G, Bréda N, Turpault M-P (2017) Local soil type variability controls the water budget and stand productivity in a beech forest. *Forest Ecology and Management* **390**, 89–103. doi:10.1016/j.foreco.2016.12.024

Kirkby M, Bracken L, Reaney S (2002) The influence of land use, soils and topography on the delivery of hillslope runoff to channels in SE Spain. Earth surface processes and landforms. *The Journal of the British Geomorphological Research Group* **27**, 1459–1473.

Klamerus-Iwan A, Link TE, Keim RF, Van Stan JT II (2020) *Storage and Routing of Precipitation Through Canopies*. Springer, Cham. doi:10.1007/978-3-030-29702-2_2

Koichiro K, Yuri T, Nobuaki T, Isamu K (2001) Generation of stemflow volume and chemistry in a mature Japanese cypress forest. *Hydrological Processes* **15**, 1967–1978. doi:10.1002/hyp.250

Kopáček J, Bače R, Hejzlar J, Kaňa J, Kučera T, Matějka K, Porcal P, Turek J (2020) Changes in microclimate and hydrology in an unmanaged mountain forest catchment after insect-induced tree dieback. *The Science of the Total Environment* **720**, 137518. doi:10.1016/j.scitotenv.2020.137518

Kunkel K (2020) Heavy downpours increasing. Climate Central, Princeton, <https://medialibrary.climatecentral.org/resources/heavy-downpours-2014>.

Lal R (2005) Forest soils and carbon sequestration. *Forest Ecology and Management* **220**, 242–258. doi:10.1016/j.foreco.2005.08.015

Larsen IJ, Macdonald LH, Brown E, Rough D, Welsh MJ, Pietraszek JH, Libohova Z, De Dios Benavides-Solorio J, Schaffrath K (2009) Causes of post-fire runoff and erosion: water repellency, cover, or soil sealing? *Soil Science Society of America Journal* **73**, 1393–1407. doi:10.2136/sssaj2007.0432

Laseter SH, Ford CR, Vose JM, Swift LW (2012) Long-term temperature and precipitation trends at the Coweeta Hydrologic Laboratory, Otto, North Carolina, USA. *Hydrology Research* **43**, 890–901.

Legates DR, Levia DF, Van Stan JT, Velasco Herrera VM (2014) Using wavelet analysis to examine bark microrelief. *Trees* **28**, 413–425. doi:10.1007/s00468-013-0959-9

Leslie IN, Heinse R (2013) Characterizing soil-pipe networks with pseudo-three-dimensional resistivity tomography on forested hillslopes with restrictive horizons. *Vadose Zone Journal* **12**, vzj2012.0200. doi:10.2136/vzj2012.0200

Levi MR, Shaw JN, Wood CW, Hermann SM, Carter EA, Feng Y (2010) Land management effects on near-surface soil properties of Southeastern US coastal plain Kandiudults. *Soil Science Society of America Journal* **74**, 258–271.

Levia DF, Herwitz SR (2005) Interspecific variation of bark water storage capacity of three deciduous tree species in relation to stemflow yield and solute flux to forest soils. *Catena* **64**, 117–137. doi:10.1016/j.catena.2005.08.001

Levia DF, Jr, Wubbena NP (2006) Vertical variation of bark water storage capacity of *Pinus strobus* L. (eastern white pine) in southern Illinois. *North-Eastern Naturalist* **13**, 131–137. doi:10.1656/1092-6194(2006)13[131:VVOBWS]2.0.CO;2

Lima LS, Coe MT, Filho BSS, Cuadra SV, Dias LCP, Costa MH, Rodrigues HO (2014) Feedbacks between deforestation, climate, and hydrology in the southwestern Amazon: implications for the provision of ecosystem services. *Landscape Ecology* **29**, 261–274. doi:10.1007/s10980-013-9962-1

Lin L, Band LE, Vose JM, Hwang T, Miniat CF, Bolstad PV (2019) Ecosystem processes at the watershed scale: influence of flowpath patterns of canopy ecophysiology on emergent catchment water and carbon cycling. *Ecohydrology* **12**, e2093. doi:10.1002/eco.2093

Liu W, Wu J, Fan H, Duan H, Li Q, Yuan Y, Zhang H (2017) Estimations of evapotranspiration in an age sequence of eucalyptus plantations in subtropical China. *PLoS One* **12**, e0174208. doi:10.1371/journal.pone.0174208

Liu L, Du J, Xu X, Kardol P, Hu D (2020) Microtopography-induced ecohydrological effects alter plant community structure. *Geoderma* **362**, 114119. doi:10.1016/j.geoderma.2019.114119

Llorens P, Poch R, Latron J, Gallart F (1997) Rainfall interception by a *Pinus sylvestris* forest patch overgrown in a Mediterranean mountainous abandoned area I. Monitoring design and results down to the event scale. *Journal of Hydrology* **199**, 331–345. doi:10.1016/S0022-1694(96)03334-3

McMahon D, Axelson J, Kocher S (2020) 'Mass tree mortality, fuels, and fire: a guide for Sierra Nevada forest landowners.' University of California Agriculture and Natural Resources Publication 8683, Davis.

Meng FR, Arain A, Li Q, Pisaric M (2019) *Forest Hydrology and Watershed.* Multidisciplinary Digital Publishing Institute, Basel.

Merriam-Webster (2020) *Evapotranspiration Definition.* Merriam-Webster Online Dictionary, <https://www.merriam-webster.com/dictionary/topography>.

Mikkelson KM, Maxwell RM, Ferguson I, Stednick JD, McCray JE, Sharp JO (2013) Mountain pine beetle infestation impacts: modeling water and energy budgets at the hill-slope scale. *Ecohydrology* **6**, 64–72. doi:10.1002/eco.278

Moore RD, Wondzell SM (2005) Physical hydrology and the effects of forest harvesting in the Pacific Northwest: a review. *Journal of the American Water Resources Association* **41**, 763–784.

National Climate Assessment (2014) *2014 National Climate Assessment.* U.S. Global Change Research Program, Washington, DC, <https://nca2014.globalchange.gov/>

National Research Council (NRC) (2008) *Hydrologic Effects of a Changing Forest Landscape.* National Academies Press, Washington, DC.

Neary DG, Ice GG, Jackson CR (2009) Linkages between forest soils and water quality and quantity. *Forest Ecology and Management* **258**, 2269–2281. doi:10.1016/j.foreco.2009.05.027

Nowacki GJ, Abrams MD (2008) The demise of fire and 'mesophication' of forests in the eastern United States. *Bioscience* **58**, 123–138. doi:10.1641/B580207

Nowacki GJ, Abrams MD (2015) Is climate an important driver of post-European vegetation change in the eastern United States? *Global Change Biology* **21**, 314–334. doi:10.1111/gcb.12663

Oishi AC, Oren R, Novick KA, Palmroth S, Katul GG (2010) Interannual invariability of forest evapotranspiration and its consequence to water flow downstream. *Ecosystems* **13**, 421–436. doi:10.1007/s10021-010-9328-3

Oishi AC, Miniat CF, Novick KA, Brantley ST, Vose JM, Walker JT (2018) Warmer temperatures reduce net carbon uptake, but do not affect water use, in a mature southern Appalachian forest. *Agricultural and Forest Meteorology* **252**, 269–282. doi:10.1016/j.agrformet.2018.01.011

Oudin L, Andréassian V, Lerat J, Michel C (2008) Has land cover a significant impact on mean annual streamflow? An international assessment using 1508 catchments. *Journal of Hydrology* **357**, 303–316. doi:10.1016/j.jhydrol.2008.05.021

Ouyang Y, Jin W, Grace JM, Obalum SE, Zipperer WC, Huang X (2019) Estimating impact of forest land on groundwater recharge in a humid subtropical watershed of the Lower Mississippi River Alluvial Valley. *Journal of Hydrology. Regional Studies* **26**, 100631. doi:10.1016/j.ejrh.2019.100631

Post WM, Kwon KC (2000) Soil carbon sequestration and land-use change: processes and potential. *Global Change Biology* **6**, 317–327. doi:10.1046/j.1365-2486.2000.00308.x

Pyle LA, Hockaday WC, Boutton T, Zygourakis K, Kinney TJ, Masiello CA (2015) Chemical and isotopic thresholds in charring: implications for the interpretation of charcoal mass and isotopic data. *Environmental Science & Technology* **49**, 14057–14064. doi:10.1021/acs.est.5b03087

Pypker TG, Levia DF, Staelens J, Van Stan JT (2011) Canopy structure in relation to hydrological and biogeochemical fluxes. In *Forest Hydrology and Biogeochemistry.* (Eds MM Caldwell,

G Heldmaier, RB Jackson, OL Lange, HA Mooney, ED Schulze and U Sommer) pp. 371–388. Springer, Dordrecht.

Renninger HJ, Clark KL, Skowronski N, Schäfer KV (2013) Effects of a prescribed fire on water use and photosynthetic capacity of pitch pines. *Trees* **27**, 1115–1127. doi:10.1007/s00468-013-0861-5

Robichaud PR (2000) Fire effects on infiltration rates after prescribed fire in Northern Rocky Mountain forests, USA. *Journal of Hydrology* **231**, 220–229. doi:10.1016/S0022-1694(00)00196-7

Roebroek CT, Melsen LA, Hoek van Dijke AJ, Fan Y, Teuling AJ (2020) Global distribution of hydrologic controls on forest growth. *Hydrology and Earth System Sciences* **24**, 4625–4639. doi:10.5194/hess-24-4625-2020

Romashchenko MI, Bohaienko VO, Matiash TV, Kovalchuk VP, Danylenko II (2020) Influence of evapotranspiration assessment on the accuracy of moisture transport modeling under the conditions of sprinkling irrigation in the south of Ukraine. *Archives of Agronomy and Soil Science* **66**, 1424–1435. doi:10.1080/03650340.2019.1674445

Royer PD, Cobb NS, Clifford MJ, Huang C-Y, Breshears DD, Adams HD, Villegas JC (2011) Extreme climatic event-triggered overstorey vegetation loss increases understorey solar input regionally: primary and secondary ecological implications. *Journal of Ecology* **99**, 714–723. doi:10.1111/j.1365-2745.2011.01804.x

Ryan KC, Knapp EE, Varner JM (2013) Prescribed fire in North American forests and woodlands: history, current practice, and challenges. *Frontiers in Ecology and the Environment* **11**(s1), e15–e24. doi:10.1890/120329

Sala A, Peters GD, McIntyre LR, Harrington MG (2005) Physiological responses of ponderosa pine in western Montana to thinning, prescribed fire and burning season. *Tree Physiology* **25**, 339–348. doi:10.1093/treephys/25.3.339

Sayer EJ (2006) Using experimental manipulation to assess the roles of leaf litter in the functioning of forest ecosystems. *Biological Reviews of the Cambridge Philosophical Society* **81**, 1–31. doi:10.1017/S1464793105006846

Schlesinger WH, Jasechko S (2014) Transpiration in the global water cycle. *Agricultural and Forest Meteorology* **189**, 115–117. doi:10.1016/j.agrformet.2014.01.011

Shakesby RA (2011) Post-wildfire soil erosion in the Mediterranean: review and future research directions. *Earth-Science Reviews* **105**, 71–100. doi:10.1016/j.earscirev.2011.01.001

Shakesby RA, Doerr SH, Walsh RPD (2000) The erosional impact of soil hydrophobicity: current problems and future research directions. *Journal of Hydrology* **231**, 178–191. doi:10.1016/S0022-1694(00)00193-1

Sidle RC, Hirano T, Gomi T, Terajima T (2007) Hortonian overland flow from Japanese forest plantations – an aberration, the real thing, or something in between? *Hydrological Processes* **21**, 3237–3247. doi:10.1002/hyp.6876

Smerdon BD, Redding T, Beckers J (2009) An overview of the effects of forest management on groundwater hydrology. *Journal of Ecosystems and Management* **10**, 22–44.

Soil Survey Staff (2019) *Soil Survey Geographic (SSURGO) Database for Clarke and Oconee Counties, Georgia*. Natural Resources Conservation Service, United States Department of Agriculture, <https://websoilsurvey.nrcs.usda.gov/>.

Staelens J, De Schrijver A, Verheyen K, Verhoest NE (2008) Rainfall partitioning into throughfall, stemflow, and interception within a single beech (*Fagus sylvatica* L.) canopy: influence of foliation, rain event characteristics, and meteorology. *Hydrological Processes: An International Journal* **22**, 33–45. doi:10.1002/hyp.6610

Stednick JD (2008) Long-term streamflow changes following timber harvesting. In *Hydrological and Biological Responses to Forest Practices: The Alsea Watershed Study.* (Ed. J Stednick) pp. 139–155. Springer, New York.

Sun G, Amatya DM, McNulty SG (2016) Forest hydrology. In *Part 7 Systems Hydrology. Handbook of Applied Hydrology*. 2nd edn. (Ed. VP Singh) pp. 85-1–85-8. McGraw Hill Education, New York.

Tromp-van Meerveld HJ, McDonnell JJ (2006a) Threshold relations in subsurface stormflow: 1. A 147-storm analysis of the Panola hillslope. *Water Resources Research* **42**, W02415. doi:10.1029/2004WR003778

Tromp-van Meerveld HJ, McDonnell JJ (2006b) Threshold relations in subsurface stormflow: 2. The fill and spill hypothesis. *Water Resources Research* **42**, W02411. doi:10.1029/2004WR003800

Uchida T, Kosugi KI, Mizuyama T (2002) Effects of pipe flow and bedrock groundwater on runoff generation in a steep headwater catchment in Ashiu, central Japan. *Water Resources Research* **38**, 24-1–24-14.

Uchida T, Tromp-van Meerveld I, McDonnell JJ (2005) The role of lateral pipe flow in hillslope runoff response: an intercomparison of non-linear hillslope response. *Journal of Hydrology* **311**, 117–133.

Van Stan JT, Van Stan JH, Levia DF (2014) Meteorological influences on stemflow generation across diameter size classes of two morphologically distinct deciduous species. *International Journal of Biometeorology* **58**, 2059–2069. doi:10.1007/s00484-014-0807-7

Wang J, Hong Y, Gourley J, Adikhari P, Li L, Su F (2010) Quantitative assessment of climate change and human impacts on long-term hydrologic response: a case study in a sub-basin of the Yellow River, China. *International Journal of Climatology* **30**, 2130–2137. doi:10.1002/joc.2023

Whipkey RZ (1965) Subsurface stormflow from forested slopes. *Hydrological Sciences Journal* **10**, 74–85.

Wullschleger SD, Hanson PJ (2006) Sensitivity of canopy transpiration to altered precipitation in an upland oak forest: evidence from a long-term field manipulation study. *Global Change Biology* **12**, 97–109.

Xiao Q, McPherson EG, Ustin SL, Grismer ME, Simpson JR (2000) Winter rainfall interception by two mature open-grown trees in Davis, California. *Hydrological Processes* **14**, 763–784. doi:10.1002/(SICI)1099-1085(200003)14:4<763::AID-HYP971>3.0.CO;2-7

Yoon JH, Wang SYS, Gillies RR, Kravitz B, Hipps L, Rasch PJ (2015) Increasing water cycle extremes in California and in relation to ENSO cycle under global warming. *Nature Communications* **6**, 8657. doi:10.1038/ncomms9657

Yu D, Li X, Coa Q, Hao R, Qiao J (2020) Impacts of climate variability and landscape pattern change on evapotranspiration in a grassland landscape mosaic. *Hydrological Processes* **34**, 1035–1051. doi:10.1002/hyp.13642

Zhang M, Liu N, Harper R, Li Q, Liu K, Wei X, Ning D, Hou Y, Liu S (2017) A global review on hydrological responses to forest change across multiple spatial scales: importance of scale, climate, forest type and hydrological regime. *Journal of Hydrology* **546**, 44–59. doi:10.1016/j.jhydrol.2016.12.040

9

Urban soils and soil water

Anita M. Thompson, Nicholas J. Balster, Edward P. Boswell

Introduction

The urbanisation of natural landscapes constitutes one of the most profound land use shifts to characterise the Anthropocene. Global data emphasise the prominence of urbanised landscapes that now include mega-cities (>10^6 people), which are projected to number over 40 by 2030 (United Nations 2015). Concurrently, urban populations will likely increase by ~1.6% annually (https://www.who.int/data/gho) and, despite considerable variability in the definition of urban land and its spatial distribution (Small *et al.* 2005; Liu *et al.* 2014), most projections indicate sustained growth of urban landscapes worldwide (United Nations 2015). Adding to this dynamic, many newly urbanised areas are less densely populated or fragmented relative to the cities from which they grew, effectively pushing their margins (and dependencies) into formerly natural and rural landscapes (York *et al.* 2011; Xu *et al.* 2018). Ameliorating the ecological impact of urban areas, the sprawling pattern of their growth and the certainty of emergent properties from these complex systems require the continued development of adaptive strategies that leverage ecosystem services, whether that be from within or outside the urban envelope (Folke *et al.* 1997; Shochat *et al.* 2006). Understanding the interdependencies between soil and water is central to mitigating the socioecological impact of urban areas.

A universal definition of urban soil continues to evolve (Morel *et al.* 2005), but its characteristics reflect a synergism between metapedogenesis (Yaalon and Yaron 1966; Craul 1992), classification and mapping (Blume and Runge 1978), edaphic impact and management of human-derived materials dating back to the late 19th century (Lehmann and Stahr 2007). Early definitions of urban soil were either too restrictive, by not considering the many characteristics transformed by human activity, or too broad, such that urban soil became lost among all anthropogenic alterations (e.g. soils of urban, industrial, traffic, mining and military areas (SUITMAs); Burghardt 1994). Subsequent modifications of the definition have better characterised the range of management challenges within complex urban systems, as well as innovations to address these challenges (Hollis 1991; Lehmann and Stahr 2007). Mirroring this evolution in terminology, various classification systems have arisen to identify and catalogue urban soil characteristics, from a hierarchical system used in the UK of four categorical levels of human-influenced soil properties (Howard 2017), to the *Soil Taxonomy: A Basic System of Soil Classification for Making and Interpreting Soil Surveys* (Soil Survey Staff 1999) of the US (also hierarchical) from which Kosse (1980) has proposed a new 'urban soil' order, namely the Anthropogenic soils, to the

Anthrosols described in the Food and Agriculture Organization of the United Nations (FAO)/UNESCO system (FAO 1988) and the Technosols described in the 2014 World Reference Base for Soil Resources (IUSS Working Group WRB 2015). However, all definitions and classification systems recognise that anthropic material either directly or indirectly affects the pedogenesis and, in turn, the unique function of urban soil.

The ecological consequences of modifying soil within urban environments among a mosaic of impervious and pervious surfaces cannot be fully understood or realised without considering the interaction of the soil with the hydrosphere. The hydrological behaviour of impervious surfaces is relatively straightforward compared with that of pervious soils, which are spatially and temporally variable due to surface sealing, unpredictable horizonation and poor structure, factors that ultimately control the timing and magnitude of run-off generation and available soil water for urban vegetation (Fletcher *et al.* 2013; Levin *et al.* 2017). In addition, impervious surfaces can reduce the amount of water available to recharge groundwater while increasing run-off volumes that overwhelm the storage capacity of available pervious soils, effectively increasing the overall 'flashiness' of run-off events (Jacobson 2011). Urban surface run-off often carries sediment, nutrients, heavy metals, microbial contaminants (e.g. faecal pathogens) and organic chemicals (e.g. pesticides and hydrocarbons; McGrane 2016). Engineered storm sewer networks, whether combined with or separate from centralised wastewater systems, also facilitate the transport of contaminants and can lead to point source impairment of receiving waters (Roy *et al.* 2014). Climatic differences across the US (as in other regions) complicate our understanding of the impacts of these modified hydrological regimes (Li *et al.* 2020a). Therefore, mitigating these effects will involve an integrated approach where soil and water are managed as concomitant systems to maintain the ecosystem services that they provide in a changing climate.

The environmental impact of urbanisation has been recognised for some time (Wilson 1967), with a few lone voices calling for intensive biophysical research of these systems; however, its unprecedented expansion outpaces quantification of ecological impacts and the identification of management practices to enhance the sustainability and resiliency of these heterogeneous systems (Pickett *et al.* 2011; Staddon *et al.* 2018). Compared with natural systems, environmental research in urban systems has occurred at a modest rate, with its contextualisation as a unique ecological system complicated by human-constructed feedbacks (built and social) being <30 years old (Bolund and Hunhammer 1999). Urban biophysical research generally includes specific technologies (e.g. engineered soil; Bassuk *et al.* 2005), evaluation of urban development strategies (Wang *et al.* 2020a), hypothesised relationships describing large-scale patterns unique to urban systems (Kühn and Klotz 2006; Groffman *et al.* 2017; Fang *et al.* 2020) and many case studies (Shao *et al.* 2020) that, although inductively valuable, often suffer from weak inference, spatial anomalies and temporal specificity such that 'on-the-ground' application is difficult to implement and often impossible to replicate.

Global climate change has hastened the need to identify innovative management practices in urban systems such as nature-based solutions (NBS; Pataki *et al.* 2011; Cohen-Shacham *et al.* 2016; Nocco *et al.* 2016; Zölch *et al.* 2017; Snep *et al.* 2020; Bouzouidja *et al.* 2021). Although definitions vary, NBS promote working with nature to meet societal challenges and address sustainability goals, many of which apply directly within the urban context (Hanson *et al.* 2020; Fig. 9.1). These practices offer an alternative to traditional engineering techniques and are often more affordable, sustainable and resource-efficient, empowering citizens to adapt, innovate and evaluate green alternatives within a more

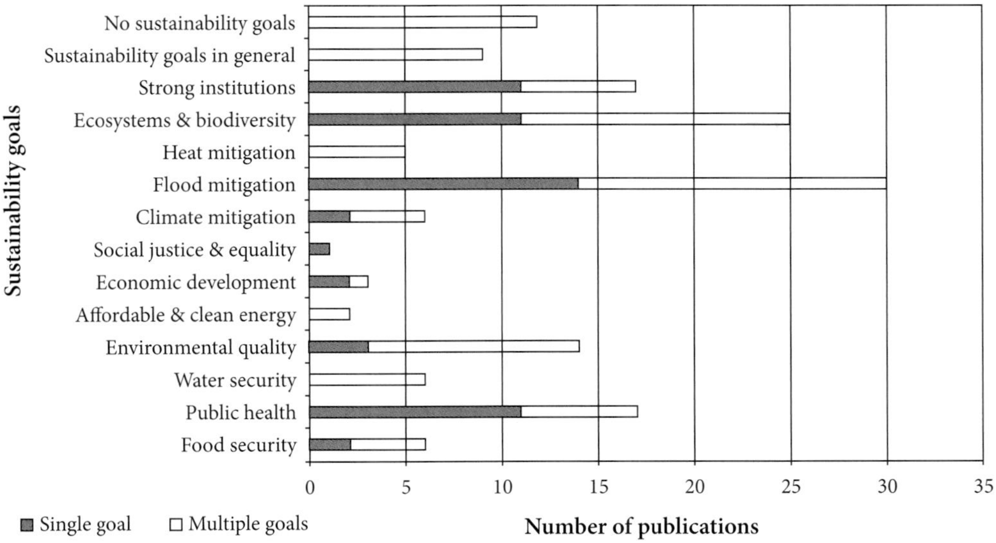

Fig. 9.1. Number of publications addressing the different sustainability goals (n = 112 publications in total). Note that a publication can consider more than one sustainability goal and the sum is therefore higher than the number of publications included in the review. Redrawn and reprinted with the permission of Elsevier from Hanson *et al.* (2020).

localised context (Maes and Jacobs 2017). NBS align with Goal 11 of the United Nations (UN) sustainable development goals, namely to 'make cities and human settlement inclusive, safe, resilient and sustainable' (UN General Assembly 2015), underscoring the need for empirical research and subsequent application of NBS (Konijnendijk *et al.* 2016).

Soil–water interactions will take centre stage as climate change over the coming decades exacerbates the environmental impact of urban systems. For example, changes in the intensity and timing of precipitation events will reveal the soil's ability to buffer the hydrological response within urban systems (Rawlins *et al.* 2015). However, whether our focus is on climate change or food security, pollution control, flood mitigation, carbon sequestration, waste management or the myriad environmental issues surrounding urbanisation, understanding the interactions between soil and water will be integral to the alignment of NBS to improve the quality of life and well-being of urban residents (Chiesura 2004; Shaw 2015; Staddon *et al.* 2018; Bouzouidja *et al.* 2021). Rawlins *et al.* (2015) provide an inventory of urban soil functions based on the ecosystem services-based categories of the Millennium Ecosystem Assessment along with qualitative assessments for: (1) the provisioning of urban soil compared to undisturbed systems; (2) the potential for urban soil enhancement; and (3) the potential impact of climate drivers on urban soil functions. In an earlier comprehensive study, Gómez-Baggethun and Barton (2013) classified key ecosystem services in urban areas to ultimately assess valuation criteria for urban planning. Among the many ecosystem services identified in these studies, soil and water hold a prominent role in the structure and function of each service (Table 9.1).

Urban soils

Humans have profoundly manipulated soil properties within urban environments, often resulting in heterogeneous mixtures and discontinuous layers of fill material and debris

Table 9.1. Classification of important ecosystem services in urban areas and underlying ecosystem functions and components

Reproduced with the permission of Elsevier from Gómez-Baggethun and Barton (2013). As noted in Gómez-Baggethun and Barton (2013), the suitability of indicators for biophysical measurement is scale dependent, and most indicators and proxies provided correspond to assessment at the plot level. $PM_{10\ \mu m}$, inhalable particles with diameters that are generally ≤10 μm

Functions and components	Ecosystem service	Examples	Examples of indicators/proxies	References
Energy conversion into edible plants through photosynthesis	Food supply	Vegetables produced by urban allotments and peri-urban areas	Production of food (tons $year^{-1}$)	Altieri *et al.* (1999)
Percolation and regulation of run-off and river discharge	Water flow regulation and run-off mitigation	Soil and vegetation percolate water during heavy and/or prolonged precipitation events	Soil infiltration capacity; percentage sealed relative to permeable surface (ha)	Villarreal and Bengtsson (2005)
Photosynthesis, shading and evapotranspiration	Urban temperature regulation	Trees and other urban vegetation provide shade, create humidity and block wind	Leaf area index; temperature decrease (°C) by tree cover × area (m^2) of plot trees cover	Bolund and Hunhammar (1999)
Absorption of sound waves by vegetation and water	Noise reduction	Absorption of sound waves by vegetation barriers, especially thick vegetation	Leaf area (m^2) and distance to roads (m); noise reduction dBA/vegetation unit (m)	Aylor (1972), Ishii (1994)
Filtering and fixation of gases and particulate matter	Air purification	Removal and fixation of pollutants by urban vegetation in leaves, stems and roots	O_3, SO_2, NO_2, CO and $PM_{10\ \mu m}$ removal (tons $year^{-1}$) multiplied by tree cover (m^2)	Chaparro and Terradas (2009)
Physical barrier and absorption on kinetic energy	Moderation of environmental extremes	Storm, floods and wave buffering by vegetation barriers; heat absorption during severe heat waves	Cover density of vegetation barriers separating built areas from the sea	Danielsen *et al.* (2005), Costanza *et al.* (2006)
Removal or breakdown of xenic nutrients	Waste treatment	Effluent filtering and nutrient fixation by urban wetlands	P, K, Mg and Ca (in mg kg^{-1}) compared to given soil/water quality standards	Vauramo and Setälä (2011)
Carbon sequestration and fixation in photosynthesis	Climate regulation	Carbon sequestration and storage by the biomass of urban shrubs and trees	CO_2 sequestration by trees (carbon multiplied by 3.67 to convert to CO_2)	Nowak (1994), McPherson (1998)
Movement of floral gametes by biota	Pollination and seed dispersal	Urban ecosystems provide habitat for birds, insects and pollinators	Species diversity and abundance of birds and bumble bees	Andersson *et al.* (2007)
Ecosystems with recreational and educational values	Recreation and cognitive development	Urban parks provide multiple opportunities for recreation, meditation and pedagogy	Surface of green public spaces (ha)/inhabitant (or every 1000 inhabitants)	Chiesura (2004)
Habitat provision for animal species	Animal sighting	Urban green space provides habitat for birds and other animals people like watching	Abundance of birds, butterflies and other animals valued for their aesthetic attributes	Blair (1996), Blair and Launer (1997)

(i.e. artefacts) within pre-existing characteristics (e.g. percentage sand) unaltered during development. These changes can significantly alter the biophysicochemical properties and functions of urban soil compared with the native soils from which they originated or the subsoils below the depth of disturbance (Pouyat *et al.* 2010). Despite the importance of these functions to urban landscapes, the number of studies focused specifically on urban soils constitutes a minute fraction of all soil publications since 1960, only seeing growth by single digits since the 1990s (Guilland *et al.* 2018; Plate 7). In general, these studies report urbanised soils as possessing high bulk densities, reduced soil organic matter contents and aggregate stability, high pH, low water-holding capacity, decreased microbial biomass and activity and often wide spatial variation in nutrient contents to support plant growth (Craul 1999; Morel *et al.* 2005; Kaye *et al.* 2006; Pouyat *et al.* 2010). Despite these limitations, urban and suburban environments still rely on soil as the primary pervious media to intercept, absorb, store and filter stormwater and snowmelt (Bullock and Gregory 1991; De Kimpe and Morel 2000; Lehmann and Stahr 2007; Pouyat *et al.* 2007). Therefore, how we interact with and manipulate this medium will influence its capacity to provide the ecosystem services (e.g. flood mitigation, contaminant remediation; Pataki *et al.* 2011; Morel *et al.* 2015), to dampen the environmental impact of urbanisation (Elmqvist *et al.* 2015) and provide a buffer against a changing climate.

Among the many soil characteristics altered by construction during urbanisation (Fig. 9.2), soil structure plays a proximate, yet paradoxical, role in the functionality of urban environments. Although soil structure is often intentionally destroyed (e.g. compaction) to provide a stable foundation for urban infrastructure, its high susceptibility to unintentional and often irreversible alteration by humans (e.g. pedestrian traffic on wet soil; Yang and Zhang 2015) can also dramatically influence its ability to support hydrological processes (e.g. infiltration). The hierarchical arrangement and subsequent stability of soil particles from physicochemical primary particle flocculation to the biological stabilisation of micro- and macroaggregates (Tisdall and Oades 1982; Six *et al.* 2004) define the three-dimensional quality of soil structure and porosity as complementary aspects critical to soil–water interactions. Soil scraping, stockpiling and replacement have been shown to reduce the proportion of macroaggregates and soil organic carbon in urban surface soils (Chen *et al.* 2014). Although there is no universally accepted way to characterise this physical organisation of particles and void space (Díaz-Zorita *et al.* 2002; Rabot *et al.* 2018), bulk density can be used as a proxy for soil structure and porosity.

Fig. 9.2. Soil disturbance during the residential construction process. (a) Destruction of soil structure by the use of heavy machinery on wet soils. (b) Rill erosion and soil loss from exposed soil surfaces during construction activities. Photographs taken by N. J. Balster.

In general, urbanisation is assumed to increase soil bulk density and often correlates with changes in hydrology, such as reductions in the rate of infiltration at the landscape scale (Woltemade 2010; Yang and Zhang 2011; Johnston *et al.* 2016). In addition, bulk densities above those detrimental to root growth (exceeding 1.75 g cm^{-3} for sands and silts and 1.4 g cm^{-3} for clays; Daddow and Warrington 1983) are commonly reported in urban landscapes (Smetak *et al.* 2007; Yang and Zhang 2011; Devigne *et al.* 2016). Although bulk density yields a first approximation of altered soil structure, it does not capture the spatial arrangement of soil pores and the degree to which these pores are connected. The seminal importance of soil structure to soil–water relationships warrants focused attention beyond quantifying changes in bulk density, because few studies investigate the direct response of soil aggregation to urbanisation, its stability within these landscapes and the degree at which processes in the built environment affect hydraulic properties.

In many cases, soil bulk density decreases from its compacted state as urban areas mature, resulting in improved physical, biological and chemical soil properties (Sharenbroch *et al.* 2005; Smetak *et al.* 2007; Johnston *et al.* 2016). These reduced bulk densities were attributed to increased soil organic matter, as well as microbial and macrofauna (e.g. earthworm) activity. Post-development rehabilitation efforts to create and maintain soil structure, such as the incorporation of compost deep into the soil, may also enhance soil hydraulic conductivity and may hold promise in creating climate-resilient urban environments (Chen *et al.* 2014; Sax *et al.* 2017). These practices suggest that NBS to enhance soil biological activity and vegetation establishment should be encouraged in urban areas in need of remediation.

A small proportion of studies deal with ecological and functional aspects of soil based on their biological communities (Guiland *et al.* 2018), with most investigating issues of the ecotoxicology, epidemiology or bioaccumulation of contaminants. However, understanding the effects of urbanisation on soil organisms as the 'soil engineers' of aggregation and structure, which, in turn, regulate soil–water interactions, is a critical component of NBS soil quality measures (Bouzouidja *et al.* 2021). Soil microbial community and functional diversity are often used to monitor variations in soil properties caused by human disturbance (Bending *et al.* 2002). Microbial biomass is generally lower in urban than rural soils, attributed to changes in soil texture, compaction and increased heavy metal content (Zhao *et al.* 2013). However, urban soils may exhibit a high degree of biological activity and have high species richness due to the active role of microorganisms (Pouyat *et al.* 2010). In contrast, in a New York City study, although microbial diversity varied across investigated soils, general bacterial communities present in soils affected by human activities did not differ from communities in other soils (Huot *et al.* 2017).

Soil pH is highlighted as a principal factor determining the diversity of microbial communities in urban (and non-urban) environments (Huot *et al.* 2017; Wang *et al.* 2017). Although earthworms are generally considered beneficial to improving soil structure (Marashi and Scullion 2003; Blouin *et al.* 2013), many species are considered invasive and may negatively affect nutrient cycling and native vegetation (Hendrix *et al.* 2008; Groffman *et al.* 2015; Craven *et al.* 2017). Earthworm populations are dynamic in urban systems as a result of abiotic and biotic factors such as soil compaction, changes in vegetation and soil moisture. For example, earthworm populations were significantly lower in younger residential soils (<10 years old) with high bulk densities (1.59 g cm^{-3}) than in older residential sites and urban parks that were >75 years old and had reduced bulk densities (1.30 g cm^{-3}; Smetak *et al.* 2007). Urban forests supported greater earthworm populations than rural forests in New York city and Baltimore, Maryland (Steinberg *et al.* 1997; Szlavecz *et al.* 2006) and native prairie vegetation supported lower earthworm densities than lawns in Madison, Wisconsin (Bajcz *et al.* 2018). Despite site-specific differences, evidence from five US, European and South African cities suggests that earthworm communities may be similar across habitat type, with this similarity driven primarily by human-mediated dispersal (Tóth *et al.* 2020).

Urban hydrology and water quality

Water quantity

The physical changes (e.g. the addition of impervious surfaces and storm sewers) and anthropogenic influences (e.g. water supply transfers) that accompany urbanisation alter surface and subsurface pathways, ultimately changing the way rainfall travels to surface waters and groundwater reservoirs. Water balance frameworks describe rainfall partitioning in urban landscapes and identify the impacts of urbanisation and the effectiveness of urban design at site and watershed scales. Replacing vegetated pervious areas with impervious roads and buildings generally reduces surface storage, infiltration and evapotranspiration, leaving surface run-off as the dominant process in the water budget (Fig. 9.3). Assessment of hydrological impacts requires differentiating between effective or directly connected impervious areas (hydraulically connected to storm sewer drainage systems) and disconnected impervious surfaces that route run-off to pervious surfaces where partial or complete infiltration can occur (Mueller and Thompson 2009). In addition, the 'urban karst', a network of constructed sewer and stormwater pipes and high-permeability trenches surrounding these and other utilities (e.g. telecommunication, drinking water, natural gas) that influence groundwater flow through leaks, cracks and fissures, has been defined as an important driver of hydrological processes in urban catchments (Kaushal and Belt 2012).

Urbanisation reduces infiltration per unit area by replacing pervious areas with impervious surfaces and limiting the infiltration capacity of the remaining soils. Soil compaction is a leading mechanism of decreased infiltration rates and hydraulic conductivities commonly found in urban soils (Gregory *et al.* 2006; Yang and Zhang 2011; Chen *et al.* 2014). However, infiltration rates vary across urban functional zones, vegetation types and time since disturbance (Yang and Zhang 2011; Fig. 9.4). For example, infiltration rates for residential lawns, estimated to cover ~23% of cities globally (Ignatieva and Hedblom 2017), built before 2000 were approximately threefold greater than for those built after 2000 (Woltemade 2010). Neglecting the impact of soil compaction would result in overestimating infiltration rates and underestimating run-off volumes and potential flood events (Gregory *et al.* 2006; Woltemade 2010).

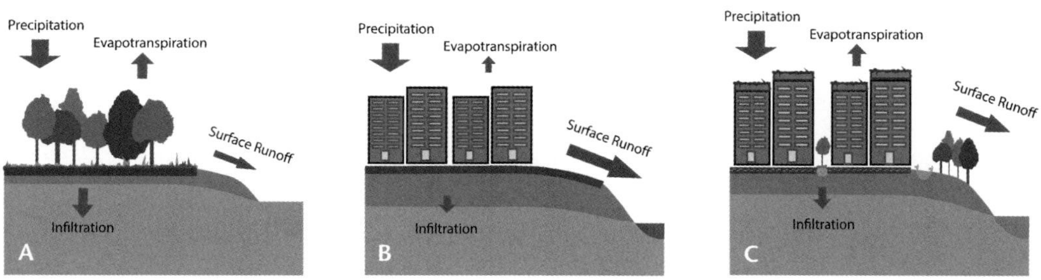

Fig. 9.3. Conceptual changes in the hydrological cycle in (a) a natural landscape, (b) an urban landscape and (c) an urban landscape with nature-based solutions. (Drawing: Yi Wang, University of Wisconsin – Madison.)

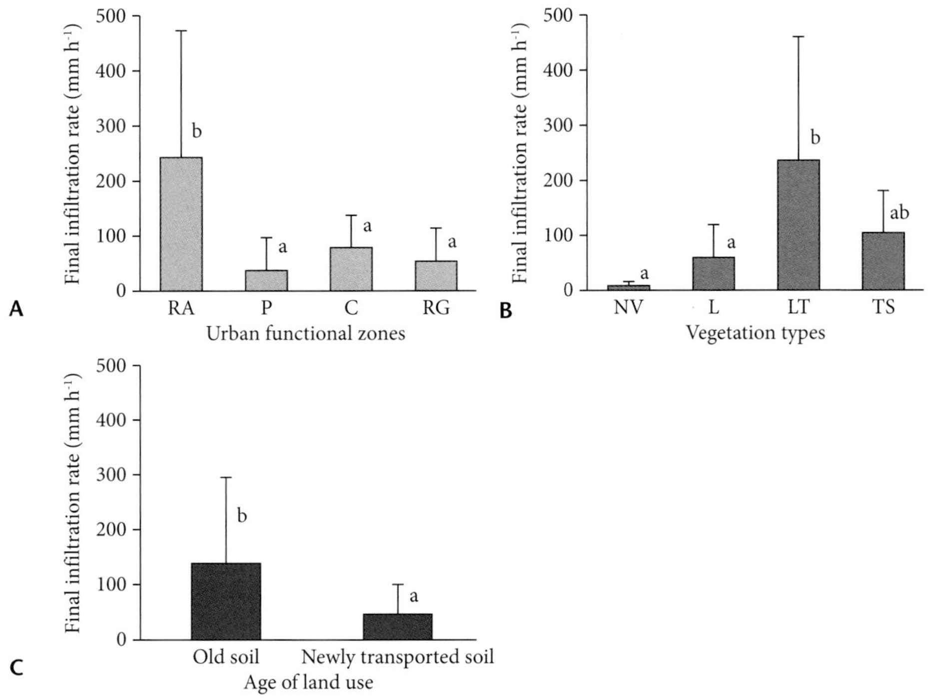

Fig. 9.4. Final infiltration rates for different (a) urban functional zones, (b) vegetation types and (c) age of land use. Data are the mean ± s.d. Different letters indicate significant differences ($P < 0.05$). RA, residential areas; P, park areas; C, campus; RG, road greenbelt; NV, no vegetation; L, lawn; LT, lawn with tree; TS, tree with shrub. Redrawn and reprinted with the permission of Springer Nature from Yang and Zhang (2011).

Soil hydraulic conductivity, as a fundamental controlling property of infiltration, is highly dynamic and influenced by the presence of soil artefacts, land use and management, natural disturbance, compaction, tillage, wetting/drying, soil organism activity and root dynamics (Zhou *et al.* 2008). Soils in parks with low-complexity habitat (components included soil, litter and vegetation) had field-saturated hydraulic conductivity values one-tenth of those in more complex habitat types, with the difference attributed to soil macropores (Ossola *et al.* 2015). Measured hydraulic conductivity in areas surrounding underground pipes was two or three orders of magnitude greater than a surrounding native alluvial soil (Sharp *et al.* 2003). Practices, especially NBS, that enhance the hydraulic conductivity of urban soils are important for mitigating stormwater run-off. For example, artificially constructed Technosols (i.e. soils created using technological materials and various artefact waste materials) offer an alternative to the consumption of natural resources and reduce construction waste streams. Technosols exhibited high porosities and high saturated hydraulic conductivities (up to 34.74 cm h^{-1}), and the physical properties (macro- and microporosity) of these artificial soils revealed high water supply for plants with available water ranging from 0.5 to 2.9 mm cm^{-1} (Yilmaz *et al.* 2018).

Land development changes surface run-off characteristics, typically increasing the volume, peak flow rate and frequency of run-off events, and can also profoundly alter watershed storage, with excavation and sealing of urban soils leading to loss of soil water storage (Rawlins *et al.* 2015; Fanelli *et al.* 2017). However, the magnitude of change varies with rainfall amount. For example, model simulations indicate that run-off volume was on average 5.5-fold greater from an urban watershed compared with a forested watershed that was 2.5-fold larger in area and that the maximum difference in run-off production occurred at more frequent low-rainfall depths (Corbett *et al.* 1997). Both the surface type and nature of connection affect the hydrological behaviour of urban landscapes, and the variable nature of pervious area contributions to run-off are difficult to predict, requiring statistical or distributed hydrology–soil–vegetation models (Fletcher *et al.* 2013). Under some circumstances, pervious areas that have been profoundly changed by urbanisation (increased bulk density and reduced vegetative cover) may be a source of run-off (Jefferson *et al.* 2017).

Groundwater and surface water recharge under many subsurface conditions relies on infiltration, and the complex nature of urban landscapes, including urban karst, may require holistic (e.g. overall water balance) approaches to estimating and managing groundwater recharge in urban environments (Lerner 2002). Generally, decreased infiltration through surface sealing and compaction increases run-off and reduces groundwater recharge and the subsequent base flow of groundwater-fed streams. However, large amounts of recharge can be generated through leaky water supply networks and storm drainage systems in many urban environments (Lerner 2002). Although engineered stormwater infiltration practices can enhance local recharge, little is known about the fate of infiltrated stormwater beyond the small site scale (Bonneau *et al.* 2017) and understanding how the below ground matrix facilitates water flow to aquifers and nearby streams and rivers is important (Archer *et al.* 2020). Low-impact development (LID) practices focus on increased infiltration by attempting to replicate natural hydrological functioning. In one Maryland LID study, stream flow and base flow increased as a small agricultural and forested watershed became urbanised, but this increase was perhaps a result of reduced evapotranspiration (ET) from the loss of 'native' vegetation, which characterised an unintended consequence of the LID (Bhaskar *et al.* 2016). Incorporation of NBS specifically targeted on the water cycle may help restore critical natural processes that control hydrology in urban environments (Nika *et al.* 2020).

Water quality

In addition to managing stormwater quantity, the quality of urban run-off must also be considered because it carries sediment, nutrients and inorganic and organic contaminants that can pollute receiving waters. During active construction, vegetation and topsoil are removed and the subsoil is exposed to accelerated rates of erosion and transport of sediment, degrading soil quality on-site (and associated ecosystem services; Table 9.1) and causing sediment-related problems off-site (Fig. 9.5).

Soil loss rates from construction sites can be 10- to 20-fold greater than those of agricultural lands and 1000- to 2000-fold greater than those of forest lands (U.S. Environmental Protection Agency 2018). Common best management practices that provide surface cover, including erosion control mats, mulches, composts, hydromulching, bonded fibre matrices and polyacrylamides, have been reported to reduce soil erosion by 62–79% compared with unprotected soil (Tyner *et al.* 2011). Russell *et al.* (2017) reviewed almost 50 studies measuring sediment yields and found yields from urbanised watersheds ranging from 120 to 27 000 t km^{-2} year^{-1} (median of 740 t km^{-2} year^{-1}), 3- to 420-fold higher than background rates from forested (median increase of ~60-fold) and agricultural (median increase of ~11-fold) watersheds. The same study found that post-development suspended sediment yields decreased but remained high compared with background rates, with median increases of approximately 6- and 3-fold compared with forest and agricultural yields respectively (Russell *et al.* 2017). These high rates after development can be attributed to channel erosion in response to altered urban flow conditions, sediment from sealed and unsealed roads (e.g. gravel roads) and imported anthropogenic particles (e.g. riprap, asphalt, brick). Russell *et al.* (2017) provide a conceptual sediment yield curve based on summarised data (Fig. 9.6) highlighting the need for implementation of erosion control practices during construction. Factors contributing to the high degree of variability in sediment yields (two to three orders of magnitude for all land use categories) include climate, watershed area and relief and the scale of urbanisation and its spatial arrangement within a watershed (Russell *et al.* 2017).

Fig. 9.5. (a) Exposed soil during the construction process and (b) off-site deposition of eroded sediment Photographs taken by N. J. Balster (a) and A. M. Thompson (b).

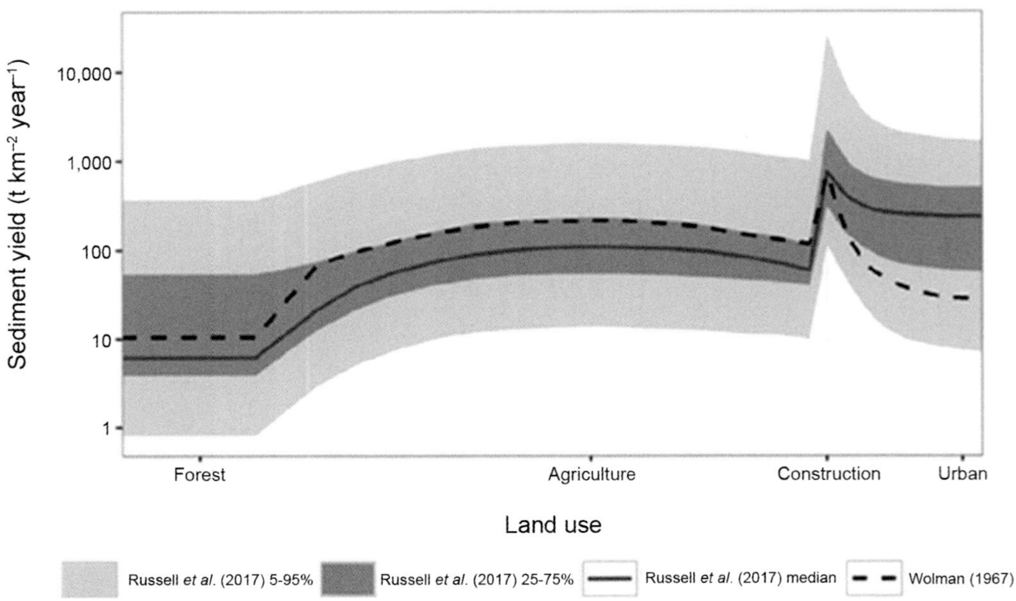

Fig. 9.6. Revised conceptual sediment yield curve based on summarised data in Russell *et al.* (2017) compared with the conceptual model of Wolman (1967). Reprinted with the permission of Elsevier from Russell *et al.* (2017).

Many natural, anthropogenic and biogenic materials contribute excess nutrients (e.g. nitrogen and phosphorus) to urban run-off, leading to eutrophication, harmful algal blooms and fish kills in receiving waters (Conley *et al.* 2009). Yang and Lusk (2018) provide a summary of concentrations, loads and potential sources of nutrients in urban waters across multiple regions in the US, and these include fertiliser applied to turfgrass and landscape plants, atmospheric deposition, pet waste, landfills, construction activities, septic systems, household products (e.g. laundry and dishwasher detergents), wastewater treatment plants, leaf litter and grass clippings, as well as landscape irrigation using reclaimed water (Carey *et al.* 2013; Yang and Lusk 2018; Wang *et al.* 2020c). Although impervious surfaces generate greater run-off volumes and can enhance nutrient export compared with previous areas, the connectivity and spatial distribution of land use types, as well as vegetation cover (e.g. the size and shape of buildings and lawns), influence water quality more so than the extent of impervious areas alone (Beck *et al.* 2016; Li *et al.* 2020b). The soil–plant–water system plays an important role in the retention and export of nutrients in urban watersheds. For example, turfgrass with dense ground cover can reduce run-off and associated nutrient losses (Linde *et al.* 1995), green roofs can function as sinks and sources of nutrient transport (National Research Council 2008) and soil mineralogy controls P removal efficiency (Carey *et al.* 2013).

Stream temperature influences the health of aquatic ecosystems, heightening the thermal impacts of urban systems, especially in cold-water fisheries. During a run-off event, thermal energy is transferred to stormwater as it flows over parking lots, streets and other impervious surfaces, leading to impairment of receiving waters (Thompson *et al.* 2008c; Zeiger and Hubbart 2015). Impervious surfaces store more heat energy, especially during summer, than vegetated areas, thus the ratio of impervious to natural surfaces serves as an important proxy to assess this impact. For example, asphalt surface temperatures in Wisconsin immediately before rainfall simulations averaged 43.6°C, whereas adjacent sod surface temperature averaged only 23.3°C, and the mean initial asphalt run-off temperature was 25°C, which was 9.5°C greater than the corresponding mean initial sod run-off temperature (Thompson *et al.* 2008a). LID strategies including porous pavements, bioretention and other NBS, such as grass swales, can mitigate the thermal impacts of urban stormwater (Li *et al.* 2019). In addition, pervious sod surfaces (those typical of residential lawns) can provide some mitigation of stormwater run-off temperature in urban watersheds (Thompson *et al.* 2008a).

Stormwater management

The scale, complexity and distribution of the built environment, and its connections to native and disturbed soils, challenge our understanding and prediction of hydrological changes. Over the past 40 years, stormwater management has transitioned from distributed storage for peak flow reduction (e.g. retention ponds) to nature-based practices that promote infiltration and ET (e.g. infiltration ponds, green roofs) to reduce run-off volumes (Petrucci *et al.* 2017). The latter requires understanding and managing the properties of urban soils and their connections to vegetation and water fluxes.

Based on available watershed-scale data from North America, East Asia, Europe and Australia and an understanding of processes driving hydrological response in watersheds, Jefferson *et al.* (2017) proposed hysteresis curves to describe how watershed function responds to increasing imperviousness and cumulative stormwater management (Fig. 9.7). These curves suggest that networks of stormwater management practices do not return watershed hydrology to predevelopment conditions and can serve as testable hypotheses

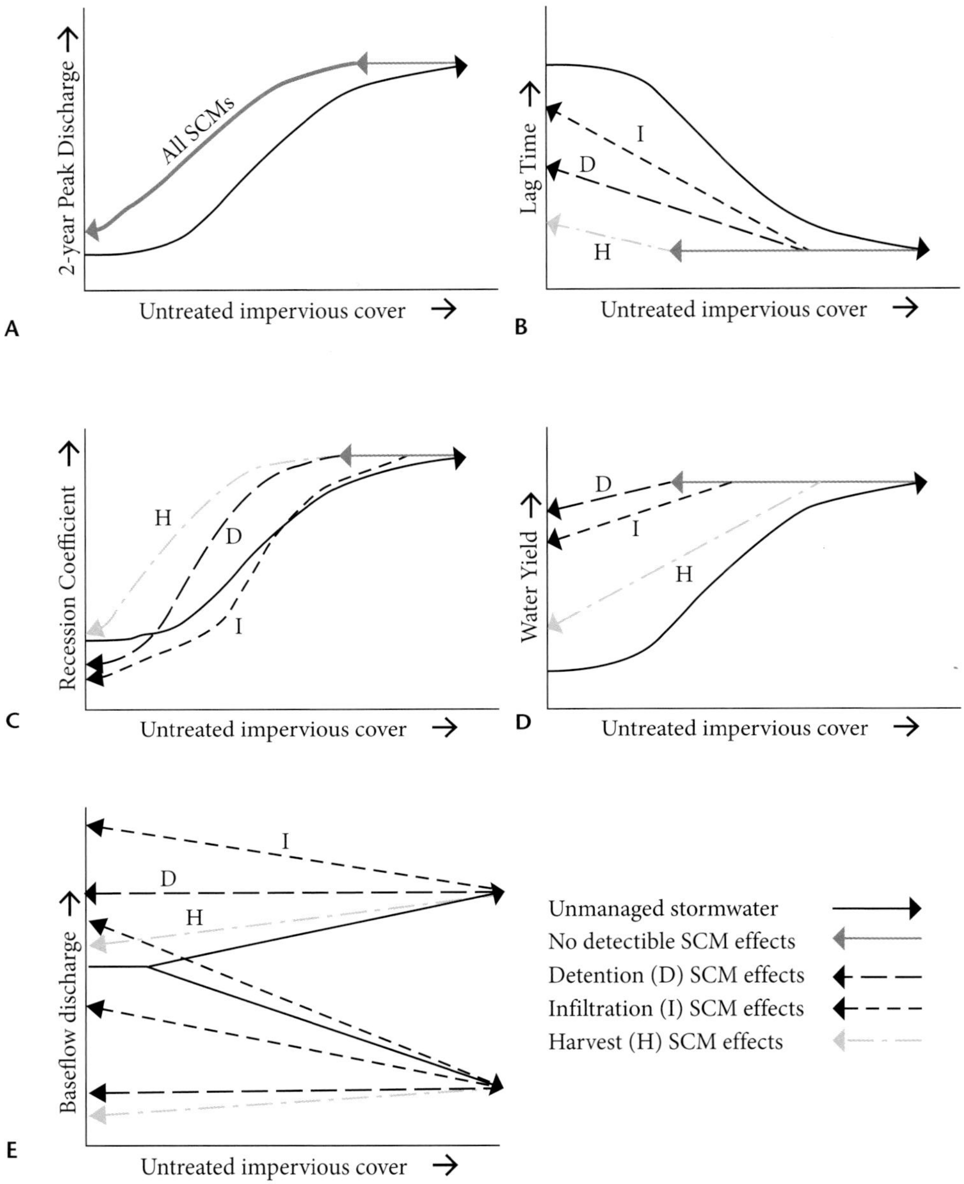

Fig. 9.7. Conceptual model of the effects of urbanisation and cumulative stormwater management on hydrological response. The solid right-pointing arrow represents the effects of urbanisation processes where stormwater is uncontrolled. The solid left-pointing represents stormwater control measure (SCM) networks that treat too little impervious area for their effects to be detected. The remaining lines represent detectible effects of SCM implementation. Long dashed lines indicate the effects of detention SCMs, short dashed lines indicate the effects of infiltration SCMs and dashed–dotted lines indicate the effects of harvest SCMs. The designed focus on peak flow mitigation across all SCM types means that their effect on peak flows is broadly similar, so is not broken out in (a). (a) Two-year peak discharge versus untreated impervious cover. (b) Lag time versus untreated impervious cover. (c) Water yield (total annual discharge) versus untreated impervious cover. (d) Discharge recession coefficient (where higher coefficients indicate faster hydrograph recessions) versus untreated impervious cover. (e) Base flow discharge versus untreated impervious cover. Redrawn and reproduced with the permission of John Wiley and Sons from Jefferson *et al.* (2017).

for future NBS research. Quantifying the cumulative effects of site-scale practices at the watershed scale is complicated by the need to separate the signal of practices from the signal of urbanisation itself, which is shaped by watershed capacitance, local climate and non-stormwater processes (e.g. vegetation changes, water import), and the varying functions of multiple stormwater management practices (Jefferson *et al.* 2017).

Urban vegetation in soil–water systems

The demographics, spatial distribution and taxonomy of urban vegetation in US cities have been quantified for some time (Schmid 1975) but only recently acknowledged as a key component in the growing trend of applying NBS to issues surrounding the urban environment (Vaverková *et al.* 2018; Snep *et al.* 2020; Martínez-Hernández *et al.* 2020). Generally, species richness increases in urban areas compared with adjacent non-urban areas largely due to a concomitant increase in exotics (Zipperer *et al.* 1997), and thus native species subsequently decrease with proximity to city centres (Kowarik 1995). More mature urban areas with larger populations often correlate with an increased number of plant species by as much as three orders of magnitude (Sukopp 1998). Although these descriptive studies are valuable to understanding the spatial heterogeneity and diversity of urban vegetation, a growing area of research has started to focus on the functional role of vegetation in urban environments (e.g. green infrastructure) and its potential use as an NBS within the soil–water matrix (Lovell and Taylor 2013; Elmqvist *et al.* 2015; Berland *et al.* 2017). Most soil–water impacts due to urbanisation are either directly or indirectly linked to stormwater issues, which can be divided into two general (not mutually exclusive) areas: quantity (e.g. pluvial flooding and run-off) and quality (e.g. point and non-point source pollution of surface and ground waters; McGrane 2016; Moore *et al.* 2017). These environmental impacts can be mitigated using a variety of NBS that fall along a continuum of plant–soil manipulations between 'green' (vegetated), 'grey' (engineered) and 'blue' (open water) structures, all suggested to improve urban resilience to climate change (Archer *et al.* 2020; Nika *et al.* 2020). In general, these NBS seek to incorporate more diverse and natural processes and to mimic the natural hydrology of the predeveloped landscape by slowing, retaining and infiltrating stormwater run-off on site.

Urban vegetation provides many ecosystem services, ranging from microclimate regulation, pollutant mitigation and hydrological buffer modification (Willis and Petrokofsky 2017) to increasing the overall liveability of built environments through public health and socioeconomic value (Kuo and Sullivan 2001; Frumkin *et al.* 2004). However, even though many plants are well adapted to urban systems (Gaston 2010), the spatial extent of vegetative cover in most urban areas worldwide has increased only in tandem with rural land acquisition, further evidence of sprawling, low-density urban expansion (Richards and Belcher 2019; Plate 6).

ET is complex in mixed-use watersheds because of the heterogeneous landscape made up of vegetation, buildings and roads (Zheng *et al.* 2020) and, in urban areas, is influenced by changes in infiltration and vegetative cover. ET losses from urban systems are beneficial for managing excess water and an important consideration in managing (i.e. irrigating) green infrastructure, especially in light of climate change. Zheng *et al.* (2020) suggest that the role of vegetation in regulating ET and water yield in urban watersheds may have been underestimated previously and that maintaining 'biological drainage' (i.e. ET) is important in controlling urban storm flow.

NBS and the concepts/approaches under its umbrella (e.g. ecosystem-based adaptation, green infrastructure and ecosystem services) are multifunctional solutions with biodiversity at their core (Nika *et al.* 2020). The performance of NBS using vegetation relies heavily on the design properties of the soil–water–plant system (Fanelli *et al.* 2017). Four practices in urban landscapes that highlight the use of vegetation within soil–water systems to address stormwater challenges include infiltration basins, bioretention systems, green roofs and vegetative buffers.

Infiltration basins

Infiltration basins are shallow, artificial, often vegetated ponds that retain stormwater and promote gradual infiltration (often constructed in coarse soils) and filtration through permeable soils. This water either recharges groundwater or leaves via ET. Infiltration systems can attenuate the frequency and magnitude of peak flow rates and decrease the volume of stormwater discharged to traditional drainage networks and downstream receiving waters (Bonneau *et al.* 2017). Enhanced hydrological function has been attributed, in part, to larger, more diverse vegetation and soil biota that can loosen compacted soils and create new flow channels (Bean and Dukes 2016).

The utility of infiltration basins as NBS centres on addressing two challenges associated with their long-term performance, namely clogging and potential contamination to underlying soil and groundwater (Dechesne *et al.* 2004), both of which reduce the capacity for groundwater recharge and remediation of surface waters. Localised infiltration can lead to water table mounding, which can reduce the thickness of the unsaturated zone available to filter pollutants. Local hydraulic conductivity has been identified as the most important factor in sizing infiltration facilities (Petrucci *et al.* 2017). Both vertical and horizontal hydraulic conductivities can be important in the design of these systems because initially vertical flow dominates but, over time, loss of hydraulic head may occur if the wetting front reaches a saturated or less-conductive soil layer, causing lateral flow to dominate (Bean and Dukes 2016). Citing basins in higher-conductivity materials or areas with greater anisotropy (ratio of horizontal to vertical hydraulic conductivity) can reduce water table mounding, which results from localised infiltration (Nimmer *et al.* 2009; Thompson *et al.* 2010). Dechesne *et al.* (2004) proposed a set of context and performance indicators for evaluating the long-term performance of infiltration basins and found the hydraulic performance indicators to be reliable, whereas the pollution performance indicators were not. Incorporating a forebay can limit sedimentation across the basin; soil modifications, such as deep tillage and the addition of compost (on a limited basis because it can be a source of nutrients) may enhance the performance of infiltration basins; and the growth and diversity of vegetation should be promoted (Bean and Dukes 2016).

Bioretention systems

Bioretention systems, vegetated depressions created in strategic locations to intercept urban run-off close to the source, are typically constructed using high-infiltration media (replacing natural soil) underlain with a gravel reservoir and underdrain (Spraakman *et al.* 2020). These small-scale distributed systems treat stormwater run-off via physical (filtration, evaporation), chemical (sorption, ion exchange, precipitation) and biological (phytoremediation, microbial-mediated transformation, transpiration) mechanisms (Shrestha *et al.* 2018). The traditional bioretention medium is a mixture of 30–60% sand, 20–30% soil from the site and 20–40% organic matter by volume (Jiang *et al.* 2019). Results from a laboratory column study of engineered soil mixtures varying in volumetric proportions of sand (30–70%), sandy or silt loam soil (0–20%) and organic compost (20–70%) suggest that although the amount of compost controls the bulk density of the mixtures, the textural class of the mineral soil may provide a more dominant control over the movement of water and long-term stability to infiltration (Thompson *et al.* 2008b). As with infiltration basins, plants are often promoted as an essential component of bioretention systems, but qualitative assessments have yielded mixed results. The main benefits of incorporating plants include increased permeability, hydraulic conductivity and removal of nitrogen, whereas the effects on sediment filtration and phosphorus removal are minor to negligible (Dagenais *et al.* 2018). Although most studies have been conducted on newly built systems, those on mature bioretention systems indicate improved hydrological and water quality performance over time resulting from soil and plant establishment (Spraakman *et al.* 2020).

Green roofs

Green roofs are typically constructed of vegetation and soil layers underlain by a filter fabric, a drainage system to remove excess water and eventually an impermeable layer to protect the roof surface. Green roofs are categorised as intensive and extensive based on the depth of the soil layer and plants they support. Intensive green roofs have a deep soil layer (>100–1200 mm) that can support larger plants and bushes, whereas extensive vegetated roofs have thin soil layers (20–150 mm) and support smaller plants generally intended to provide full coverage of the roof surface (Czemiel Berndtsson 2010). The planting medium is typically a mixture of graded materials, including composted organic matter and granular mineral material (Hill *et al.* 2016). In their survey of extensive green roof media, Hill *et al.* (2016) found no significant differences in green roof depths in systems ranging from <3 to 17 years, and that higher organic matter content corresponded to increased water-holding capacity, lower bulk density and higher permeability.

Green roofs can be used as an NBS in a variety of urban environments. In dense urban areas where availability of land space is limited, green roofs can attenuate stormwater run-off and improve water quality and wildlife habitat while reducing temperature, noise and pollution (Czemiel Berndtsson 2010; Nawaz *et al.* 2015). The vegetative component mediates these soil–water interactions through increased retention (via changes in antecedent soil moisture) and subsequent ET of stormwater (Czemiel Berndtsson 2010; Nawaz *et al.* 2015). Rainfall retained in green roofs varies considerably and depends on the thickness of the soil substrate, its water content and precipitation characteristics, and may depend on roof age, vegetation cover and slope (Czemiel Berndtsson 2010). An average rainfall retained for extensive green roofs has been reported to range from 45% to 78% (Czemiel Berndtsson 2010) and from 15% to 83% (Nawaz *et al.* 2015).

Vegetative buffers

Vegetative buffers are planted areas set aside in the environment to help reduce soil erosion and prevent contaminants found in stormwater run-off from entering sensitive areas such as surface waters. These NBS provide a natural alternative to traditional engineered structures designed to hold existing soil in place or trap sediment in surface run-off (Elmqvist *et al.* 2015; Hubble *et al.* 2017). These buffers can take a variety of forms. For example, vegetated strips and residential yards can intercept and treat stormwater before it runs off-site into more sensitive areas (Fig. 9.8); or, improving stream bank integrity with vegetation can improve slope stability via increases in soil shear strength (Müller *et al.* 2020) and decrease erosion into urban streams (Gray and Sotir 1996). This use of vegetation is key in changing rainfall–run-off relationships (Schueler 1992), especially given the complexity of identifying the origin of soil erosion or pollutants (e.g. from parking lots, deposition, rooftop

Fig. 9.8. Residential yard planted with vegetated buffers. Photograph taken by N. J. Balster.

particulate) to focus mitigation (Thorpe and Harrison 2008). For example, polycyclic aromatic hydrocarbons were the dominant source of run-off pollution in the sediments of a stormwater catchment pond in Minnesota, USA, with 67% originating from a combination of driveways, sidewalks, parking lots, roads, urban trails and even playgrounds (Crane 2014). Vegetative buffers can also be a source of contaminants that must be considered in their design and distribution within urban areas. For example, urban forests have been shown to be a source of increased dissolved organic carbon (Zhao *et al.* 2019) and residential lawns can deliver nutrient loads up to fourfold greater than adjacent non-turf areas (Shaver *et al.* 2007), as well as suspended solids (Gromaire *et al.* 2015) and pesticides (Metcalfe *et al.* 2016). Thus, balancing the soil–water relationships of vegetative buffers will vary with the biophysical features of the location, changes in climate (e.g. intensity and frequency of heavy rainfall events) and the addition of emergent pollutants. However, the advantages of vegetative buffers and other green infrastructure, especially when factoring in sociohealth benefits, warrant continued research and close assessment of innovative practices.

While cities increasingly embrace greening their environments (Endreny 2018), the time has come for more experimental investigations to better match vegetation type, extent and diversity with services formally provided through more traditionally engineered structures. The use of rain gardens (small depressions within residential yards designed to infiltrate roof water on site) exemplifies this effort. In a controlled experiment, three vegetation types (shrubs, native prairie and turfgrass) replicated among 12 residential-sized rain gardens were found to significantly alter the hydrological and N budgets by changing drainage, ET, soil moisture and the transport of N relative to one another and to non-vegetated controls (Nocco *et al.* 2016) via plant-induced changes in antecedent soil moisture (Johnston *et al.* 2020). Similarly, urban trees, and by extension urban forests, have been shown to reduce stormwater run-off via increased canopy interception (Carlyle-Moses and Gash 2011) and ET, likely the largest flux of water within heavily vegetated urban environments (Peters *et al.* 2011). Similarly, these effects vary by species even within the same genus. Van Stan *et al.* (2015) measured ~6% greater interception loss from a canopy of a tulip tree (*Liriodendron tulipifera*) compared with an American beech (*Fagus grandifolia*) in north-eastern Maryland, USA, and Livesley *et al.* (2014) measured interception differences as much as 13% between two *Eucalyptus* species in Victoria, Australia. Rates of transpiration (e.g. a difference of 172.8 kg tree^{-1} day^{-1} between a Canary Island pine and a London plane tree in Los Angeles, CA, USA; Pataki *et al.* 2011) and infiltration (e.g. a difference of 0.13 cm min^{-1} between canopies of *Robinia pseudoacacia* L. and *Tilia cordata* Mill. in Munich, Germany; Rahman *et al.* 2019) also varied by species.

In a 3-year field study, differences in hydrological regime resulted in different levels of ecosystem services in three experimental wetlands for treating urban stormwater (Doherty *et al.* 2014). Three parallel wetlands had the same size, shape, elevation, slope and soils, were planted with the same species (native wet prairie herbs) and received similar surface inflows. However, they drained differently because of variation in subsurface clay thickness. The fastest-draining wetland provided ecosystem services of flow attenuation, stormwater retention, plant diversity support, erosion resistance and water quality improvement at their highest levels, but had lowest net primary productivity (NPP). The ponded wetland facilitated invasion and dominance by cattail (*Typha* spp.) and provided the highest level of NPP but the lowest levels of the other five services (e.g. P was mobilised and exported; Doherty *et al.* 2014). In the wetlands, epibenthic mats that developed in the fastest- and intermediate-draining wetlands played a disproportionate role in soil stabilisation compared with vascular plants (Prellwitz and Thompson 2014).

These findings demonstrate the utility of vegetation as a powerful and versatile NBS to alter soil–water relationships within our built environments (Gaston *et al.* 2013). To better quantify the variety of ecosystem services vegetation provides, land managers could focus on improving the complexity of habitat patches within existing ecosystems by managing soils, litter and vegetation holistically (Ossola *et al.* 2015) and coupling these relationships with biochemical models to help scale these practices to other municipalities and settings (see Fletcher *et al.* 2013). Additional studies that provide direct quantification of multiple ecosystem services to identify trade-offs and bundled services are needed to provide guidance at the site scale and within a larger watershed context (Fig. 9.9).

Summary

Soils support a wide variety of critical ecosystem services, many of which centre on the interactions between soil and water. Urbanisation has profoundly modified these interactions, often through changes in soil structure, diminishing ecosystem services that buffer the degradation of fluvial systems. Changes in climate add even greater uncertainty to these relationships. In addition to predictions of hotter and drier conditions, there is growing interest in quantifying winter changes in urban environments (Valtanen *et al.* 2017; Boswell *et al.* 2020a; Masson *et al.* 2020). For example, while we are elucidating the important role of these processes (e.g. freeze–thaw cycling) on soil aggregation in natural landscapes (Sorensen *et al.* 2018; Boswell *et al.* 2020b; Wang *et al.* 2020b), there is a general lack of understanding of how these effects will manifest in urban areas. How we address challenges within the soil–water–climate nexus and evaluate future NBS in urban systems will determine the liveability of our built environments and the sustainability of the natural environment they affect.

Using vegetation to manage soil–water systems within urban landscapes offers an exciting and relatively new alternative for mitigating the environmental impacts of urbanisation. Compared with traditional engineering techniques, these NBS use an integrated

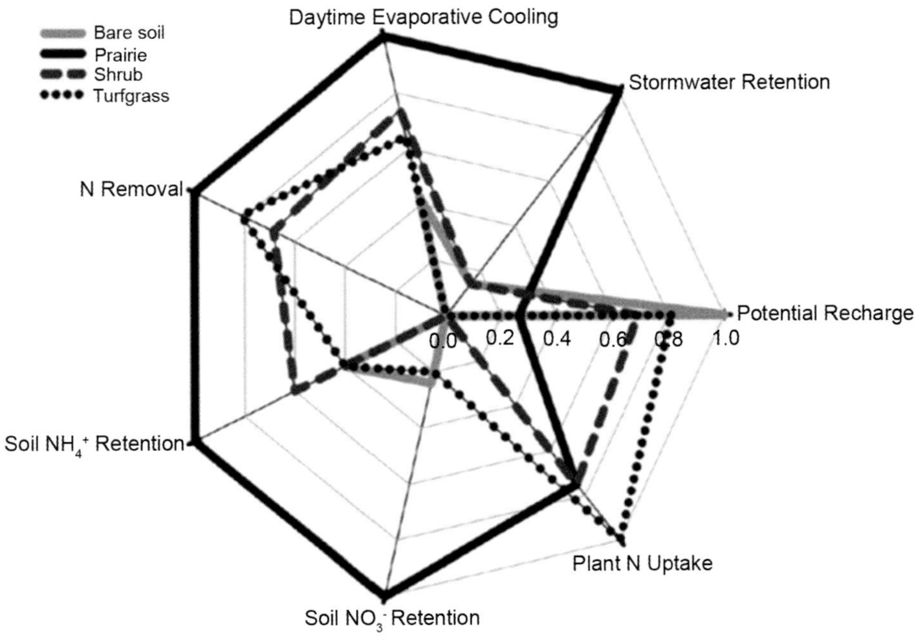

Fig. 9.9. Relative provisioning of seven ecosystem services from four vegetation treatments based on analogous measured parameters. Total drainage, evapotranspiration and the change in storage were used as ecosystem service analogues for groundwater recharge, daytime evaporative cooling and stormwater retention respectively. N removal efficiency, soil NO_3^- and NH_4^+ storage, and percentage total Kjeldahl nitrogen (N) in plant tissue were used as ecosystem service analogues for N removal, soil N retention and plant N uptake respectively. All data were normalised to the vegetation treatment that had the maximum mean for each measured parameter. The spatial arrangement of ecosystem services is arbitrary, thus polygon shapes are not unique to these normalised data. Linear interpolation between and across ecosystem services is inappropriate in this context. Reproduced with the permission of Springer Nature from Nocco *et al.* (2016).

approach that seeks to mimic the natural hydrology of the predeveloped landscape while maintaining multiple ecosystem services. Understanding and manipulating soil and water interactions will be increasingly important as climate change exacerbates the environmental impact of urban systems. Several recent studies (Bean and Dukes 2016; Dagenais *et al.* 2018; Yang and Lusk 2018; Spraakman *et al.* 2020) call for future research to focus on the role of vegetation in the management of soil–water relationships across a range of landscapes, climate and environmental conditions.

References

Altieri MA, Companioni N, Cañizares K, Murphy C, Rosset P, Bourque M, Nicholls CI (1999) Greening of the 'barrios': urban agriculture for food security in Cuba. *Agriculture and Human Values* **16**, 131–140.

Andersson E, Barthel S, Ahrné K (2007) Measuring social-ecological dynamics behind the generation of ecosystem services. *Ecological Applications* **17**, 1267–1278.

Archer NAL, Bell RAB, Utcher AS, Bricker SH (2020) Infiltration efficiency and subsurface water processes of a sustainable drainage system and consequences to flood management. *Journal of Flood Risk Management* **13**, 1–17. doi:10.1111/jfr3.12629

Aylor D (1972) Noise reduction by vegetation and ground. *Journal of the Acoustical Society of America* **51**, 197–205.

Bajcz AW, Johnston MR, Boswell EP, Balster NJ (2018) Earthworm densities correlate with aboveground plant biomass and vegetation type across residential properties in Madison, Wisconsin, USA. *Pedobiologia* **69**, 41–44. doi:10.1016/j.pedobi.2018.05.001

Bassuk N, Grabosky J, Trowbridge P (2005) *Using CU-Structural Soil in the Urban Environment*. Urban Horticulture Institute, Cornell University, Ithaca.

Bean EZ, Dukes MD (2016) Evaluation of infiltration basin performance on coarse soils. *Journal of Hydrologic Engineering* **21**, 04015050. doi:10.1061/(ASCE)HE.1943-5584.0001258

Beck SM, McHale MR, Hess GR (2016) Beyond impervious: urban land-cover pattern variation and implications for watershed management. *Environmental Management* **58**, 15–30. doi:10.1007/s00267-016-0700-8

Bending GD, Turner MK, Jones J (2002) Interactions between crop residue and soil organic matter quality and the functional diversity of soil microbial communities. *Soil Biology & Biochemistry* **34**, 1073–1082. doi:10.1016/S0038-0717(02)00040-8

Berland A, Shiflett SA, Shuster WD, Garmestani AS, Goddard HC, Herrmannd DL, Hopton ME (2017) The role of trees in urban stormwater management. *Landscape and Urban Planning* **162**, 167–177. doi:10.1016/j.landurbplan.2017.02.017

Bhaskar AS, Hogan DM, Arch SA (2016) Urban base flow with low impact development. *Hydrological Processes* **30**, 3156–3171. doi:10.1002/hyp.10808

Blair RB (1996) Land use and avian species diversity along an urban gradient. *Ecological Applications* **6**, 506–519.

Blair RB, Launer AE (1997) Butterfly diversity and human land use: species assemblages along an urban gradient. *Biological Conservation* **80**, 113–125.

Blouin M, Hodson ME, Delgado EA, Baker G, Brussaard L, Butt KR, Brun JJ (2013) A review of earthworm impact on soil function and ecosystem services. *European Journal of Soil Science* **64**, 161–182.

Blume HP, Runge M (1978) Genese und Oekologie innerstaedtischer boeden aus Bauschutt [Development and ecology of inner-city soils from rubble]. *Journal of Plant Nutrition and Soil Science* **141**, 727–740.

Bolund P, Hunhammar S (1999) Ecosystem services in urban areas. *Ecological Economics* **29**, 293–301.

Bonneau J, Fletcher TD, Costelloe JF, Burns MJ (2017) Stormwater infiltration and the 'urban karst' – a review. *Journal of Hydrology* **552**, 141–150. doi:10.1016/j.jhydrol.2017.06.043

Boswell EP, Bajcz A, Thompson AM, Balster NJ (2020*a*) Novel determination of effective freeze–thaw cycles as drivers of ecosystem change. *Journal of Environmental Quality* **49**, 314–323. doi:10.1002/jeq2.20053

Boswell EP, Balster NJ, Bajcz A, Thompson AM (2020*b*) Soil aggregation returns to a setpoint despite seasonal response to snow manipulation. *Geoderma* **357**, 113954.

Bouzouidja R, Béchet B, Hanzlikova J, Sněhota M, Le Guern C, Capiaux H, Jean-Soro L, Claverie R, Joime S, Schwartz C, *et al.* (2021) Simplified performance assessment methodology for addressing soil quality of nature-based solutions. *Journal of Soils and Sediments* **21**, 1909–1927. doi:10.1007/s11368-020-02731-y

Bullock P, Gregory P (1991) *Soils in Urban Environment*. Blackwell Scientific Publications, Oxford.

Burghardt W (1994) Soils in urban and industrial environments. *Plant Nutrition and Soil Science* **157**, 205–214.

Carey RO, Hochmuth GJ, Martinez CJ, Boyer TH, Dukes MD, Toor GS, Cisar JL (2013) Evaluating nutrient impacts in urban watersheds: challenges and research opportunities. *Environmental Pollution* **173**, 138–149. doi:10.1016/j.envpol.2012.10.004

Carlyle-Moses DE, Gash JHC (2011) Rainfall interception loss by forest canopies. In *Forest Hydrology and Biogeochemistry: Synthesis of Past Research and Future Directions*. Ecological Studies Series 216. (Eds DF Levia, DE Carlyle-Moses and T Tanaka) pp. 407–423. Springer, Heidelberg.

Chaparro L, Terradas J (2009) *Ecological services of urban forest in Barcelona*. Institut Municipal de Parcs i Jardins Ajuntament de Barcelona, Àrea de Medi Ambient, Barcelona, <https://www.itreetools.org/documents/302/Barcelona%20Ecosystem%20Analysis.pdf>.

Chen Y, Day SD, Wick AF, Mcguire KJ (2014) Influence of urban land development and subsequent soil rehabilitation on soil aggregates, carbon, and hydraulic conductivity. *The Science of the Total Environment* **494–495**, 329–336. doi:10.1016/j.scitotenv.2014.06.099

Chiesura A (2004) The role of urban parks for the sustainable city. *Landscape and Urban Planning* **68**, 129–138. doi:10.1016/j.landurbplan.2003.08.003

Cohen-Shacham E, Walters G, Janzen C, Maginnis S (2016) *Nature-Based Solutions to Address Global Societal Challenges*. IUCN, Gland.

Conley DJ, Paerl HW, Howarth RW, Boesch DF, Seitzinger SP, Havens KE, Lancelot C, Likens GE (2009) Controlling eutrophication: nitrogen and phosphorus. *Science* **323**, 1014–1015. doi:10.1126/science.1167755

Corbett CW, Wahlb M, Portera DE, Edwards D, Moised C (1997) Nonpoint source runoff modelling: a comparison of a forested watershed and an urban watershed on the South Carolina coast. *Journal of Experimental Marine Biology and Ecology* **213**, 133–149. doi:10.1016/S0022-0981(97)00013-0

Costanza R, Wilson M, Troy A, Voinov A, Liu S, D'Agostino J (2006) *The Value of New Jersey's Ecosystem Services and Natural Capital*. Gund Institute for Ecological Economics, University of Vermont for New Jersey Department of Environmental Protection, <http://archives.pdx.edu/ds/psu/8357>.

Crane JL (2014) Source apportionment and distribution of polycyclic aromatic hydrocarbons, risk considerations, and management implications for urban stormwater pond sediments in Minnesota, USA. *Archives of Environmental Contamination and Toxicology* **66**, 176. doi:10.1007/s00244-013-9963-8

Craul PJ (1992) *Urban Soil in Landscape Design*. Wiley, New York.

Craul PJ (1999) *Urban Soils: Applications and Practices*. John Wiley and Sons, New York.

Craven D, Thakur MP, Cameron EK, Frelich LE, Beauséjour R, Blair RB, Blossey B, Burtis J, Choi A, Dávalos A, *et al.* (2017) The unseen invaders: introduced earthworms as drivers of change in plant communities in North American forests (a meta-analysis). *Global Change Biology* **23**, 1065–1074. doi:10.1111/gcb.13446

Czemiel Berndtsson J (2010) Green roof performance towards management of runoff water quantity and quality: a review. *Ecological Engineering* **36**, 351–360. doi:10.1016/j.ecoleng.2009.12.014

Daddow RL, Warrington GE (1983) Growth-limiting soil bulk densities as influenced by soil texture. USDA Forest Service Watershed Systems Development Group Report WSDG-TN-00005, Fort Collins.

Dagenais D, Brisson J, Fletcher TD (2018) The role of plants in bioretention systems; does the science underpin current guidance? *Ecological Engineering* **120**, 532–545. doi:10.1016/j.ecoleng.2018.07.007

Danielsen F, Sorensen MK, Olwig MF, Selvam V, Parish F, Burgess ND, Hiraishi T, Karunagaran VM, Rasmussen MS, Hansen LB, *et al.* (2005) The Asian tsunami: a protective role for coastal vegetation. *Science* **310**, 643.

Dechesne M, Barraud S, Bardin J (2004) Indicators for hydraulic and pollution retention assessment of stormwater infiltration basins. *Journal of Environmental Management* **71**, 371–380. doi:10.1016/j.jenvman.2004.04.005

De Kimpe CR, Morel JL (2000) Urban soil management: a growing concern. *Soil Science* **165**, 31–40.

Devigne C, Mouchon P, Vanhee B (2016) Impact of soil compaction on soil biodiversity – does it matter in urban context? *Urban Ecosystems* **19**, 1163–1178. doi:10.1007/s11252-016-0547-z

Díaz-Zorita M, Perfect E, Grove JH (2002) Disruptive methods for assessing soil structure. *Soil & Tillage Research* **64**, 3–22. doi:10.1016/S0167-1987(01)00254-9

Doherty J, Miller J, Prellwitz S, Thompson AM, Loheide S, Zedler J (2014) Bundles and tradeoffs among six wetland services were associated with hydrologic regime. *Ecosystems* **17**, 1026–1039. doi:10.1007/s10021-014-9775-3

Elmqvist T, Setälä H, Handel SN, van der Ploeg S, Aronson J, Blignaut JN, Gómez-Baggethun E, Nowak DJ, Kronenberg J, De Groot R (2015) Benefits of restoring ecosystem services in urban areas. *Current Opinion in Environmental Sustainability* **14**, 101–108. doi:10.1016/j.cosust.2015.05.001

Endreny TA (2018) Strategically growing the urban forest will improve our world. *Nature Communications* **9**, 1160–1163. doi:10.1038/s41467-018-03622-0

Fanelli R, Prestegaard K, Palmer M (2017) Evaluation of infiltration- based stormwater management to restore hydrological processes in urban headwater streams. *Hydrological Processes* **31**, 3306–3319. doi:10.1002/hyp.11266

Fang C, Cui X, Deng X, Liang L (2020) Urbanization and eco-environment coupling circle theory and coupler regulation. *Journal of Geographical Sciences* **30**, 1043–1059. doi:10.1007/s11442-020-1769-9

Fletcher TD, Andrieu H, Hamel P (2013) Understanding, management and modelling of urban hydrology and its consequences for receiving waters: a state of the art. *Advances in Water Resources* **51**, 261–279. doi:10.1016/j.advwatres.2012.09.001

Folke C, Jansson A, Larsson J, Costanza R (1997) Ecosystem appropriation by cities. *Ambio* **26**, 167–172.

Food and Agriculture Organization of the United Nations (FAO) (1988) 'Soil map of the world: revised legend.' World Soil Resources Report 60, Final Draft, FAO, Rome.

Frumkin K, Frank L, Jackson R (2004) Urban sprawl and public health: designing, planning and building for healthy communities. MIT Press, Cambridge.

Gaston KJ (Ed.) (2010) *Urban Ecology. Ecological Reviews.* Cambridge University Press, New York.

Gaston KJ, Ávila-Jiménez ML, Edmondson JL (2013) Managing urban ecosystems for goods and services. *Journal of Applied Ecology* **50**, 830–840. doi:10.1111/1365-2664.12087

Gómez-Baggethun E, Barton DN (2013) Classifying and valuing ecosystem services for urban planning. *Ecological Economics* **86**, 235–245. doi:10.1016/j.ecolecon.2012.08.019

Gray D, Sotir R (1996) *Biotechnical and Soil Bioengineering Slope Stabilization: A Practical Guide for Erosion Control.* Wiley, New York.

Gregory JH, Dukes MD, Jones PH, Miller GL (2006) Effect of urban soil compaction on infiltration rate. *Journal of Soil and Water Conservation* **61**, 117–124.

Groffman PM, Fahey TJ, Fisk MC, Yavitt JB, Sherman RE, Bohlen PJ, Maerz JC (2015) Earthworms increase soil microbial biomass carrying capacity and nitrogen retention in northern hardwood forests. *Soil Biology & Biochemistry* **87**, 51–58. doi:10.1016/j.soilbio.2015.03.025

Groffman PM, Avolio M, Cavender-Bares J, Bettez ND, Grove JM, Hall SJ, Hobbie SE, Larson KL, Lerman SB, Locke DH, *et al.* (2017) Ecological homogenization of residential macrosystems. *Nature Ecology & Evolution* **1**, 0191. doi:10.1038/s41559-017-0191

Gromaire MC, Van de Voorde A, Lorgeoux C, Chebbo G (2015) Benzalkonium runoff from roofs treated with biocide products – *in situ* pilot-scale study. *Water Research* **81**, 279–287. doi:10.1016/j.watres.2015.05.060

Guilland C, Maron PA, Damas O, Ranjard L (2018) Biodiversity of urban soils for sustainable cities. *Environmental Chemistry Letters* **16**, 1267–1282. doi:10.1007/s10311-018-0751-6

Hanson HL, Wickenberg B, Olsson JA (2020) Working on the boundaries – how do science use and interpret the nature-based solution concept? *Land Use Policy* **90**, 104302. doi:10.1016/j.landusepol.2019.104302

Hendrix PF, Callaham MA Jr, Drake J, Huang CY, James SW, Snyder BA, Zhang W (2008) Pandora's box contained bait: the global problem of introduced earthworms. *Annual Review of Ecology, Evolution, and Systematics* **39**, 593–613. doi:10.1146/annurev.ecolsys.39.110707.173426

Hill J, Drake J, Sleep B (2016) Comparisons of extensive green roof media in Southern Ontario. *Ecological Engineering* **94**, 418–426. doi:10.1016/j.ecoleng.2016.05.045

Hollis JM (1991) The classification of soils in urban areas. In *Soils in the Urban Environment.* (Eds P Bullock and PJ Gregory) pp. 5–27. Wiley, Blackwell Scientific Publications, Oxford.

Howard J (2017) *Anthropogenic Soils*. Springer International Publishing, Cham.

Hubble T, Clarke S, Stokes A, Phillips C (2017) 4th International Conference on soil bio- and eco-engineering (SBEE2016) 'The Use of Vegetation to Improve Slope Stability'. *Ecological Engineering* **109**, 141–144. doi:10.1016/j.ecoleng.2017.11.003

Huot H, Joyner J, Córdoba A, Shaw RK, Wilson MA, Walker R, Muth TR, Cheng Z (2017) Characterizing urban soils in New York City: profile properties and bacterial communities. *Journal of Soils and Sediments* **17**, 393–407. doi:10.1007/s11368-016-1552-9

Ignatieva M, Hedblom M (2017) An alternative green carpet: how can we move to sustainable lawns in a time of climate change? *Science* **362**, 148–149.

Ishii M (1994) Measurement of road traffic noise reduced by the employment of low physical barriers and potted vegetation. *Proceedings. Inter-Noise* **94**, 595–598.

IUSS Working Group WRB (2015) *World Reference Base for Soil Resources 2014. International soil classification System for Naming Soils and Creating Legends for Soil Maps. Update 2015.* World Soil Resources Reports No. 106. FAO, Rome, <https://www.fao.org/3/i3794en/I3794en.pdf>.

Jacobson CR (2011) Identification and quantification of the hydrological impacts of imperviousness in urban catchments: a review. *Journal of Environmental Management* **92**, 1438–1448. doi:10.1016/j.jenvman.2011.01.018

Jefferson AJ, Bhaskar AS, Hopkins KG, Fanelli R, Avellaneda PM, Mcmillan SK (2017) Stormwater management network effectiveness and implications for urban watershed function: a critical review. *Hydrological Processes* **31**, 4056–4080. doi:10.1002/hyp.11347

Jiang C, Li J, Li H, Li Y (2019) An improved approach to design bioretention system media. *Ecological Engineering* **136**, 125–133. doi:10.1016/j.ecoleng.2019.06.014

Johnston MR, Balster NJ, Zhu J (2016) Impact of residential prairie gardens on the physical properties of urban soil in Madison, Wisconsin. *Journal of Environmental Quality* **45**, 45–52. doi:10.2134/jeq2015.02.0093

Johnston MR, Balster NJ, Thompson AM (2020) Vegetation alters soil water drainage and retention of replicate rain gardens. *Water (Basel)* **12**, 3151. doi:10.3390/w12113151

Kaushal SS, Belt KT (2012) The urban watershed continuum: evolving spatial and temporal dimensions. *Urban Ecosystems* **15**, 409–435. doi:10.1007/s11252-012-0226-7

Kaye JP, Groffman PM, Grimm NB, Baker LA, Pouyat RV (2006) A distinct urban biogeochemistry? *Trends in Ecology & Evolution* **21**, 192–199. doi:10.1016/j.tree.2005.12.006

Konijnendijk CC, Ferrini F, Fini A (2016) Introduction. In *Routledge Handbook of Urban Forestry.* (Eds F Ferrini, CC Konijnendijk and A Fini) pp. 1–13. Routledge, Abington.

Kosse A (1980) Anthrosols: proposals for a new soil order. In *Agronomy Abstracts.* Vol. 182. ASA, CSSA, and SSSA, Madison.

Kowarik I (1995) On the role of alien species in urban flora and vegetation. In *Plant Invasions – General Aspects and Special Problems.* (Eds P Pysek, K Prach, M Rejmánek and PM Wade) pp. 85–103. SPB Academic, Amsterdam.

Kühn I, Klotz S (2006) Urbanization and homogenization – comparing the floras of urban and rural areas in Germany. *Biological Conservation* **127**, 292–300. doi:10.1016/j.biocon.2005.06.033

Kuo FE, Sullivan WC (2001) Environment and crime in the inner city – does vegetation reduce crime? *Journal of Environment and Behavior* **33**, 343–367. doi:10.1177/00139160121973025

Lehmann A, Stahr K (2007) Nature and significance of anthropogenic urban soils. *Journal of Soils and Sediments* **7**, 247–260. doi:10.1065/jss2007.06.235

Lerner DN (2002) Identifying and quantifying urban recharge: a review. *Hydrogeology Journal* **10**, 143–152. doi:10.1007/s10040-001-0177-1

Levin MJ, Kim K-HJ, Morel JL, Burghardt W, Charzynski P, Shaw RK (2017) *Soils within cities: Global Approaches to their Sustainable Management.* Schweizerbart Science Publishers, Stuttgart.

Li J, Gong Y, Li X, Yin D, Shi H (2019) Urban stormwater runoff thermal characteristics and mitigation effect of low impact development measures. *Journal of Water and Climate Change* **10**, 53–62. doi:10.2166/wcc.2018.145

Li C, Sun G, Caldwell PV, Cohen E, Fang Y, Zhang Y (2020a) Impacts of urbanization on watershed water balances across the conterminous United States. *Water Resources Research* **56**, 1–19. doi:10.1029/2019WR026574

Li Y, Boswell E, Thompson AM (2020b) Correlations between land use and stream nitrate–nitrite concentrations in the Yahara River Watershed in south-central Wisconsin. *Journal of Environmental Management* **278**, 111535.

Linde DT, Watschke TL, Jarret AR, Borger JA (1995) Surface runoff assessment from creeping bentgrass and perennial ryegrass turf. *Agronomy Journal* **87**, 176–182. doi:10.2134/agronj1995.00021962008700020007x

Livesley SJ, Baudinette B, Glover D (2014) Rainfall interception and stem flow by eucalypt street trees – the impacts of canopy diversity and bark type. *Urban Forestry & Urban Greening* **13**, 192–197. doi:10.1016/j.ufug.2013.09.001

Lovell ST, Taylor JR (2013) Supplying urban ecosystem services through multifunctional green infrastructure in the United States. *Landscape Ecology* **28**, 1447–1463. doi:10.1007/s10980-013-9912-y

Liu A, Ha C, Zhou Y, Wu J (2014) How much of the world's land has been urbanized really: a hierarchical framework for avoiding confusion. *Landscape Ecology* **29**, 763–771. doi:10.1007/s10980-014-0034-y

Maes J, Jacobs S (2017) Nature-based solutions for Europe's sustainable development. *Conservation Letters* **10**, 121–124. doi:10.1111/conl.12216.

Marashi ARA, Scullion J (2003) Earthworm casts form stable aggregates in physically degraded soils. *Biology and Fertility of Soils* **37**, 375–380.

Martínez-Hernández V, Meffe R, Hernandez-Martín J, González AA, de Santiago-Martín A, Bustamante I (2020) Sustainable soil amendments to improve nature-based solutions for wastewater treatment and resource recovery. *Journal of Environmental Management* **261**, 110255. doi:10.1016/j.jenvman.2020.110255

Masson V, Lemonsu A, Hidalgo J, Voogt J (2020) Urban climates and climate change. *Annual Review of Environment and Resources* **45**, 411–444. doi:10.1146/annurev-environ-012320-083623

McGrane SJ (2016) Impacts of urbanization on hydrological and water quality dynamics, and urban water management: a review. *Hydrological Sciences Journal* **61**, 2295–2311. doi:10.1080/02626667.2015.1128084

McPherson EG (1998) Atmospheric carbon dioxide reduction by Sacramento's urban forest. *Journal of Arboriculture* **24**, 215–223.

Metcalfe CD, Sultana T, Li H, Helm PA (2016) Current-use pesticides in urban watersheds and receiving waters of western Lake Ontario measured using polar organic chemical integrative samplers (POCIS). *Journal of Great Lakes Research* **42**, 1432–1442. doi:10.1016/j.jglr.2016.08.004

Moore TL, Rodak CM, Vogel JR (2017) Urban stormwater characterization, control, and treatment. *Water Environment Research* **89**, 1876–1927. doi:10.2175/106143017X15023776270692

Morel JL, Schwartz C, Florentin L, de Kimpe C (2005) Urban soils. In *Encyclopedia of Soils in the Environment*. (Ed. D Hillel) pp. 202–208. Elsevier, New York.

Morel JL, Chenu C, Lorenz K (2015) Ecosystem services provided by soils of urban, industrial, traffic, mining, and military areas (SUITMAs). *Journal of Soils and Sediments* **15**, 1659–1666. doi:10.1007/s11368-014-0926-0

Mueller G, Thompson AM (2009) The ability of urban residential lawns to disconnect impervious area from municipal sewer systems. *Journal of the American Water Resources Association* **45**, 1116–1126. doi:10.1111/j.1752-1688.2009.00347.x

Müller A, Osterlund H, Marsalek J, Viklander M (2020) The pollution conveyed by urban runoff: a review of sources. *The Science of the Total Environment* **709**, 136215. doi:10.1016/j.scitotenv.2019.136125

National Research Council (2008) *Urban Stormwater Management in the United States*. National Academy Press. Washington, DC.

Nawaz R, McDonald A, Postoyko S (2015) Hydrological performance of a full-scale extensive green roof located in a temperate climate. *Ecological Engineering* **82**, 66–80. doi:10.1016/j.ecoleng.2014.11.061

Nika CE, Gusmaroli L, Ghafourian M, Atanasova N, Buttliglieri G, Katsou E (2020) Nature-based solutions as enablers of circularity in water systems: a review on assessment methodologies, tools and indicators. *Water Research* **183**, 115988. doi:10.1016/j.watres.2020.115988.

Nimmer M, Thompson AM, Misra D (2009) Water table mounding beneath stormwater infiltration basins. *Journal of Environmental and Engineering Geoscience* **15**, 67–79. doi:10.2113/gseegeosci.15.2.67

Nocco MA, Rouse SE, Balster NJ (2016) Vegetation type alters water and nitrogen budgets in a controlled, replicated experiment on residential-sized rain gardens planted with prairie, shrub, and turfgrass. *Urban Ecosystems* **19**, 1665–1691. doi:10.1007/s11252-016-0568-7

Nowak DJ (1994) Atmospheric carbon dioxide reduction by Chicago's urban forest. In *Chicago's Urban Forest Ecosystem: Results of the Chicago Urban Forest Climate Project*. General Technical Report NE-186. (Eds EG McPherson, DJ Nowak and RA Rowntree) pp. 83–94. U.S. Department of Agriculture, Forest Service, Northeastern Forest Experiment Station, Radnor.

Ossola A, Kristin A, John S (2015) Habitat complexity influences fine scale hydrological processes and the incidence of stormwater runoff in managed urban ecosystems. *Journal of Environmental Management* **159**, 1–10. doi:10.1016/j.jenvman.2015.05.002

Pataki DE, Carreiro MM, Cherrier J, Grulke NE, Jennings V, Pincetl S, Zipperer WC (2011) Coupling biogeochemical cycles in urban environments: ecosystem services, green solutions, and misconceptions. *Frontiers in Ecology and the Environment* **9**, 27–36. doi:10.1890/090220

Peters EB, Hiller RV, McFadden JP (2011) Seasonal contributions of vegetation types to suburban evapotranspiration. *Biogeosciences* **116**(G1), 1–16.

Petrucci G, De Bondt K, Claeys P (2017) Toward better practices in infiltration regulations for urban stormwater management. *Urban Water Journal* **14**, 546–550. doi:10.1080/1573062X.2016.1176224

Pickett STA, Cadenasso ML, Grove JM, Boone CG, Groffman PM, Irwin E, Kaushal SS, Marshall V, McGrath PB, Nilon CH, *et al.* (2011) Urban ecological systems: scientific foundations and a decade of progress. *Journal of Environmental Management* **92**, 331–362. doi:10.1016/j.jenvman.2010.08.022

Pouyat RV, Pataki DE, Belt KT, Groffman PM, Hom J, Band LE (2007) Effects of urban land-use change on biogeochemical cycles. In *Terrestrial Ecosystems in a Changing World*. (Eds JG Canadell, DE Pataki and LF Pitelka) pp. 45–56. Springer, New York.

Pouyat RV, Szlavecz K, Yesilonis ID, Groffman PM, Schwarz K (2010) Chemical, physical, and biological characteristics of urban soils. In *Urban Ecosystem Ecology*. Agronomy Monograph 55. (Eds J Aitkenhead-Peterson and A Volder) pp. 119–152. American Society of Agronomy, Crop Science Society of America, Soil Science Society of America, Madison.

Prellwitz S, Thompson AM (2014) Biota and hydrology influence soil stability in constructed wetlands. *Ecological Engineering* **64**, 360–366. doi:10.1016/j.ecoleng.2014.01.010

Rabot E, Wiesmeier M, Schlüter S, Vogel HJ (2018) Soil structure as an indicator of soil functions: a review. *Geoderma* **314**, 122–137. doi:10.1016/j.geoderma.2017.11.009

Rahman MA, Moser A, Anderson M, Zhang C, Rötzer T, Pauleit S (2019) Comparing the infiltration potentials of soils beneath the canopies of two contrasting urban tree species. *Urban Forestry & Urban Greening* **38**, 22–32. doi:10.1016/j.ufug.2018.11.002

Rawlins BG, Harris J, Price S, Bartlett M (2015) A review of climate change impacts on urban soil functions with examples and policy insights from England, UK. *Soil Use and Management* **31**, 46–61. doi:10.1111/sum.12079

Richards DR, Belcher RN (2019) Global changes in urban vegetation cover. *Remote Sensing* **12**, 23. doi:10.3390/rs12010023

Roy AH, Rhea LK, Mayer AL, Shuster W D, Beaulieu JJ, Hopton ME, Amand AS (2014) How much is enough? Minimal responses of water quality and stream biota to partial retrofit stormwater management in a suburban neighborhood. *Public Library of Science* **9**, e85011. doi:10.1371/journal.pone.0085011

Russell KL, Vietz GJ, Fletcher TD (2017) Global sediment yields from urban and urbanizing watersheds. *Earth-Science Reviews* **168**, 73–80. doi:10.1016/j.earscirev.2017.04.001

Sax MS, Bassuk N, van Es H, Rakow D (2017) Long-term remediation of compacted urban soils by physical fracturing and incorporation of compost. *Urban Forestry & Urban Greening* **24**, 149–156. doi:10.1016/j.ufug.2017.03.023

Scharenbroch BC, Lloyd JE, Johnson-Maynard JL (2005) Distinguishing urban soils with physical, chemical, and biological properties. *Pedobiologia* **49**, 283–296. doi:10.1016/j.pedobi.2004.12.002

Schmid JA (1975) 'Urban vegetation. A review and Chicago case study.' Research paper no. 161, University of Chicago, Department of Geography. Chicago.

Schueler TR (1992) Design of storm water wetland systems: guidelines for creating diverse and effective stormwater wetland systems in the Mid-Atlantic region. Anacostia Restoration Team, Department of Environmental Programs, Metropolitan Washington Council of Governments, No. 92710, Washington, DC.

Shao Z, Ding L, Li D, Altan O, Huq ME, Li C (2020) Exploring the relationship between urbanization and ecological environment using remote sensing images and statistical data: a case study in the Yangtze River Delta, China. *Sustainability* **12**, 5620. doi:10.3390/su12145620

Sharp JM Jr, Krothe JN, Mather JD, Garcia-Fresca B, Stewart CA (2003) Effects of urbanization on groundwater systems. In *Earth Science in the City: A Reader.* Vol. 56. (Eds G Heiken, R Fakundiny and J Sutter) pp. 257–278. American Geophysical Union, Washington, DC.

Shaver E, Horner R, Skupien J, May C, Ridley G (2007) Fundamentals of *Urban Runoff Management: Technical and Institutional Issues.* 2nd edn. The North American Lake Management Society (NALMS) Madison.

Shaw RK (2015) Soils in urban areas: characterization, management, challenges. *Soil Science* **180**, 135. doi:10.1097/SS.0000000000000138

Shochat E, Warren PS, Faeth SH, McIntyre NE, Hope D (2006) From patterns to emerging processes in mechanistic urban ecology. *Trends in Ecology & Evolution* **21**, 186–191. doi:10.1016/j.tree.2005.11.019

Shrestha P, Hurley SE, Wemple BC (2018) Effects of different soil media, vegetation, and hydrologic treatments on nutrient and sediment removal in roadside bioretention systems. *Ecological Engineering* **112**, 116–131. doi:10.1016/j.ecoleng.2017.12.004

Six J, Bossuyt H, Degryze S, Denef K (2004) A history of research on the link between (micro) aggregates, soil biota, and soil organic matter dynamics. *Soil & Tillage Research* **79**, 7–31. doi:10.1016/j.still.2004.03.008

Small C, Pozzi F, Elvidge CD (2005) Spatial analysis of global urban extent from DMSP-OLS night lights. *Remote Sensing of Environment* **96**, 277–291. doi:10.1016/j.rse.2005.02.002

Smetak KM, Johnsn-Maynard JL, Lloyd JE (2007) Earthworm population density and diversity in different-aged urban systems. *Applied Soil Ecology* **37**, 161–168. doi:10.1016/j.apsoil.2007.06.004

Snep RPH, Voeten Joris GWF, Mol G, Van Hattum T (2020) Nature based solutions for urban resilience: a distinction between no-tech, low-tech and high-tech solutions. *Frontiers in Environmental Science* **8**, 599060. doi:10.3389/fenvs.2020.599060

Soil Survey Staff (1999) *Soil Taxonomy: A Basic System of Soil Classification for Making and Interpreting Soil Surveys*. 2nd edn. U.S. Department of Agriculture Handbook 436, Natural Resources Conservation Service, Washington, DC.

Sorensen PO, Finzi AC, Giasson MA, Reinmann AB, Sanders-DeMott R, Templer PH (2018) Winter soil freeze–thaw cycles lead to reductions in soil microbial biomass and activity not compensated for by soil warming. *Soil Biology & Biochemistry* **116**, 39–47. doi:10.1016/j.soilbio.2017.09.026

Spraakman S, Van Seters T, Drake J, Passeport E (2020) How has it changed? A comparative field evaluation of bioretention infiltration and treatment performance post-construction and at maturity. *Ecological Engineering* **158**, 106036. doi:10.1016/j.ecoleng.2020.106036

Staddon C, Ward S, De Vito L, Zuniga-Teran A, Gerlak AK, Schoeman Y, Hart A, Booth G (2018) Contributions of green infrastructure to enhancing urban resilience. *Environment Systems & Decisions* **38**, 330–338. doi:10.1007/s10669-018-9702-9

Steinberg DA, Pouyat RV, Parmelee RW, Groffman PM (1997) Earthworm abundance and nitrogen mineralization rates along an urban–rural land use gradient. *Soil Biology and Biochemistry* **29**, 427–430.

Sukopp H (1998) Urban ecology – scientific and practical aspects. In *Urban Ecology*. (Eds J Breuste, H Feldman and O Uhlmann) pp. 3–16. Springer-Verlag, Berlin.

Szlavecz K, Placella S, Pouyat RV, Groffman PM, Csuzdi C, Yesilonis ID (2006) Invasive earthworms and N-cycling in remnant forest patches. *Applied Soil Ecology* **32**, 54–62. doi:10.1016/j.apsoil.2005.01.006

Thompson AM, Kim K, Vandermuss AJ (2008*a*) Thermal characteristics of stormwater runoff from asphalt and sod surfaces. *Journal of the American Water Resources Association* **44**, 1325–1336. doi:10.1111/j.1752-1688.2008.00226.x

Thompson AM, Paul AC, Balster NJ (2008*b*) Physical and hydraulic properties of engineered soil media for bioretention basins. *Transactions of the ASABE* **51**, 499–514. doi:10.13031/2013.24391

Thompson AM, Wilson T, Norman J, Gemechu A, Roa-Espinosa A (2008*c*) Modeling the effect of summertime heating on urban runoff temperature. *Journal of the American Water Resources Association* **44**, 1548–1563. doi:10.1111/j.1752-1688.2008.00259.x

Thompson AM, Nimmer M, Misra D (2010) Effects of variations in hydrogeological parameters on water-table mounding in sandy loam and loamy sand soils beneath stormwater infiltration basins. *Hydrogeology Journal* **18**, 501–508. doi:10.1007/s10040-009-0532-1

Thorpe A, Harrison RM (2008) Sources and properties of non-exhaust particulate matter from road traffic: a review. *The Science of the Total Environment* **400**, 270–282. doi:10.1016/j.scitotenv.2008.06.007

Tisdall J, Oades J (1982) Organic matter and water-stable aggregates in soils. *Journal of Soil Science* **33**, 141–163. doi:10.1111/j.1365-2389.1982.tb01755.x

Tóth Z, Szlavecz K, Epp Schmidt DJ, Hornung E, Setälä H, Yesilonis ID, Kotze DJ, Dombos M, Pouyat R, Mishra S, *et al.* (2020) Earthworm assemblages in urban habitats across biogeographical regions. *Applied Soil Ecology* **151**, 103530. doi:10.1016/j.apsoil.2020.103530

Tyner JS, Yoder DC, Chomicki BJ, Tyagi A (2011) A review of construction site best management practices for erosion control. *Transactions of the ASABE* **54**, 441–450. doi:10.13031/2013.36447

United Nations (UN) (2015) 'World population prospects: the 2015 revision, key findings and advance tables.' Working Paper ESA/P/WP.241, UN, Department of Economic and Social Affairs, Population Division, New York.

United Nations (UN) General Assembly (2015) *Resolution Adopted by the General Assembly on 25 September 2015 [Without Reference to a Main Committee (A/70/L.1)] 70/1. Transforming Our World: The 2030 Agenda for Sustainable Development*. UN, New York, <https://www.refworld.org/docid/57b6e3e44.html>.

U.S. Environmental Protection Agency (USEPA) (2018) 'Stormwater Phase II Final Rule. Construction Site Runoff Control Minimum Control Measure.' Office of Water (4203). Fact Sheet 2.6. USEPA, Washington, DC.

Valtanen M, Sillanpää N, Setälä H (2017) A large-scale lysimeter study of stormwater biofiltration under cold climatic conditions. *Ecological Engineering* **100**, 89–98. doi:10.1016/j.ecoleng.2016.12.018

Van Stan JT II, Levia DF Jr, Jenkins RB (2015) Forest canopy interception loss across temporal scales: Implications for urban greening initiatives. *The Professional Geographer* **67**, 41–51. doi:10.1080/00330124.2014.888628

Vauramo S, Setälä H (2011) Decomposition of labile and recalcitrant litter types under different plant communities in urban soils. *Urban Ecosystems* **14**, 59–70.

Vaverková MD, Radziemska M, Bartoň S, Cerdà A, Koda E (2018) The use of vegetation as a natural strategy for landfill restoration. *Land Degradation & Development* **29**, 3674–3680. doi:10.1002/ldr.3119

Villarreal EL, Bengtsson L (2005) Response of a Sedum green-roof to individual rain events. *Ecological Engineering* **25**, 1–7.

Wang J, Zhou L, Ping H, Li G (2020a) The impact of urbanization and LID technology on hydrological effect. *Coastal Research* **104**, 14–22. doi:10.2112/JCR-SI104-004.1

Wang QI, Lv W, Li B, Zhou Y, Jiang L, Piao S, Wang Y, Zhang L, Meng F, Liu P, *et al.* (2020b) Annual ecosystem respiration is resistant to changes in freeze–thaw periods in semi-arid permafrost. *Global Change Biology* **26**, 2630–2641. doi:10.1111/gcb.14979

Wang Y, Thompson AM, Selbig W (2020c) Leachable phosphorus from senesced Green Ash and Norway Maple leaves in urban watersheds. *The Science of the Total Environment* **743**, 140662. doi:10.1016/j.scitotenv.2020.140662

Willis KJ, Petrokofsky G (2017) The natural capital of city trees. *Science* **356**, 374–376. doi:10.1126/science.aam9724

Wilson KV (1967) A preliminary study of the effects of urbanization on floods in Jackson, Mississippi. *U.S. Geological Survey Professional Paper* **575-D**, D259–D261.

Wolman MG (1967) A cycle of sedimentation and erosion in urban river channels. *Geografiska Annaler. Series A. Physical Geography* **49**, 385–395. doi:10.1080/04353676.1967.11879766

Woltemade CJ (2010) Impact of residential soil disturbance on infiltration rate and stormwater runoff. *Journal of the American Water Resources Association* **46**, 700–711. doi:10.1111/j.1752-1688.2010.00442.x

Xu C, Haase D, Pauleit S (2018) The impact of different urban dynamics on green space availability: a multiple scenario modeling approach for the region of Munich, Germany. *Ecological Indicators* **93**, 1–12. doi:10.1016/j.ecolind.2018.04.058

Yaalon DH, Yaron B (1966) Framework for man-made soil changes: an outline of metapedogenesis. *Soil Science* **102**, 272–277. doi:10.1097/00010694-196610000-00010

Yang Y, Lusk MG (2018) Nutrients in urban stormwater runoff: current state of the science and potential mitigation options. *Current Pollution Reports* **4**, 112–127. doi:10.1007/s40726-018-0087-7

Yang J, Zhang G (2011) Water infiltration in urban soils and its effects on the quantity and quality of runoff. *Journal of Soils and Sediments* **11**, 751–761. doi:10.1007/s11368-011-0356-1

Yang JL, Zhang GL (2015) Formation, characteristics and eco-environmental implications of urban soils – a review. *Soil Science and Plant Nutrition* **61**, 30–46. doi:10.1080/00380768.2015.1035622

Yilmaz D, Cannavo P, Séré G, Vidal-Beaudet L, Legret M, Damas O, Peyneau PE (2018) Physical properties of structural soils containing waste materials to achieve urban greening. *Journal of Soils and Sediments* **18**, 442–455. doi:10.1007/s11368-016-1524-0

York AM, Shreshtha M, Boone CG, Zhang S, Harrington JA Jr, Prebyl TJ, Swann A, Agar M, Antolin MF, Nolen B, *et al.* (2011) Land fragmentation under rapid urbanization: a cross-site analysis of southwestern cities. *Urban Ecosystems* **14**, 429–455. doi:10.1007/s11252-011-0157-8

Zeiger SJ, Hubbart JA (2015) Urban stormwater temperature surges: a central US watershed study. *Hydrology* **2**, 193–209. doi:10.3390/hydrology2040193

Zhao D, Li F, Yang Q, Wang R, Song Y, Tao Y (2013) The influence of different types of urban land use on soil microbial biomass and functional diversity in Beijing, China. *Soil Use and Management* **29**, 230–239. doi:10.1111/sum.12034

Zhao C, Gao SJ, Zhou L, Li X, Wang CC (2019) Dissolved organic matter in urban forestland soil and its interactions with typical heavy metals: a case of Daxing District, Beijing. *Environmental Science and Pollution Research International* **26**, 2960–2973. doi:10.1007/s11356-018-3860-7

Zheng Q, Hao L, Huang X, Sun L, Sun G (2020) Effects of urbanization on watershed evapotranspiration and its components in southern China. *Water (Basel)* **12**, 645. doi:10.3390/w12030645

Zhou X, Lin HS, White EA (2008) Surface soil hydraulic properties in four soil series under different land uses and their temporal changes. *Catena* **73**, 180–188. doi:10.1016/j.catena.2007.09.009

Zipperer WC, Foresman TW, Sisinni SM, Pouyat RV (1997) Urban tree cover: an ecological perspective. *Urban Ecosystems* **1**, 229–246. doi:10.1023/A:1018587830636

Zölch T, Henze L, Keilholz P, Pauleit S (2017) Regulating urban surface runoff through nature-based solutions – an assessment at the micro-scale. *Environmental Research* **157**, 135–144. doi:10.1016/j.envres.2017.05.023

10

Maintaining wetland ecosystem services in a changing climate

W. Carter Johnson, G. R. Guntenspergen

Introduction

Value of wetlands

There is increasing interest in understanding the many ecosystem services provided by wetlands (Scott and Jones 1995) at global, national and local scales (Zedler and Kircher 2005; Colloff *et al.* 2016; Trettin *et al.* 2020). These services are the result of the natural processes and properties of wetland systems that directly and indirectly benefit humans. The Millennium Ecosystem Assessment (2005) recognised four categories of ecosystem services: (1) provisioning services (e.g. fisheries support, timber production, direct food production); (2) regulating services (e.g. water quality improvement, river flooding mitigation, carbon sequestration); (3) supporting services (e.g. chemical transformers, water storage); and (4) cultural services (e.g. ecotourism, landscape aesthetics). Coastal and inland wetlands contribute disproportionately by area to the estimated US$16–54 trillion per year of the monetary value of these services (Costanza *et al.* 1997) and account for 43% of the value of all ecosystems on Earth, even though wetlands only cover 6% of the Earth's surface. Wetlands are arguably among the most valuable ecosystems on the planet (Mitsch and Gosselink 2015).

Coastal intertidal wetlands (mangrove and salt marsh) are among the most valuable of all ecosystems globally (Fig. 10.1; Barbier *et al.* 2011; Costanza *et al.* 2014; de Groot *et al.* 2012). Ecosystem services provided by mangroves (e.g. Lee *et al.* 2014) and salt marshes (e.g. Craft *et al.* 2009; Shepard *et al.* 2011; Barbier 2019) have been summarised into seven broad categories: protection from storms and flooding; fisheries habitat and production; habitat for threatened species; cultural services and values; biogeochemistry services; nutrient cycling; and carbon sequestration (Kelleway *et al.* 2017). Salt marshes and mangroves protect coastal regions from storms (Alongi 2008; Gedan *et al.* 2011; Shepard *et al.* 2011; Temmerman *et al.* 2013), sequester carbon (Mcleod *et al.* 2011; Duarte *et al.* 2013; Rogers *et al.* 2019), transform nutrients (Boorman 1999; Valiela *et al.* 2002) and provide the organic matter and nursery grounds that support commercial fisheries (Barbier and Strand 1998; Aburto-Oropeza *et al.* 2008). Mangrove forests provide a wide range of ecosystem services conservatively estimated to be worth US$194 000 per hectare per year (Lovelock *et al.* 2015).

Fig. 10.1. A coastal wetland on the Atlantic Coast of the US near Fire Island, New York, 26 August 2020. Photograph taken by Charles Flagg, Stony Brook University.

Inland wetlands differ from coastal wetlands by the biotic and abiotic properties that control their ecology (Batzer and Baldwin 2012). Most inland wetlands are not exposed to salt water, and the interaction of hydrological and geomorphic processes, landform and surface and subsurface waters differ from those of coastal wetlands. Brinson (1993) classified the many types of freshwater wetlands into four broad categories based on geomorphic setting, the water source and its transport and hydrodynamics: depressional (dependent on atmospheric or groundwater sources of water); extensive peatland (where peat substrates dominate the movement and storage of water); riverine (which form as linear strips throughout the landscape with predominately unidirectional flow); and fringe wetlands (which occur in estuaries where tidal forces dominate or in lakes where water moves in and out of the wetland from the effects of wind, waves and seiches). Hydrology and geomorphology combined with other biotic and abiotic properties (e.g. vegetation, soil texture, soil pH) differentiate the ecosystem functions of these diverse inland wetland types (Smith *et al.* 1995).

Inland wetland types vary in the hydroperiod, water chemistry, soil properties, landscape position and biotic assemblages that are the underpinnings of ecosystem services (Brinson and Eckles 2011). Brinson (1993) expressed the relative importance of water source (i.e. precipitation, groundwater discharge and lateral surface flow) to major wetland types (bog, riverine etc.) and how hydrology contributes to wetland functions. Studies of wetland ecosystem services (Zedler and Kircher 2005; De Steven and Lowrance 2011; Fennessy and Craft 2011; Vasić *et al.* 2020) emphasise the hydrological, biogeochemical and biotic processes that provide the 'services' to maintain sustainable ecosystems and provide human benefits. Certain categories of ecosystem services provided by these wetlands stand out as having global significance and value, namely biodiversity support, water quality improvement, flood abatement and carbon management (Zedler and Kircher 2005).

Different wetland types vary in their relative contribution to providing ecosystem services (Fennessy and Craft 2011). The value of riparian and floodplain forest, depressional wetlands, bogs and fens and vernal pool ecosystem services may range from high to low when comparing water quality maintenance, habitat, hydrological or carbon sequestration delivery.

Vasić *et al.* (2020) provide a comprehensive review of ecosystem services supplied by depressional wetlands in agricultural landscapes. Vasić *et al.* (2020) identified eight ecosystem services from studies in Europe, New Zealand and North America (i.e. flood control, improved water quality, nursery populations and habitats, native pollinator populations, climate change mitigation, nutrient cycling (biotic remediation of wastes), local regulation of air temperature and humidity and landscape aesthetics) and focused on four: flood control (Gleason *et al.* 2008; van der Kamp and Hayashi 2009), improving water quality (Neely and Baker 1989; Detenbeck *et al.* 2002), biodiversity (Swanson and Dubbert 1989; Withey and van Kooten 2014) and carbon management (Downing *et al.* 2008; Phillips and Beeri 2008; Tangen *et al.* 2015).

Wetland vulnerability to climate change

Considerable work has assessed the resilience of coastal wetlands to climate change (Kirwan and Megonigal 2013) and the resulting impact on the ecosystem services they have historically provided. The response of wetland function to climate change also depends on the interaction with non-climate-related stresses, such as land cover changes, which have historically been responsible for the loss of wetland numbers and function (Burkett and Kusler 2000). Tidal wetland conversion to open water through climate-mediated sea level rise and increased flooding frequency is currently occurring at an accelerated pace (Lovelock *et al.* 2015; Saintilan *et al.* 2020). Vegetation response to rising seas and saltwater intrusion can result in mangroves migrating upslope into salt marsh habitat and having a significant effect on below-ground carbon stocks (Kelleway *et al.* 2017; Simpson *et al.* 2017).

The response of depressional wetlands to climate change is influenced by changes in major components of the water budget: precipitation, temperature and water source (Brinson 2006). Both temperature and precipitation patterns have changed in recent decades, and these changes are projected to intensify into the future. Depressional wetlands are inextricably linked to the hydrological regime of the site and so are sensitive to changes in environmental conditions and management regimes. Wetlands dependent on precipitation are highly vulnerable to a drying climate, but, in contrast, wetlands that receive water from regional groundwater flow systems for their source of water are more resilient (Winter 2000).

Warmer summers and greater weather extremes in mid-continental climates may lead to higher summer evapotranspiration rates, resulting in accelerated drying of small depressional basins and longer drought periods for larger wetlands (Johnson *et al.* 2004, 2005, 2010). Altered hydrology can reverberate in many ways, ranging from a loss of wildlife to a shortage of water for livestock to reduction in water quality services. Changes in the hydrological regime of these wetlands control the anaerobic processes in the soil that are responsible for water quality improvement and carbon sequestration (Trettin *et al.* 2020). Climate-induced changes in habitat can have major effects on wetland biodiversity.

Amphibians are typically the most sensitive wetland taxa to habitat alteration by climate change (Corn 2005; Sievers *et al.* 2018), but drying of depressional wetlands can disrupt landscape-scale connectivity among wetlands and may prohibit animals, including amphibians and water-dependent birds, from moving to more favourable locations or being able to complete their life cycle (McIntyre *et al.* 2019; Allen *et al.* 2020).

Mitigation measures

Potential mitigation measures to reduce the impact of climate change would differ markedly between coastal and depressional wetlands. There are few options for protecting coastal wetlands globally from increased temperature (save from less burning of fossil fuel), changes in precipitation or rapidly rising sea levels, although various management measures could provide temporary relief. Tidal wetlands often do not keep pace vertically with increases in sea level (Lovelock *et al.* 2015; Peteet *et al.* 2018) because their sediment supply is captured by dams across rivers and levees against tides (Weston 2014). Applying sediment to these sediment-deficient marshes may provide a subsidy raising the elevation capital of these marshes (Ford *et al.* 1999; Cahoon *et al.* 2019), allowing them to persist. In some locations where the topography of the landscape is suitable, sea level rise can lead to coastal wetlands migrating landward at a rate that exceeds seaward displacement until they reach anthropogenic or physical barriers (Enwright *et al.* 2016; Kirwan and Gedan 2019).

The fact that depressional wetlands are bounded by land frequently managed for farming, grazing or forestry opens up opportunities for mitigation and restoration (Zedler 2003; Brinson and Eckles 2011; Vasić *et al.* 2020). It goes without saying that the most direct mitigative approach to maintaining wetland services is to protect remaining wetlands and restore degraded wetlands (Bedford 1999; Zedler 2003). However, researchers have identified, almost by accident, land management actions that can lower the risk of climate change on the water budget of depressional wetlands. These mitigation approaches are identified and evaluated in the case study (below) for prairie pothole wetlands.

Case studies: potential for climate change mitigation in the Prairie Pothole Region

Ecology of prairie pothole wetlands

The Prairie Pothole Region (PPR) is located in the centre of North America and is divided roughly in half by the border between the US and Canada (Plate 8). The word 'pothole' refers to the small size (<1 ha) and circular shape of many of these depressional wetlands, likened to the rain-filled potholes that form on poorly maintained roads. Approximately 25 million wetlands were left in the 780 000-km^2 PPR during the melting of the last episode of Pleistocene glaciation, ~12 000 BP (Prairie Pothole Joint Venture 1998). The wetlands range widely in area, shape and depth (Fig. 10.2). Some dry up quickly nearly every year, whereas others hold surface water in most years. This hydroperiod of wetlands (i.e. the length of time of inundation and the portion of year the wetland holds ponded water) is used to characterise prairie pothole wetlands as temporary, seasonal, semipermanent or permanent ponds (Stewart and Kantrud 1971). The glacial topography ranges from flat, ancient lakebeds with lacustrine soils to rugged moraines with coarse-textured soils of all combinations of sand, pebbles and rocks intermixed with heavy clay. The region historically was dominated by two types of grassland or prairie: tall and mixed (both tall and

Fig. 10.2. Image of prairie wetlands in a grassland matrix in the Prairie Pothole Region near Clear Lake, South Dakota. Note the wide variety of sizes, shapes and disposition of wetlands. Photograph taken by Craig Novotny, South Dakota State University.

mid-height grasses) that follow the strong moisture gradient from east (~56 cm mean annual precipitation) to west (~25 cm mean annual precipitation) in the rain shadow of the Rocky Mountains (Leitch 1989). The PPR was included in the group of 11 of the world's largest wetland areas, highlighting their conservation importance at the global scale (Keddy *et al.* 2009)

An incredibly diverse biota lives in and near prairie pothole wetlands. Plant life ranges from tiny algae to submerged aquatics (pond weeds) to emergents (cattails, bulrushes). Algae form the energy base for hundreds of species of invertebrate animals, especially crustaceans, aquatic beetles, clams and snails (Kantrud *et al.* 1989). In turn, invertebrates are the food base for ducks, one of the major groups of vertebrates in prairie wetlands (Weller 1999). Other wetland-dependent wildlife in the PPR include muskrats, beaver, fish, frogs, salamanders, blackbirds, herons, terns, sedge wrens and egrets. These organisms are distributed unevenly across the different wetland types depending on hydroperiod and water quality (from fresh to saline and eutrophic to oligotrophic). Both hydroperiod and water quality are strongly affected by climate (Mushet *et al.* 2015; Johnson and Poiani 2016).

Of all the groups of organisms that are associated with prairie pothole wetlands, waterfowl have been studied the most. A most astounding fact is that the PPR produces at least half of North America's ducklings in an average year yet constitutes only 10% of the continent's breeding habitat (Kantrud *et al.* 1989). The numbers and geographic location of waterfowl are monitored by many organisations, both public (e.g. USA Fish and Wildlife Service, Canadian Wildlife Service) and private (e.g. Ducks Unlimited). People who live in the PPR, including birders and waterfowl hunters, and those who live where waterfowl born in the PPR overwinter are concerned with the variation in waterfowl population sizes due to weather cycles (deluge or drought), climate variability and climate change (Sauchyn *et al.* 2010). Several studies (discussed below) have been conducted recently to determine how sensitive the prairie wetland ecosystem is to climate change and how any negative consequences discovered could be curbed through adjustments to the management of wetlands and their watersheds.

Prairie wetland water budgets

All ecosystems are influenced by climate; prairie pothole wetlands are no exception. Studies have shown that water levels and the presence or absence of associated wetland organisms are affected by precipitation, run-off from snowmelt and rainstorm deluges, groundwater recharge, evaporation and transpiration, relative humidity and wind speed (Weller 1981; van der Valk 1989; Johnson *et al.* 2004; Galatowitsch 2012; van der Valk and Mushet 2016). Simulation models of PPR wetlands must account for these and other variables to understand their relative importance in affecting wetland functioning (Fig. 10.3). Models such as WETLANDSCAPE (Johnson and Poiani 2016) track the water regime of wetlands quite well when compared to observed drought (1930s Dust Bowl) and deluge (1990s floods) periods (Fig. 10.4a). They also match up well with long-term field measurements (Fig. 10.4b).

How sensitive would PPR wetlands be to a future climate? Varying precipitation and temperature within the ranges projected by the Intergovernmental Panel on Climate Change (IPCC; 2007) would subject PPR wetlands to temperatures several degrees (3–4°C) warmer and precipitation ranging from 10% less to 10% more than that of the 20th century. The key question that could only be addressed by modelling (e.g. Johnson *et al.* 2010; Liu

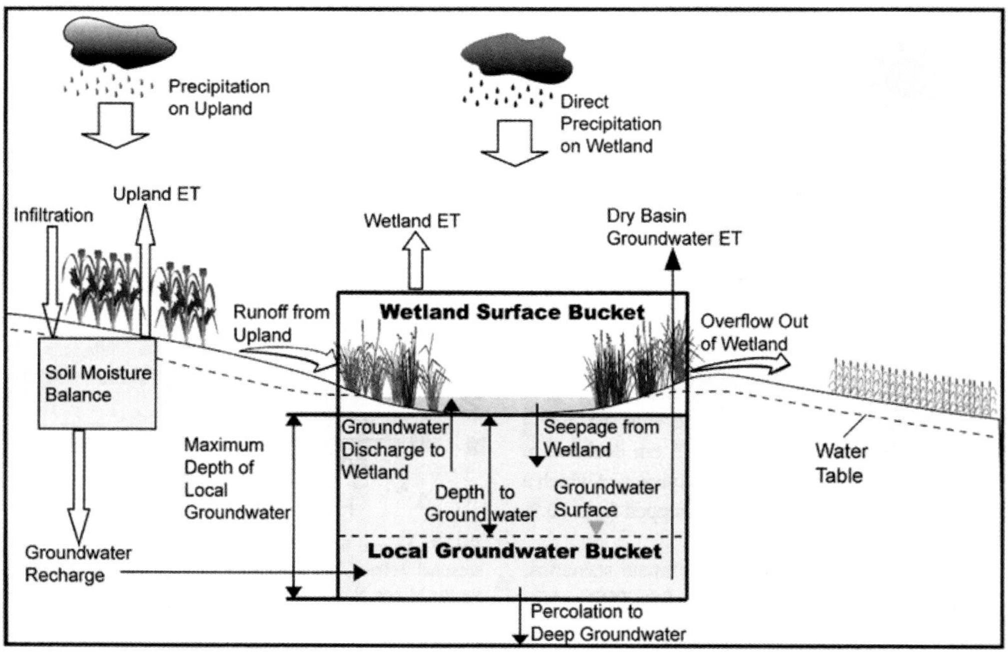

Fig. 10.3. Generalised wetland water budget with a double-bucket submodel used in numerical model analyses. ET, evapotranspiration. Reproduced with the permission of Springer Nature from Johnson and Poiani (2016).

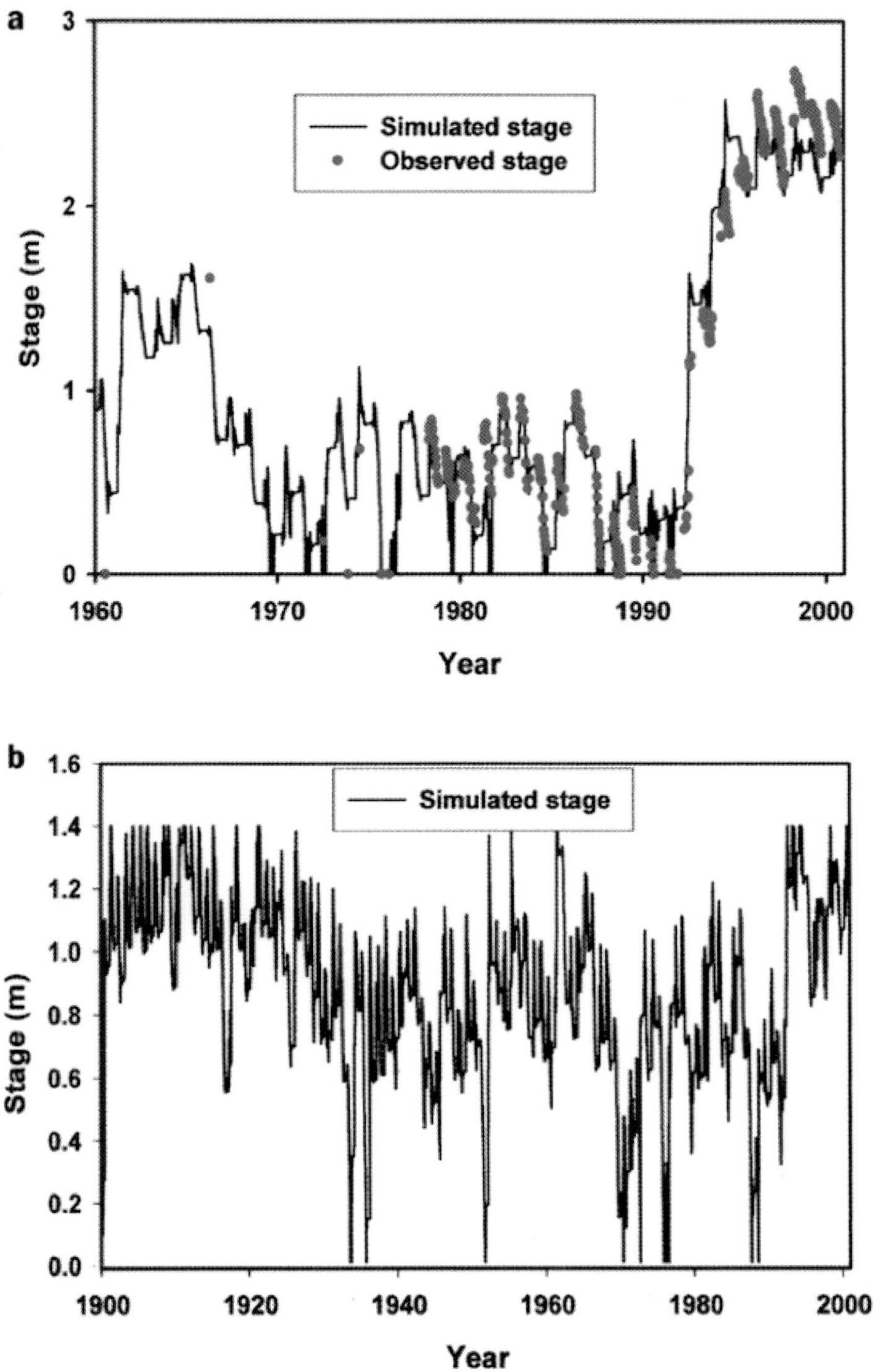

Fig. 10.4. (a) Comparison of simulated (solid line) and observed (grey dots) wetland stage levels for the period 1960–2002 and (b) stage levels simulated for the period of record (101 years) for the Jamestown, North Dakota, USA, weather station, showing simulated wet and dry periods. Reproduced with the permission of Oxford University Press from Johnson *et al.* (2005).

and Schwartz 2011; McKenna *et al.* 2018) was to determine whether this magnitude of climate change would have little, moderate or major impact on wetland functioning. WETLANDSCAPE (Johnson *et al.* 2010) uses a combination of downscaled climate data from global circulation models and a 'coding' approach that varies precipitation and temperature by percentage changes in combination to make projections about the response of prairie pothole wetlands across a range of climate scenarios. Initial 100-year simulations using WETLANDSCAPE varied temperature and precipitation by using three different climate scenarios (+2°C, +4°C and +4°C and +10% precipitation; Johnson *et al.* 2010) and indicated major changes in the water regime of the three wetland types (semipermanent, seasonal, temporary) that make up the wetland complex (Stewart and Kantrud 1971; Johnson *et al.* 2004). In most cases wetlands became drier, became dry more often and experienced shortened hydroperiods. Wetlands also shifted from wetter to drier types (e.g. from semipermanent to seasonal flood types). However, increasing precipitation by ~20% restored more normal water regimes.

Serious consequences for waterfowl breeding and for key measures of wetland productivity, which affect all wetland organisms, could occur under future climate scenarios (Fig. 10.5; Johnson *et al.* 2005; Mushet 2016). Should more precipitation than projected accompany the warmer temperature, the seriousness of climate change to wetland function would diminish. If precipitation remained similar to current amounts and actual

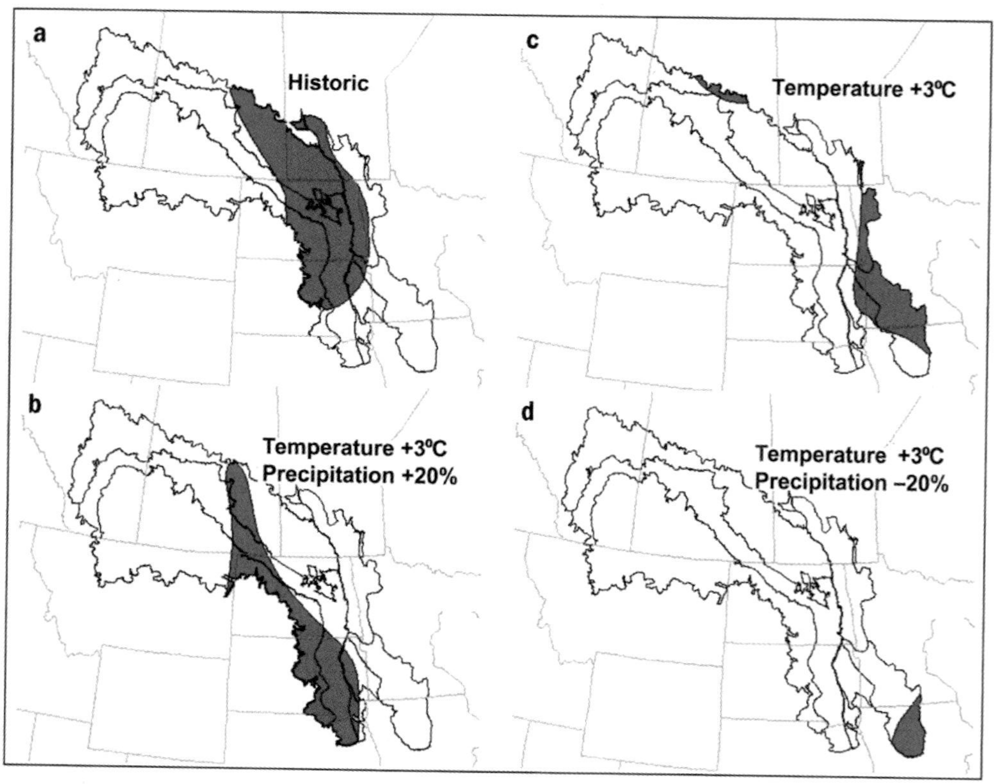

Fig. 10.5. Simulated location of the most favourable wetland conditions for breeding waterfowl in the Prairie Pothole Region under (a) historic and (b–d) alternative future scenarios. Note the significant geographic shifts caused by rather modest changes in temperature and precipitation. Reproduced with the permission of Oxford University Press from Johnson *et al.* (2005).

temperature becomes higher than projected, mitigative measures would be essential to continue the historic role that the PPR has played in maintaining the millions of ducks and other water birds in North America. So, what measures should be considered to soften the impact of climate change?

Mitigation measures

Conservation and wetland protection

Prairie pothole wetlands have been vulnerable to drainage and modification since Euro-American settlement began ~150 years ago (Prince 1997). Wetland loss has already affected water bird numbers in major ways due to near-complete drainage of prairie pothole wetlands in Iowa and Minnesota, USA. Drainage has been less severe in North and South Dakota, USA, ranging from 30% to 50% (Dahl 1990), but continues to occur. Some primary mitigative measures would be to substantially increase efforts to protect the remaining wetlands from drainage via easements from government and non-government organisations, strengthening wetland legal protection on farms offered by the US *Farm Bill* and funding restoration of drained wetlands and the natural cover in their watersheds (Doherty *et al.* 2012). The more wetlands and the more diverse types we have, the faster that dependent organisms can bounce back from weather extremes.

Land management

Should wetlands become drier in the future, what measures could be adopted to improve the water regime? Or how can we 'squeeze' more water into wetlands from the precipitation that falls as rain and snow in wetland watersheds? Modifying land cover may be one answer. An unexpected effect of changing land use in wetland landscapes occurred at the St. Denis National Wildlife Area in southern prairie Canada (Saskatchewan) and spurred research into the study of the effects of land management on wetland habitat (van der Kamp *et al.* 2003). After changing land cover from cultivated crops, such as wheat, to smooth brome (a perennial, cool-season grass), formerly wet prairie pothole wetlands dried up (van der Kamp *et al.* 1999, 2003). Although the land use change at the St. Denis National Wildlife Area was intended to improve wildlife habitat, it unintentionally devastated water bird nesting and foraging habitat by removing surface water (van der Kamp *et al.* 1999, 2003).

How did this happen? The smooth brome affected several components of the wetland water budget. First, the tall, stout grass captured blowing snow in the uplands that otherwise would have drifted into wetland basins. Second, the soil macropores from dead roots and invertebrate tunnelling, characteristic of perennial prairie, allowed more soil water to percolate into the soil, during both summer and winter, compared with the annually tilled cropland, where much of the rain and snow ran off into the wetland. Third, run-off from snowmelt and rainstorms was slowed by the dense vegetation and was able to infiltrate soil and reach the groundwater (van der Kamp *et al.* 1999, 2003). Another possibility is that the very productive smooth brome may have depleted soil moisture more than wheat or other similar crops via higher evapotranspiration, water that may otherwise have supplied the wetland with groundwater.

The conundrum is that restoring prairie in wetland landscapes can be a 'double-edged sword'. On the one side, restoring grassland on cropland reduces soil erosion and wetland sedimentation, recharges groundwater, increases soil carbon content and contributes to prairie biodiversity. Although all are environmental improvements, on the other side of the sword, wetlands dried out, reducing their value for use by livestock and water-dependent wild nature, most notably waterfowl (van der Kamp *et al.* 1999, 2003).

Enter land management. Might there be methods to manage prairie pothole wetland landscapes that possess both functional wetlands and biodiverse surrounding prairie? The study in the St. Denis National Wildlife Area (van der Kamp *et al.* 2003) hinted at possibilities that include burning, mowing and grazing of the prairie. Might any or all of these management methods lessen the impact of a warmer climate on water levels in PPR wetlands? Intrigued by the Canadian study, Voldseth *et al.* (2007, 2009) weighed in specifically on this issue by incorporating farming practices into existing digital models (Johnson *et al.* 2005, 2010) to project the effects of a range of future climates on the water budgets of PPR wetlands. The model (WETSIM 3.2) was a single-basin hydrological and wetland vegetation model parameterised and calibrated to a semi-permanent wetland in South Dakota. The model simulated watershed surface processes, watershed ground water, wetland surface processes and wetland vegetation dynamics. Components of a cropping systems model, namely the Environmental Policy Integrated Climate (EPIC) model (Williams 1995) developed in the US, were used in WETSIM to estimate run-off and evapotranspiration under various land management strategies and crop cover types. The land use practices examined in this example were: unmanaged grassland; grassland with moderately heavy grazing; and cultivated crops. Land cover was further subdivided into native grass, brome grass, grazed grass, row crop and small grain because changes in upland cover type and above-ground vegetative structure affect soil structure, soil organic matter and root channel macropores, all of which can have an effect on infiltration, run-off and evapotranspiration demand (Voldseth *et al.* 2007).

The modelling results (Voldseth *et al.* 2007, 2009) match up well with the observations at St. Denis, but also offer suggestions of ways to mitigate for the effects of climate change on wetlands (Fig. 10.6). Over a 41-year simulation period, the model wetland that went dry least often (22%) was grazed grass in the historic period (late 20th century), whereas brome and native grass categories in the warmest future went dry most often (85%). As in St. Denis, the model wetland became drier when surrounded by brome grass than by small grain (one-half vs one-third of the time respectively during the historic period). The wetland in a watershed of cropland was always wetter than unmanaged grass. In contrast, a wetland surrounded by grazed grass was, in all scenarios, wetter than in any other land use category. The climate scenarios caused major reductions in wetland wetness of all land uses, showing again the vulnerability of wetlands to climate when surrounded by diverse land uses. Dryness increased by 150–250% across climate scenarios. Increasing precipitation by 10% reduced dryness by ~10% in all land uses.

Discussion

These two case studies lead to several conclusions relevant to mitigating the negative consequences of climate change on prairie pothole wetlands. Clearly, the utilisation of grassland in these studies has the most potential to ameliorate wetland dryness expected by a warmer climate. Why is this so? Grazing reduces transpiration by reducing the leaf area of grassland vegetation, and therefore more moisture remains in the soil to potentially find its way to wetlands through subsurface drainage. Hoof action by grazing animals also closes some soil macropores and compacts the surface soil, thereby increasing run-off into the wetland. Mowing also reduces leaf area, but is less likely than grazing to close surface connected macropores. Fire during the growing season also reduces leaf area and transpiration and increases run-off. Management of grassland also affects snow input to wetlands. Managed grassland allows more snow to reach the wetland basin and to be trapped by the

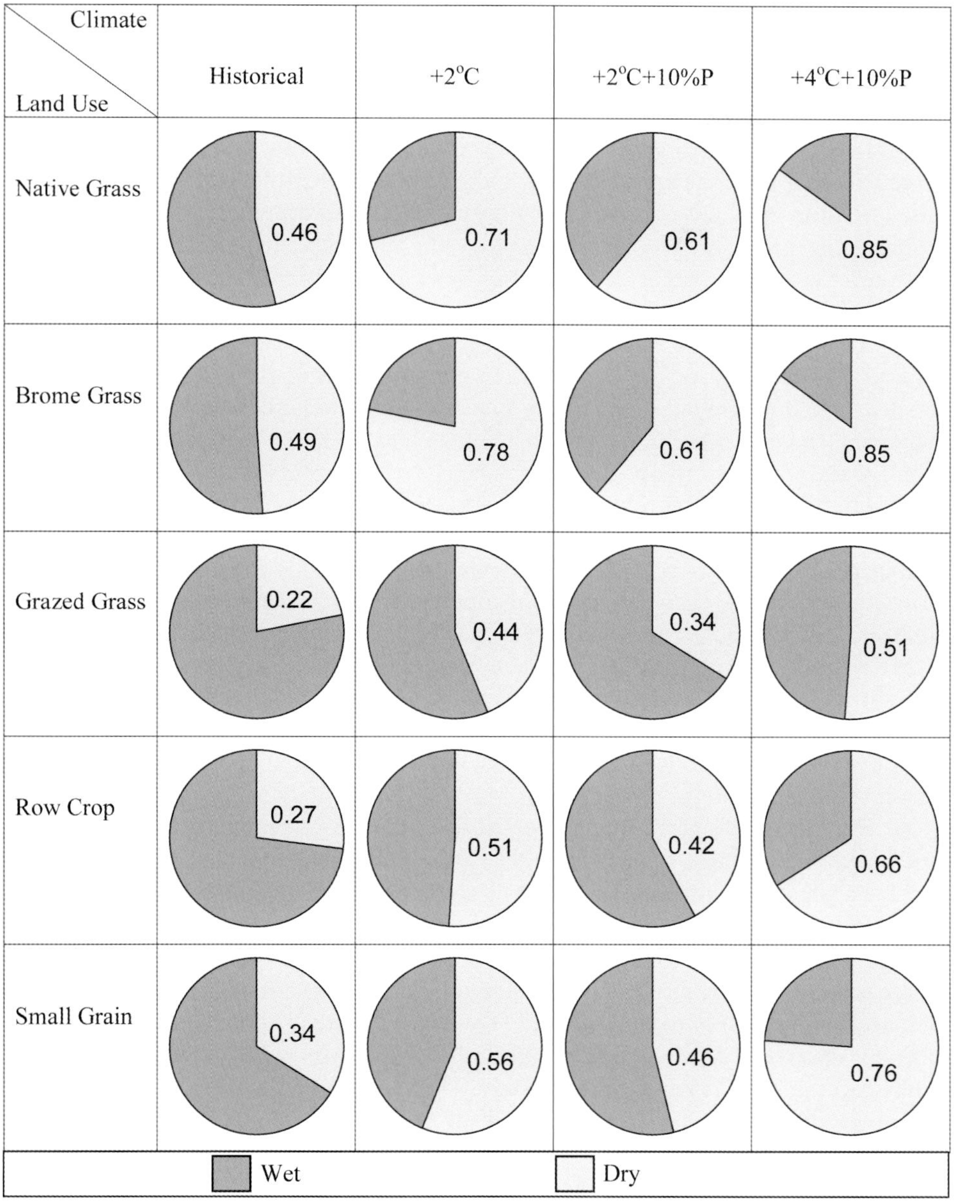

Fig. 10.6. Numerical model calculations of the proportion of years that were simulated (41 years) that the model wetland went dry (light grey) during the growing season. The mean annual values were based on a 41-year simulation. Reproduced with the permission of Springer Nature from Voldseth *et al.* (2009).

taller, stiff vegetation. Unmowed or ungrazed grassland captures snow and keeps it from drifting into the wetland. Snow (snowmelt) is the lifeblood of PPR wetlands, making up ~85% of total surface run-off. These relationships have been discussed in Voldseth *et al.* (2007, 2009).

But here we encounter another conundrum. Thinning grass cover to improve wetland functioning in a warmer climate may also diminish habitat for key wildlife species, where increasing habitat was often the goal of converting cropland to grassland in the first place. Clearly, to have a net positive outcome, a land management program is needed that both optimises management practices, including a mixing of grazing, burning and haying, and enhances wildlife populations across the landscape that depend solely on functional wetlands or on both nearby grassland and wetlands. The success of management programs will require adaptive management over longer periods of time, at the scale of years to decades (Renton *et al.* 2015). Interchanging practices could involve periods of grassland rest, managing parts of the watershed differently and choosing practices that are more efficacious in certain stages of weather cycles (wet to dry). Indications are that wetlands in the currently drier parts of the PPR, more vulnerable to a warmer climate, such as at St. Denis, could benefit the most from managed grassland (Voldseth *et al.* 2007, 2009).

Most PPR wetlands occur on privately owned working farms. Those in grassland watersheds are now grazed and/or hayed, so could not benefit much more in a warmer climate from the practices laid out above (Voldseth *et al.* 2009). Wetlands are managed for wildlife by maintaining tall and dense grassland in their watersheds (Voldseth *et al.* 2009), such as wetlands on wildlife refuges, federal- and state- or provincial-owned wildlife areas (e.g. Wetland Production Areas in the US), wetlands on private lands restricted by no haying or grazing easements and those owned by private conservation agencies (e.g. The Nature Conservancy, Ducks Unlimited). Federal grassland easements adjacent to wetlands could be modified to allow management to maximise wildlife production by balancing the different methods to manage wetlands in watersheds subjected to increasing dryness. Conservation and best management practices for prairie pothole wetlands share similarity with recommendations made more generally for wetlands: prescribed grazing, wetland restoration, upland land cover prescriptions, upland water management and buffer strips (Brinson and Eckles 2011).

Research alerts managers and restorationists to the mixed outcomes of converting prairie pothole landscapes from cropland to grassland. The water budget of the wetland is bound to be negatively affected, especially in dry regions, even though many other changes will be positive and welcome. The shift from farmland to grassland is a major environmental change. An abundance of ecosystem services accompanies this shift, including sediment capture, groundwater recharge and discharge, stock water processing and purification, habitat and forage for plants and animals, including livestock, and climate protection and carbon storage. These services occur at multiple scales, from global to local (Johnson 2019).

Inland wetlands in other parts of the world share their vulnerability to climate change with prairie pothole wetlands. Projections of a drier future climate in many wetland regions focus attention on ways to mitigate for the effects of drying and reduced wetland functioning. For example, reduced ecosystem services are expected in the Murray–Darling Basin in Australia, the result of less water available to support wetlands and dryland rivers (Lavorel *et al.* 2014). Some wetlands may respond to climate change by adapting, persisting or transforming into different ecological states. Landscape-scale approaches could help us understand the biophysical processes underpinning ecosystem services needed by managers and decision makers in an area that covers 14% of the Australian landmass (Lavorel *et al.* 2014). Management approaches that could be used include changes in upland crops to reduce water demand, alterations to grazing practices, accounting for possible novel ecosystem services that may develop from systems transformed positively by climate change, reductions in management costs to maintain high-value wetlands by reconnecting them to river flow and recognising that some wetlands can collectively provide ecosystem services even though they may exist in a moderately degraded state (Colloff *et al.* 2016).

Conclusions

1. The Earth's wetland ecosystems are especially vulnerable to climate change. The fate of coastal wetlands is closely tied to increasing atmospheric temperature causing the melting of glaciers and thermal expansion, leading to rising ocean levels. In contrast, inland wetlands in terrestrial landscapes are especially sensitive to components of the global hydrological cycle, particularly changes in the amount and periodicity of precipitation and rates of evapotranspiration.
2. These differences argue for different approaches to mitigation as a means to lessen the severity of climate change. The restoration of both historically impacted ecosystems and those yet to be affected is critical to maintaining essential ecosystem services. This challenge is especially daunting for coastal wetlands given that all of them around the globe will be affected by sea level rise. The restoration of depressional wetlands requires tailored landscape and regional approaches given that the expected changes to the current climate will differ greatly geographically.
3. Research in the PPR of central North America has identified strategic land management practices that can potentially mitigate for the negative consequences of a drier future climate. An experiment in St. Denis, Saskatchewan, Canada, quite unexpectedly revealed the close relationship between land cover (crops vs grassland) in wetland watersheds and water levels in prairie pothole wetlands. Through the use of management practices such as fire, grazing and haying, more surface water can be delivered to the wetland basin. Proper intermixing of these practices over years has the potential to maintain vital ecosystem services, especially the provisioning of wildlife, in both the wetlands and surrounding grassland.
4. A multiple-pronged approach could be considered to mitigate for the effects of climate change associated with depressional wetlands. This approach could include legal protection of remaining wetlands from drainage and damage, restoration of drained wetlands and the adoption of land management practices that offset the negative consequences of a drier climate. Improving ecosystem functions in existing wetlands and restoring ecosystem function in degraded wetlands can promote system resilience, enabling wetland services to more quickly rebound and recover from losses encountered during extreme hydrological periods.

References

Aburto-Oropeza O, Ezcurra E, Danemann G, Valdez V, Murray J, Sala E (2008) Mangroves in the Gulf of California increase fishery yields. *Proceedings of the National Academy of Sciences of the United States of America* **105**, 10456–10459. doi:10.1073/pnas.0804601105

Allen C, Gonzales R, Parrott L (2020) Modelling the contribution of ephemeral wetlands to landscape connectivity. *Ecological Modelling* **419**, 108944. doi:10.1016/j.ecolmodel.2020.108944

Alongi DM (2008) Mangrove forests: resilience, protection from tsunamis, and responses to global climate change. *Estuarine, Coastal and Shelf Science* **76**, 1–13. doi:10.1016/j.ecss.2007.08.024

Barbier EB (2019) The value of coastal wetland ecosystem services. In *Coastal Wetlands: An Integrated Ecosystem Approach.* (Eds G Perillo, E Wolanski, DR Cahoon and C Hopkinson) pp. 947–964. Elsevier Publishing, Amsterdam.

Barbier EB, Strand I (1998) Valuing mangrove–fishery linkages: a case study of Campeche, Mexico. *Environmental and Resource Economics* **12**, 151–166. doi:10.1023/A:1008248003520

Barbier EB, Hacker SD, Kennedy C, Koch EW, Stier AC, Silliman BR (2011) The value of estuarine and coastal ecosystem services. *Ecological Monographs* **81**, 169–193. doi:10.1890/10-1510.1

Batzer DP, Baldwin AH (Eds) (2012) *Wetland Habitats of North America: Ecology and Conservation concerns*. University of California Press, Berkeley.

Bedford BL (1999) Cumulative effects on wetland landscapes: links to wetland restoration in the United States and southern Canada. *Wetlands* **19**, 775–788. doi:10.1007/BF03161784

Boorman LA (1999) Salt marshes – present functioning and future change. *Mangroves and Salt Marshes* **3**, 227–241. doi:10.1023/A:1009998812838

Brinson M (1993) 'A hydrogeomorphic classification for wetlands.' Technical Report WRP-DE-4, United States Army Engineer Waterways Experiment Station, Vicksburg.

Brinson M (2006) Consequences for wetlands of a changing global environment. In *Ecology of Freshwater and Estuarine Wetlands*. (Eds DP Batzer and RR Sharitz) pp. 436–461. University of California Press, Berkeley.

Brinson M, Eckles SD (2011) United States Department of Agriculture conservation program and practice effects on wetland ecosystem services: a synthesis. *Ecological Applications* **21**(Supplement), S116–S127. doi:10.1890/09-0627.1

Burkett V, Kusler J (2000) Climate change: potential impacts and interactions in wetlands of the United States. *Journal of the American Water Resources Association* **36**, 313–320. doi:10.1111/j.1752-1688.2000.tb04270.x

Cahoon DR, Lynch JC, Roman CT, Schmit JP, Skidds DE (2019) Evaluating the relationship among wetland vertical development, elevation capital, sea-level rise, and tidal marsh sustainability. *Estuaries and Coasts* **42**, 1–15. doi:10.1007/s12237-018-0448-x

Colloff MJ, Lavorel S, Wise RM, Dunlop M, Overton IC, Williams KJ (2016) Adaptation services of floodplains and wetlands under transformational climate change. *Ecological Applications* **26**, 1003–1017. doi:10.1890/15-0848

Corn PS (2005) Climate change and amphibians. *Animal Biodiversity and Conservation* **1**, 59–67.

Costanza R, d'Arge R, de Groot RS, Farber S, Grasso M, Hannon B, Limburg K, Naeem S, O'Neill RV, Paruel J, *et al.* (1997) The value of the world's ecosystem service and natural capital. *Nature* **387**, 253–260. doi:10.1038/387253a0

Costanza R, Groot R, Sutton P, der Ploeg S, Anderson SA, Kubiszewski I, Farber S, Tuner RK (2014) Changes in the global value of ecosystem services. *Global Environmental Change* **26**, 152–158. doi:10.1016/j.gloenvcha.2014.04.002

Craft CJ, Clough J, Ehman J, Joye S, Park D, Pennings S, Guo H, Machmuller M (2009) Forecasting the effects of accelerated sea level rise on tidal marsh ecosystem services. *Frontiers in Ecology and the Environment* **7**, 73–78. doi:10.1890/070219

Dahl TE (1990) 'Wetland losses in the United States 1780s to 1980s.' United States Department of the Interior, Fish and Wildlife Service, Washington, DC.

de Groot R, Brander L, van der Ploeg S, Costanza R, Bernard F, Braat L, Christie M, Crossman N, Ghermandi A, Hein L, *et al.* (2012) Global estimates of the value of ecosystems and their services. *Ecosystem Services* **1**, 50–61. doi:10.1016/j.ecoser.2012.07.005

De Steven D, Lowrance R (2011) Agricultural conservation practices and wetland ecosystem services in the wetland-rich Piedmont-Coastal Plain region. *Ecological Applications* **21**(Supplement), S3–S17. doi:10.1890/09-0231.1

Detenbeck NE, Elonen CM, Taylor DL, Cotter AM, Puglisi FA, Sanville WD (2002) Effects of agricultural activities and best management practices on water quality of seasonal prairie pothole wetlands. *Wetlands Ecology and Management* **10**, 335–354. doi:10.1023/A:1020397103165

Doherty KE, Ryba AJ, Stemler CI, Niemuth ND, Meeks WA (2012) Conservation planning in an era of change: state of the U.S. Prairie Pothole Region. *Wildlife Society Bulletin* **37**, 546–563.

Downing JA, Cole JJ, Middelburg JJ, Striegl RG, Duarte CM, Kortelainen P, Prairie YT, Laube KA (2008) Sediment organic carbon burial in agriculturally eutrophic impoundments over the last century. *Global Biogeochemical Cycles* **22**, GB1018. doi:10.1029/2006GB002854

Duarte CM, Losada IJ, Hendriks IE, Mazarrasa I, Marba N (2013) The role of coastal plant communities for climate change mitigation and adaptation. *Nature Climate Change* **3**, 961–968. doi:10.1038/nclimate1970

Enwright NM, Griffith KT, Osland MJ (2016) Barriers to and opportunities for landward migration of coastal wetlands with sea-level rise. *Frontiers in Ecology and the Environment* **14**, 307–316. doi:10.1002/fee.1282

Fennessy S, Craft C (2011) Agricultural conservation practices increase wetland ecosystem services in the Glaciated Interior Plains. *Ecological Applications* **21**(Supplement), S49–S64. doi:10.1890/09-0269.1

Ford MA, Cahoon DR, Lynch JC (1999) Restoring marsh elevation in a rapidly subsiding salt marsh by thin layer deposition of dredged material. *Ecological Engineering* **12**, 189–205. doi:10.1016/S0925-8574(98)00061-5

Galatowitsch S (2012) Northern Great Plains wetlands. In *Wetland Habitats of North America. Ecology and Conservation Concerns.* (Eds D Batzer and A Baldwin) pp. 283–298. University of California Press, Berkeley.

Gedan KB, Kirwan ML, Wolanski E, Barbier EB, Silliman BR (2011) The present and future role of coastal wetland vegetation in protecting shorelines: answering recent challenges to the paradigm. *Climatic Change* **106**, 7–29. doi:10.1007/s10584-010-0003-7

Gleason RA, Tangen BA, Laubhan MK, Kermes KE, Euliss NH (2008) 'Estimating water storage capacity of existing and potentially restorable wetland depressions in a sub basin of the Red River of the North.' U.S. Geological Survey (USGS) Open-File Report no. 2007–1159, USGS, Reston.

Intergovernmental Panel on Climate Change (2007) *Climate Change 2007: The Physical Science Basis*. Cambridge University Press, Cambridge.

Johnson WC (2019) Ecosystem services provided by prairie wetlands in northern rangelands. *Rangelands* **41**, 44–48. doi:10.1016/j.rala.2018.12.003

Johnson WC, Poiani KA (2016) Climate change effects on prairie pothole wetlands: findings from a 25-year numerical modeling project. *Wetlands* **36** (Supplement 2), S273–S286. doi:10.1007/s13157-016-0790-3

Johnson WC, Boettcher SE, Poiani KA, Guntenspergen GR (2004) Influence of weather extremes on the water levels of glaciated prairie wetlands. *Wetlands* **24**, 385–398. doi:10.1672/0277-5212(2004)024[0385:IOWEOT]2.0.CO;2

Johnson WC, Millett BV, Gilmanov T, Voldseth RA, Guntenspergen GR, Naugle DE (2005) Vulnerability of northern prairie wetlands to climate change. *Bioscience* **55**, 863–872. doi:10.1641/0006-3568(2005)055[0863:VONPWT]2.0.CO;2

Johnson WC, Werner B, Guntenspergen GR, Voldseth RA, Millett B, Naugle D, Tulbure M, Carroll RWH, Tracy J, Olawsky C (2010) Prairie wetland complexes as landscape functional units in a changing climate. *Bioscience* **60**, 128–140. doi:10.1525/bio.2010.60.2.7

Kantrud HA, Krapu GL, Swanson GA (1989) 'Prairie basin wetlands of the Dakotas: a community profile.' Biological Report 85 (7.28), United States Department of the Interior, Fish and Wildlife Service, Washington, DC.

Keddy PA, Fraser LH, Solomeshch AI, Junk WJ, Campbell DR, Arroyo MTK, Alho CJR (2009) Wet and wonderful: the world's largest wetlands are conservation priorities. *Bioscience* **59**, 39–51. doi:10.1525/bio.2009.59.1.8

Kelleway JJ, Cavanaugh K, Rogers K, Feller I, Doughty C, Saintilan N (2017) Review of the ecosystem service implications of mangrove encroachment into salt marshes. *Global Change Biology* **23**, 3967–3983. doi:10.1111/gcb.13727

Kirwan ML, Gedan KB (2019) Sea-level driven land conversion and the formation of ghost forests. *Nature Climate Change* **9**, 450–457. doi:10.1038/s41558-019-0488-7

Kirwan ML, Megonigal JP (2013) Tidal wetland stability in the face of human activity and sea level rise. *Nature* **504**, 53–60. doi:10.1038/nature12856

Lavorel S, Colloff MJ, McIntyre S, Doherty MD, Murphy HT, Metcalfe DJ, Dunlop M, Williams RJ, Wise RM, Williams KJ (2014) Ecological mechanisms underpinning climate adaptation services. *Global Change Biology* **21**, 12–31. doi:10.1111/gcb.12689

Lee SY, Primavera JH, Dahdouh-Guebas F, McKee K, Bosire JO, Cannicci S, Diele K, Fromard F, Koedam N, Marchand C, *et al.* (2014) Ecological role and services of tropical mangrove ecosystems: a reassessment. *Global Ecology and Biogeography* **23**, 726–743. doi:10.1111/geb.12155

Leitch JA (1989) Politicoeconomic overview of Prairie Potholes. In *Northern Prairie Wetlands.* (Ed. AG van der Valk) pp. 2–15. Iowa State University Press, Ames.

Liu G, Schwartz FW (2011) An integrated observational and model-based analysis of the hydrologic response of prairie pothole systems to variability in climate. *Water Resources Research* **47**, W02504. doi:10.1029/2010WR009084

Lovelock CE, Cahoon DR, Friess DA, Guntenspergen GR, Krauss KW, Reef R, Triet T (2015) The vulnerability of Indo-Pacific mangrove forests to sea-level rise. *Nature* **526**, 559–563. doi:10.1038/nature15538

McIntyre N, Liu G, Gorzo J, Wright C, Guntenspergen G, Schwartz F (2019) Simulating the effects of climate variability on waterbodies and wetland-dependent birds in the Prairie Pothole Region. *Ecosphere* **10**, e02711. doi:10.1002/ecs2.2711

McKenna OP, Mushet DM, Scherff EJ, McLean KI, Mills CT (2018) The Pothole Hydrology-Linked Systems Simulator (PHyLiSS) – Development and application of a systems model for prairie-pothole wetlands: United States Geological Survey (USGS) Report 2018–1165, USGS, Reston. doi:10.3133/ofr20181165.

Mcleod E, Chmura GL, Bouillon S, Dalm R, Bjork M, Duarte CM, Lovelock CE, Schlesinger WH, Silliman BR (2011) A blueprint for blue carbon: toward an improved understanding of the role of vegetated coastal habitats in sequestering CO_2. *Frontiers in Ecology and the Environment* **9**, 552–560. doi:10.1890/110004

Millennium Ecosystem Assessment (2005) *Ecosystems and Human Well-being: Current State and Trends.* Island Press, Washington, DC.

Mitsch WJ, Gosselink JG (2015) *Wetlands.* 5th edn. John Wiley and Sons, New York.

Mushet D (Ed.) (2016) Midcontinent Prairie-Pothole wetlands and climate change. *Wetlands* **36**(Supplement 2), S223–S459.

Mushet DM, Goldhaber MB, Mills CT, McLean KI, Aparicio VM, McCleskey B, Holloway JM, Stockwell CA (2015) 'Chemical and biotic characteristics of prairie lakes and large wetlands in south-central North Dakota – effects of a changing climate.' U.S. Geological Survey (USGS) Scientific Investigations Report no. 2015–5126. USGS, Reston. doi:10.3133/sir20155126

Neeley R, Baker J (1989) Nitrogen and phosphorous dynamics and fate in agricultural runoff. In *Northern Prairie Wetlands.* (Ed. A van der Valk) pp. 92–131. Iowa State University Press, Ames.

Peteet DM, Nichols J, Kenna T, Chang C, Browne J, Reza M, Kovari S, Liberman L, Stern-Protz S (2018) Sediment starvation destroys New York City marshes' resistance to sea level rise. *Proceedings of the National Academy of Sciences of the United States of America* **115**, 10281–10286. doi:10.1073/pnas.1715392115

Phillips RL, Beeri O (2008) The role of hydropedologic vegetation zones in greenhouse gas emissions for agricultural wetland landscapes. *Catena* **72**, 386–394. doi:10.1016/j.catena.2007.07.007

Prairie Pothole Joint Venture (1998) *The Prairie Pothole Region: Land of Extremes...Landscape of Change.* United States Fish and Wildlife Service, Denver.

Prince H (1997) *Wetlands of the American West. A Historical Geography of Changing Attitudes.* University of Chicago Press, Chicago.

Renton DA, Mushet DM, DeKeyser ES (2015) Climate change and prairie pothole wetlands: mitigating water-level and hydroperiod effects through upland management. United States Department of the Interior, U.S. Geological Survey (USGS) Scientific Investigations Report no. 2015–5004, USGS, Reston. doi:10.3133/sir20155004

Rogers K, Kelleway JJ, Saintilan N, Megonigal JP, Adams JB, Holmquist JR, Lu M, Schile-Beers L, Zawadzki A, Mazumder D, *et al.* (2019) Wetland carbon storage controlled by millennial-scale variation in relative sea-level rise. *Nature* **567**, 91–95. doi:10.1038/s41586-019-0951-7

Saintilan N, Khan NS, Ashe E, Kelleway JJ, Rogers K, Woodroffe CD, Horton BP (2020) Thresholds of mangrove survival under rapid sea level rise. *Science* **368**, 1118–1121. doi:10.1126/science.aba2656

Sauchyn D, Diaz H, Kulshreshtha S (2010) 'The new normal. The Canadian prairies in a changing climate.' Canadian Plains Research Center, University of Regina, Saskatchewan.

Scott DA, Jones TA (1995) Classification and inventory of wetlands: a global overview. *Vegetatio* **118**, 3–16. doi:10.1007/BF00045186

Shepard CC, Crain CM, Beck MW (2011) The protective role of coastal marshes: a systematic review and meta-analysis. *PLoS One* **6**, e27374. doi:10.1371/journal.pone.0027374

Sievers MR, Hale KM, Parris KM, Swearer SE (2018) Impacts of human induced environmental change in wetlands on aquatic animals. *Biological Reviews of the Cambridge Philosophical Society* **93**, 529–554. doi:10.1111/brv.12358

Simpson LT, Osborne TZ, Duckett LJ, Feller IC (2017) Carbon storage along a climate induced coastal wetland gradient. *Wetlands* **37**, 1023–1035. doi:10.1007/s13157-017-0937-x

Smith RD, Ammann A, Bartoldus C, Brinson MM (1995) 'An approach for assessing wetland functions using hydrogeomorphic classification, reference wetlands, and functional indices.' United States Army Engineer Waterways Experiment Station, Technical Report WRP-DE-9. Vicksburg.

Stewart RE, Kantrud HE (1971) Classification of natural ponds and lakes in the Glaciated Prairie Region.' United States Bureau of Sport Fisheries and Wildlife. Resource Publication 92, Bureau of Sport Fisheries and Wildlife, Washington.

Swanson GA, Duebbert HF (1989) Wetland habitats of waterfowl in the prairie pothole region. In *Northern Prairie Wetlands.* (Ed. AG van der Valk) pp. 228–267. Iowa State University Press, Ames.

Tangen BA, Finocchiaro R, Gleason RA (2015) Effects of land use on greenhouse gas fluxes and soil properties of wetland catchments in the Prairie Pothole Region of North America. *The Science of the Total Environment* **533**, 391–409. doi:10.1016/j.scitotenv.2015.06.148

Temmerman S, Meire P, Bouma TJ, Herman PMJ, Ysebaert T, De Vriend HJ (2013) Ecosystem-based coastal defense in the face of global change. *Nature* **504**, 79–83. doi:10.1038/nature12859

Trettin CC, Kolka RK, Marsh AS, Bansal S, Lilleskov EA, Megonigal P, Stelk MJ, Lockaby G, D'Amore DV, MacKenzie RA, *et al.* (2020) Wetland and hydric soils. In *Forest and Rangeland Soils of the United States Under Changing Conditions: A Comprehensive Science Synthesis.* (Eds R Pouyat, DS Page-Dumroese, T Patel-Weynand and LH Geiser) pp. 99–125. Springer, Cham.

Valiela I, Cole ML, Mcclelland J, Hauxwell J, Cebrian J, Joye SB (2002) Role of salt marshes as part of coastal landscapes. In *Concepts and Controversies in Tidal Marsh Ecology.* (Eds M Weinstein and DA Kreeger) pp. 22–36. Springer, Dordrecht.

van der Kamp G, Hayashi M (2009) Groundwater–wetland ecosystem interaction in the semiarid glaciated plains of North America. *Hydrogeology Journal* **17**, 203–214. doi:10.1007/s10040-008-0367-1

van der Kamp G, Stolte WJ, Clark RG (1999) Drying out of small prairie wetlands after conversion of their catchments from cultivation to permanent brome grass. *Hydrological Sciences Journal* **44**, 387–397. doi:10.1080/02626669909492234

van der Kamp G, Hayashi M, Gallen D (2003) Comparing the hydrology of grassed and cultivated catchments in the semi-arid Canadian prairies. *Hydrological Processes* **17**, 559–575. doi:10.1002/hyp.1157

van der Valk A (Ed.) (1989) *Northern Prairie Wetlands.* Iowa State University Press, Ames.

van der Valk A, Mushet D (2016) Interannual water-level fluctuations and the vegetation of prairie potholes: potential impacts of climate change. *Wetlands* **36**(Supplement 2), 397–406. doi:10.1007/s13157-016-0850-8

Vasić F, Paul C, Strauss V, Helming K (2020) Ecosystem services of kettle holes in agricultural landscapes. *Agronomy (Basel)* **10**, 1326. doi:10.3390/agronomy10091326

Voldseth RA, Johnson WC, Gilmanov T, Guntenspergen GR, Millett BV (2007) Model estimates of land-use effects on water levels of northern prairie wetlands. *Ecological Applications* **17**, 527–540. doi:10.1890/05-1195

Voldseth RA, Johnson WC, Guntenspergen GR, Gilmanov T, Millett B (2009) Adaptation of farming practices could buffer effects of climate change on northern prairie wetlands. *Wetlands* **29**, 635–647. doi:10.1672/07-241.1

Weller MW (1981) *Freshwater Marshes. Ecology and Wildlife Management*. University of Minnesota Press, Minneapolis.

Weston N (2014) Declining sediments and rising seas: an unfortunate convergence for tidal wetlands. *Estuaries and Coasts* **37**, 1–23. doi:10.1007/s12237-013-9654-8

Williams JR (1995) The EPIC model. In *Computer Model of Watershed Hydrology*. (Ed. V. Singh) pp. 909–1000. Water Resources Publication, Littleton.

Winter TC (2000) The vulnerability of wetlands to climate change: a hydrologic landscape perspective. *Journal of the American Water Resources Association* **36**, 305–311. doi:10.1111/j.1752-1688.2000.tb04269.x

Withey P, van Kooten GC (2014) Wetlands retention and optimal management of waterfowl habitat under climate change. *Journal of Agricultural and Resource Economics* **9**, 1–18.

Zedler J (2003) Wetlands at your service. *Frontiers in Ecology and the Environment* **1**, 65–72. doi:10.1890/1540-9295(2003)001[0065:WAYSRI]2.0.CO;2

Zedler J, Kircher S (2005) Wetland resources: status, trends, ecosystem services, and restorability. *Annual Review of Environment and Resources* **30**, 39–74. doi:10.1146/annurev.energy.30.050504.144248

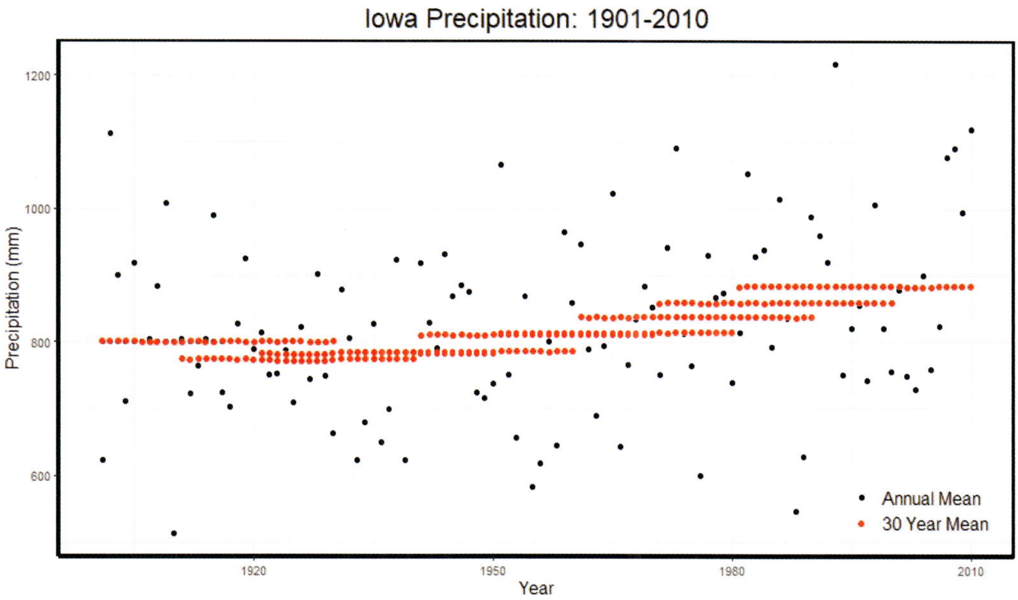

Plate 1. Changes in 30-year normal precipitation from 1900 to 2010 for Iowa.

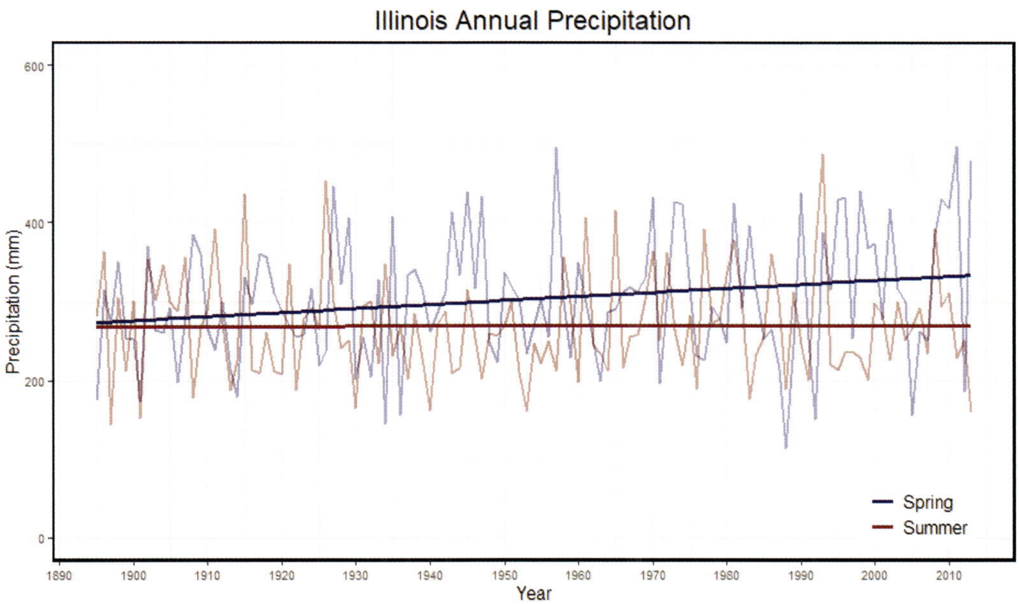

Plate 2. Spring (April–June) and summer (July–September) precipitation for Illinois since 1985.

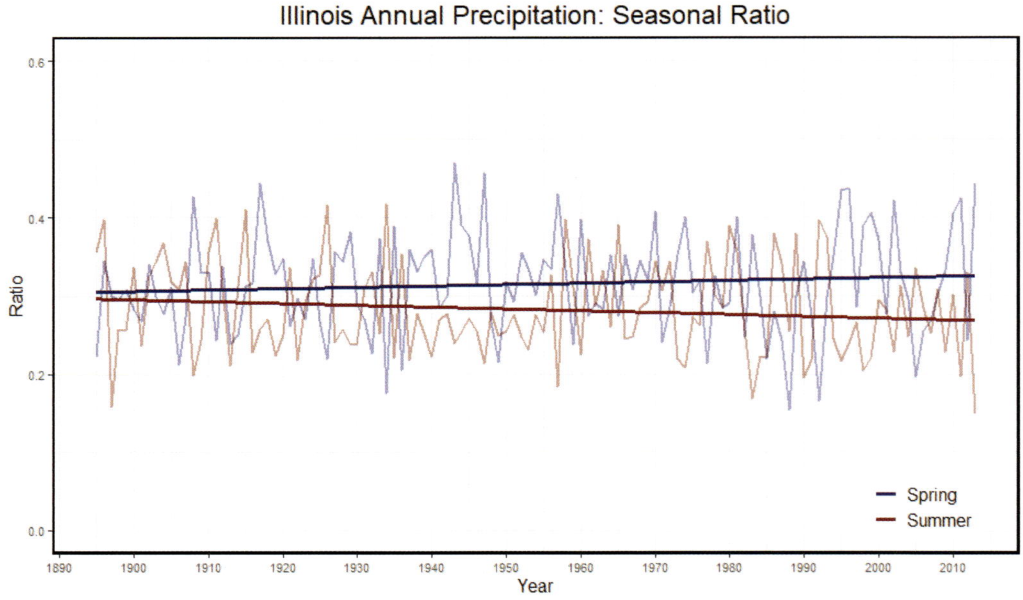

Plate 3. Fraction of the total annual precipitation in the spring (April–June) and summer (July–September) period for Illinois.

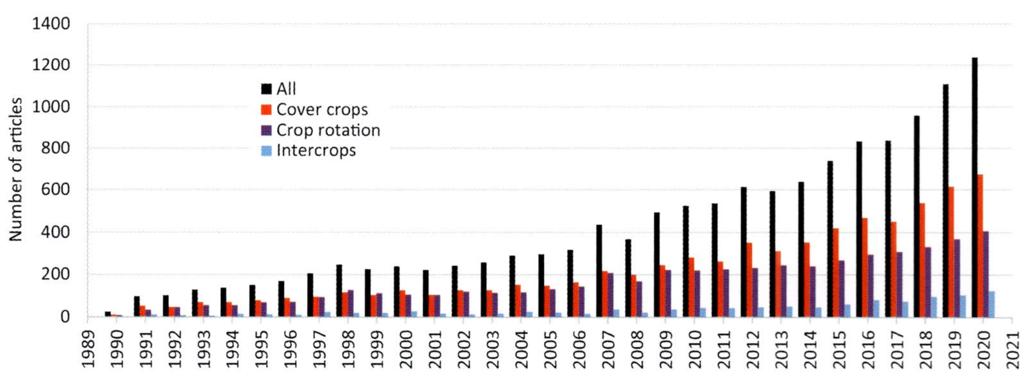

Plate 4. Number of peer-reviewed journal articles published, by year, on crop diversity and soil water. Query results from Clarivate Analytics' Web of Science Core Collection accessed on 10 February 2021. The search terms included 'crop diversity', 'cover crop', 'crop rotation' or 'intercrop' in combination with 'soil water', 'soil moisture', 'soil hydrology' or 'soil hydraulic' in article titles, abstracts and keywords. The sum of articles on cover crops, crop rotation and intercrops is greater than the total because some articles had multiple terms present (e.g. 'intercropping of a winter rye cover crop').

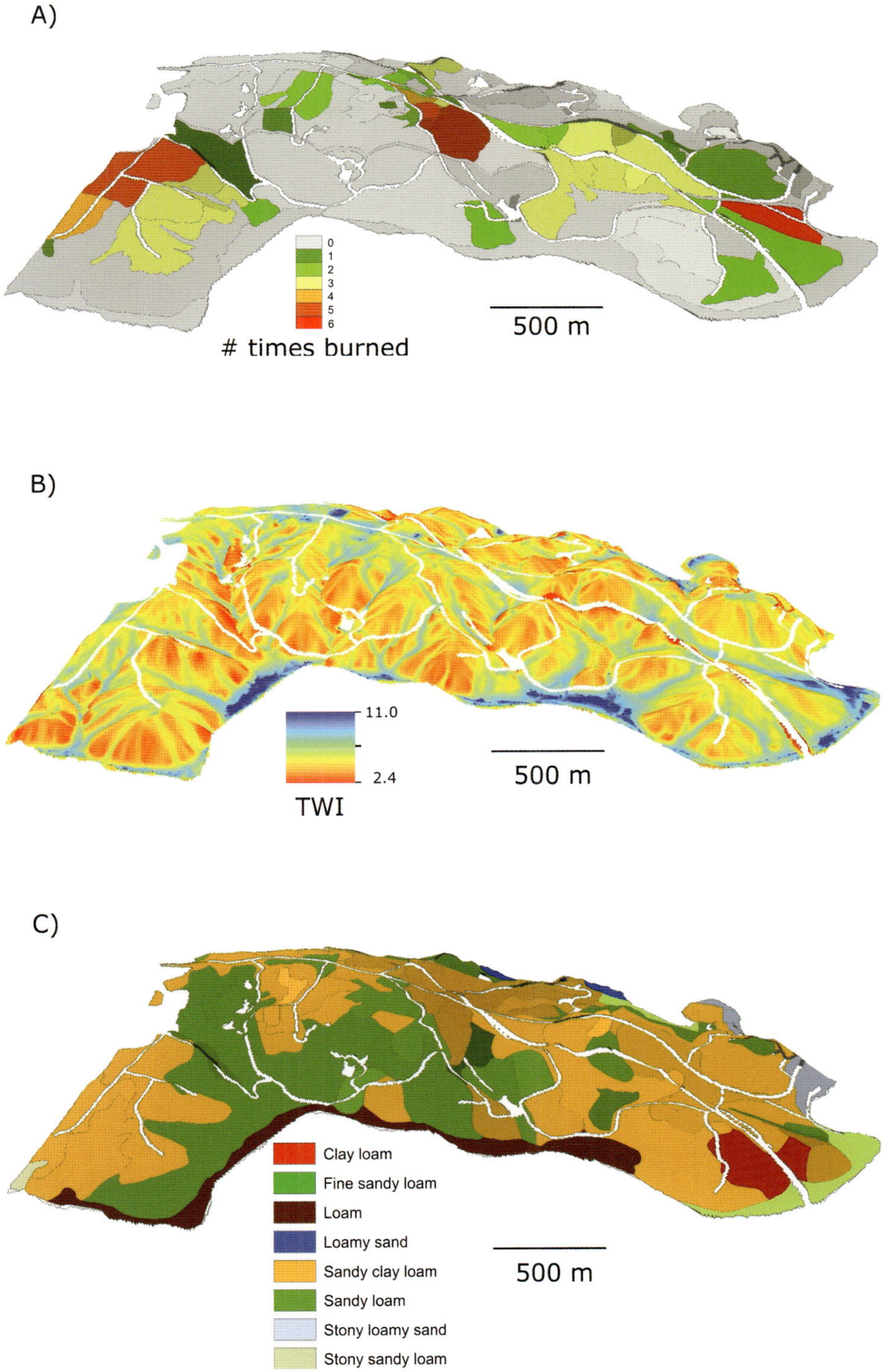

Plate 5. Whitehall Forest landscape showing (a) the number of times it has burned between 2011 and 2019, (b) the topographic wetness index (TWI) and (c) the soil texture class of surface soils from the Soil Survey Geographic Database (SSURGO; Soil Survey Staff 2019).

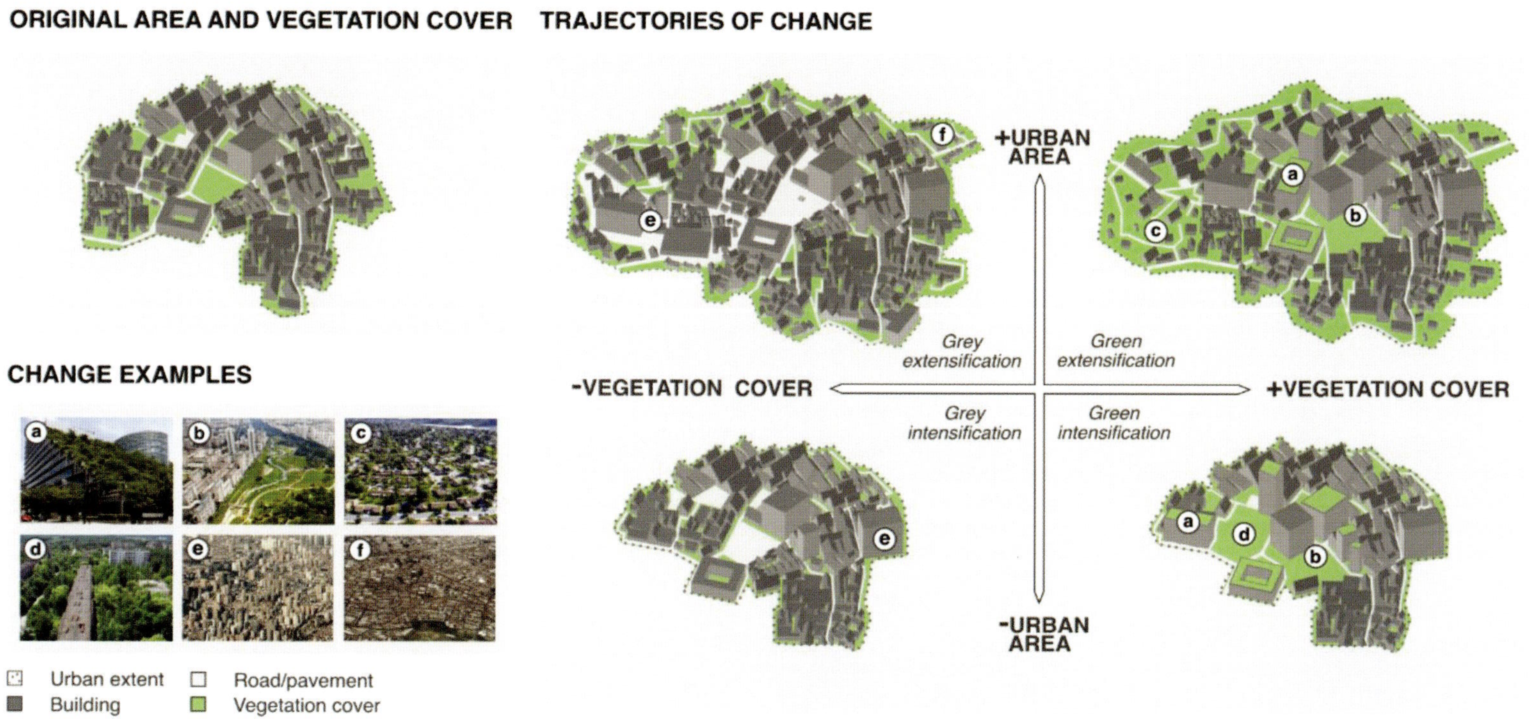

Plate 6. Urban areas can show different trajectories of change in green cover and total area over time, from those that: decline in the proportion of vegetation cover and simultaneously expand in area (grey extensification); decline in the proportion of vegetation cover and become smaller (grey intensification); increase in vegetation cover and expand in extent (green extensification); or increase in vegetation cover and become smaller (green intensification). These macro-scale trajectories are the net result of varied location mechanisms, including (a) intentional addition of vegetation on top of buildings, (b) intentional removal of buildings to create open parks, (c) suburban sprawl creating low-density urban areas with high vegetation cover, (d) the growth of spontaneous vegetation after land abandonment, (e) the loss of vegetation following densification in urban cores and (f) suburban sprawl to create high-density urban areas with low vegetation cover. Reproduced with the permission of MDPI from Richards and Belcher (2019). Image sources are given in the Acknowledgement section of Richards and Belcher (2019).

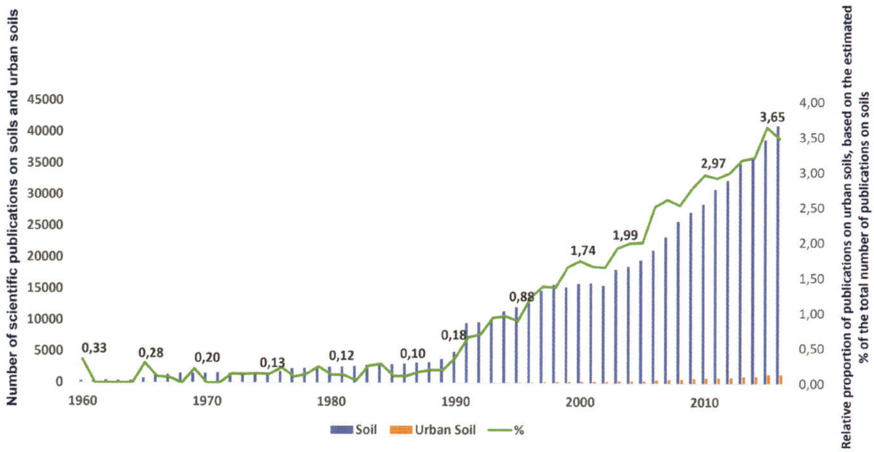

Plate 7. Evolution in the number of scientific publications on soils and urban soils published annually in the international scientific literature for the period 1960–2016. The green curve depicts the change in the relative proportion of publications on urban soils, based on the estimated percentage of the total number of publications on soils. This figure highlights the weak consideration of the biology of urban soils in the literature. Reprinted with permission from Springer Nature from Guilland *et al.* (2018).

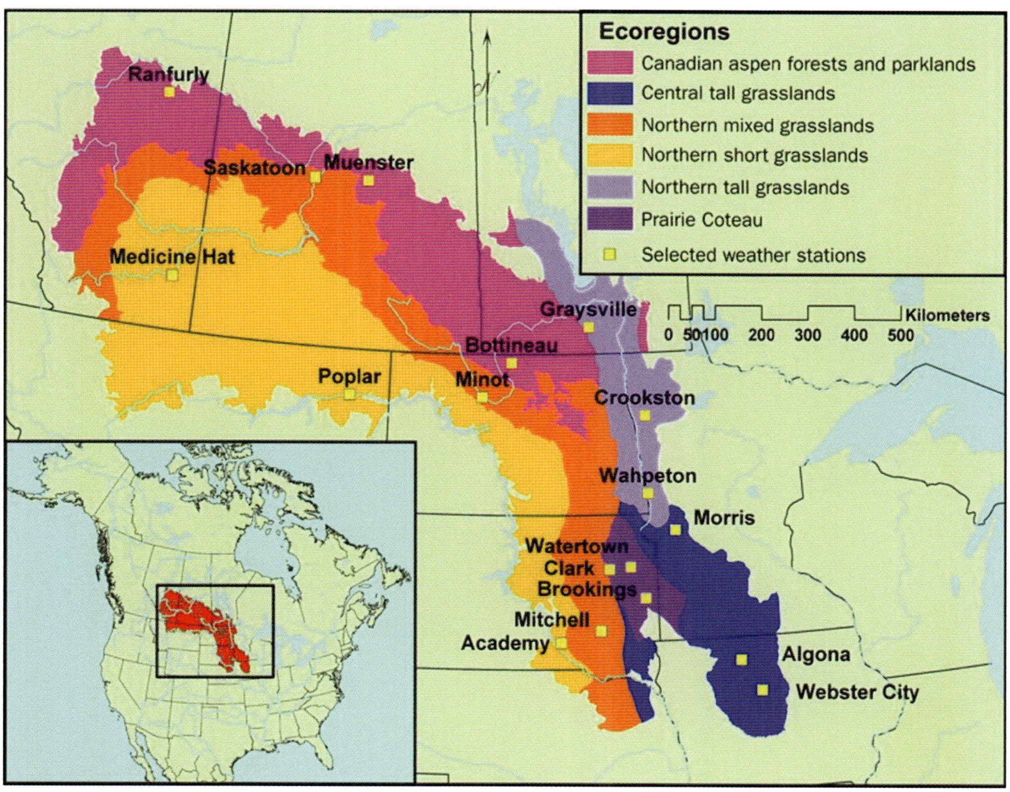

Plate 8. The Prairie Pothole Region of central North America (inset) and six ecoregion subdivisions. Yellow symbols represent weather stations (three per ecoregion) used in model simulations. Reproduced with the permission of Oxford University Press from Johnson *et al.* (2005).

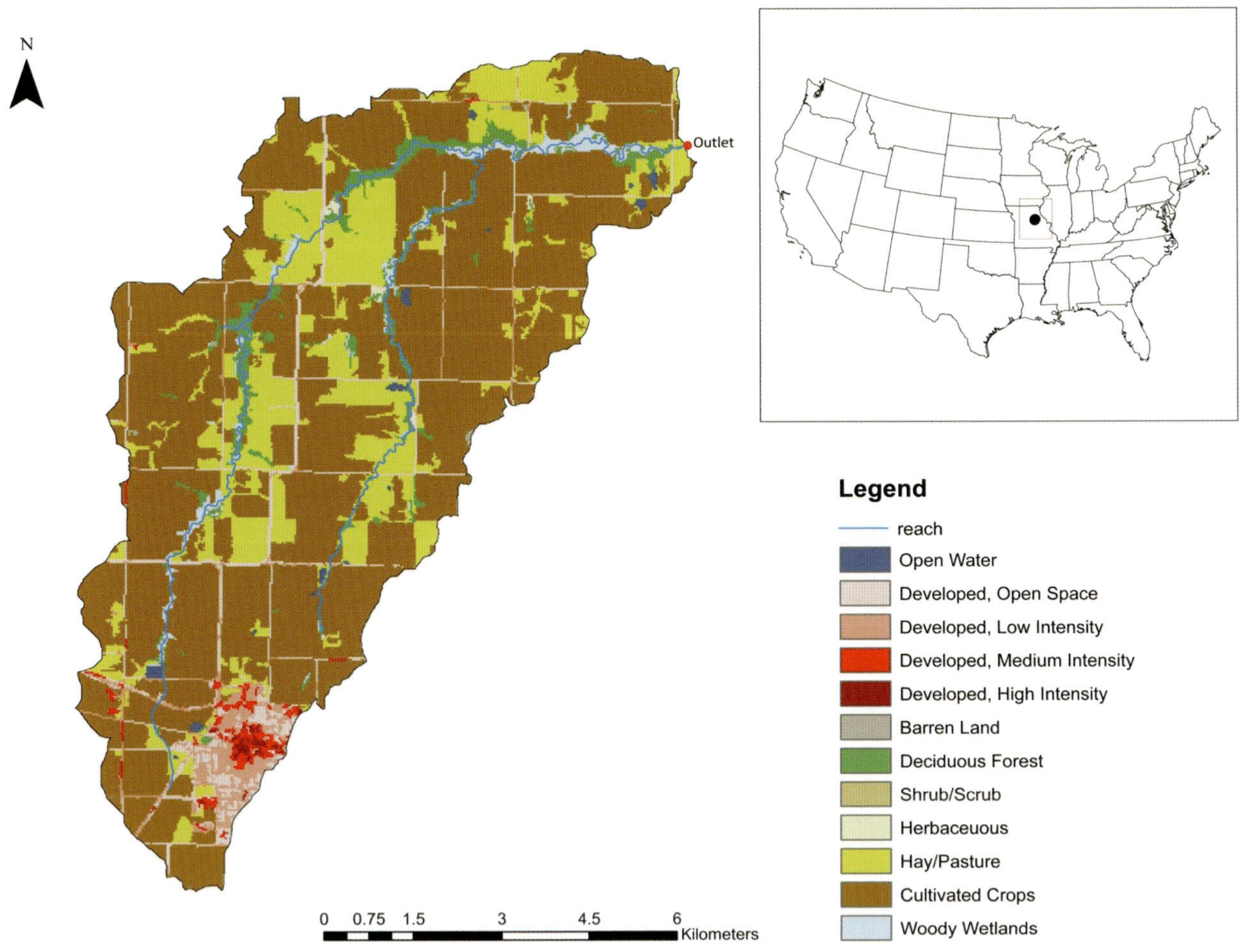

Plate 9. Location and land use (US Department of Agriculture National Land Cover Dataset 2016; https://www.mrlc.gov) of the Goodwater Creek Experimental Watershed. Inset, map showing the location of the watershed in Missouri, USA (adapted from Gautam *et al.* 2021).

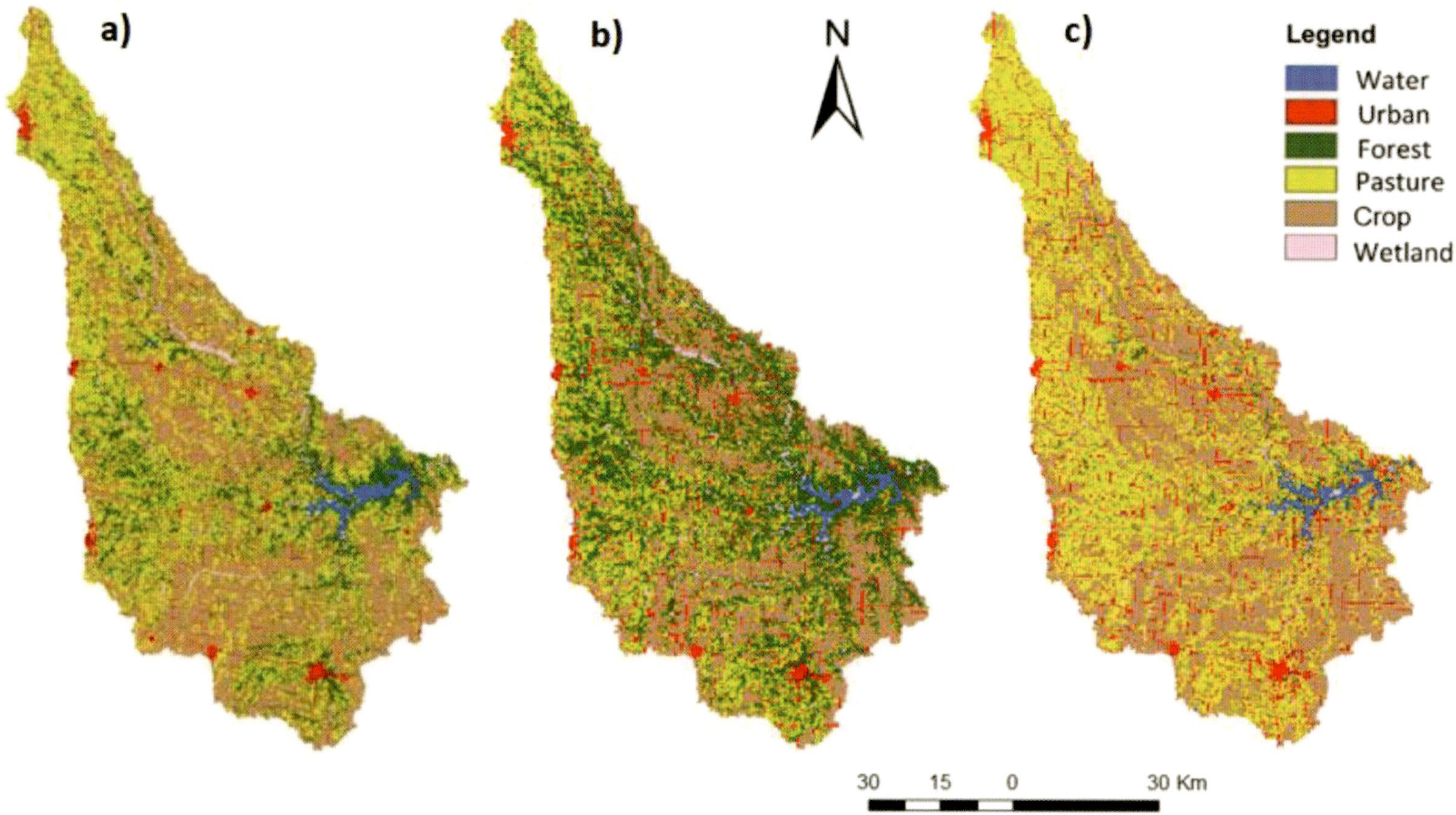

Plate 10. Land use in the Salt River Basin under (a) 2011 status according to the National Land Cover Dataset (https://data.nal.usda.gov/dataset/national-land-cover-database-2011-nlcd-2011), (b) the 2040–59 best case scenario and (c) the 2040–59 worst case scenario (adapted from Phung *et al.* 2019).

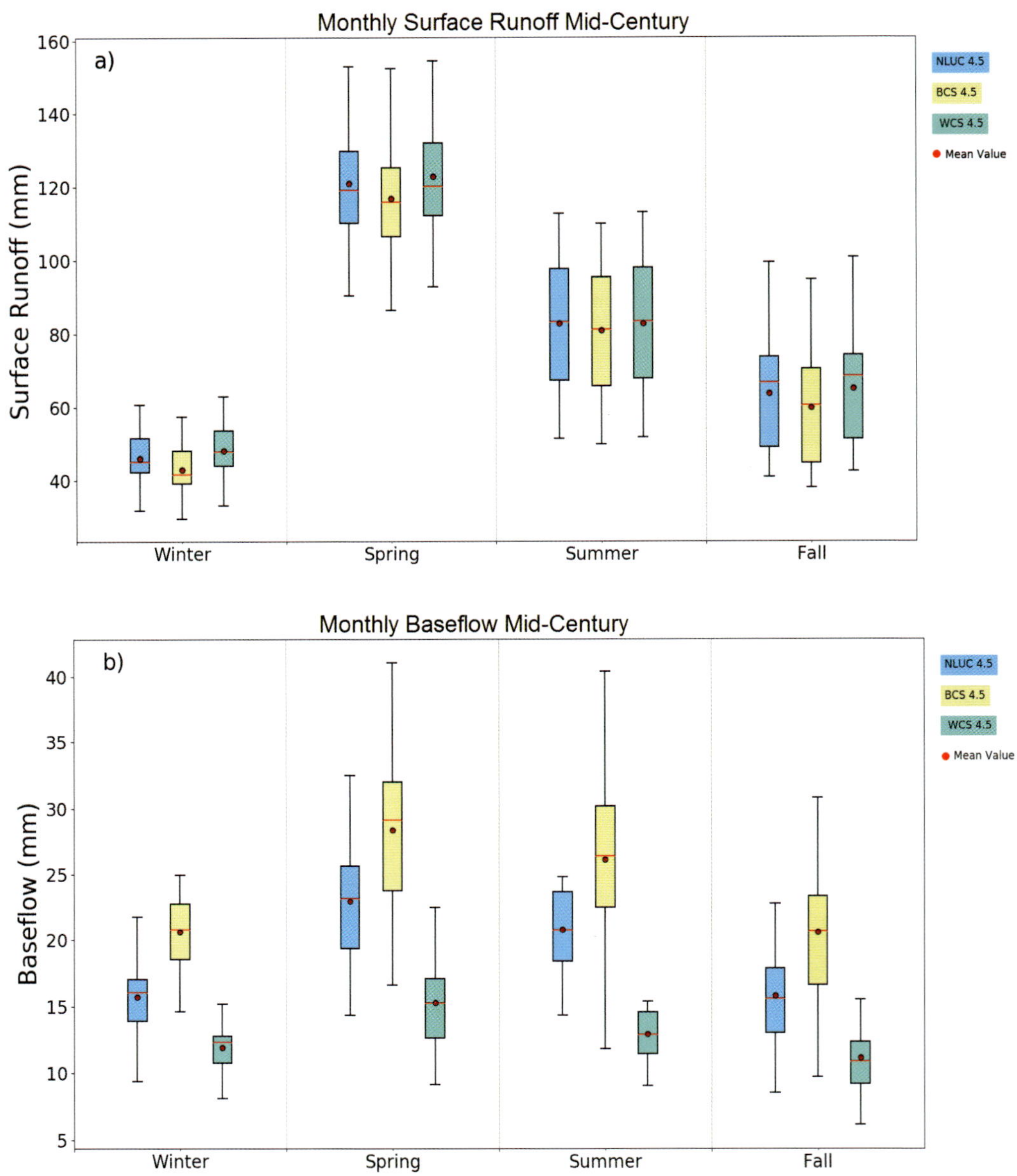

Plate 11. Box plots of seasonal (a) surface run-off and (b) base flow from ensemble model runs based on representative concentration pathway (RCP) 4.5 for mid-century (2040–59) given no land use change (NLUC), best-case scenario (BCS) and worst-case scenario (WCS). The boxes show the interquartile range, with median values indicated by horizontal lines and mean values indicated by filled circles; whiskers show the range.

11

Soil water management and climate fluctuations: modelling approach

Claire Baffaut, Christine Costello, Sagar Gautam, Quang Phung, Allen Thompson

Introduction

Soil water management entails a range of agricultural practices, infrastructure decisions and social and economic policies that ensure or optimise the availability of water to meet the goals defined by a community (Iglesias and Garrote 2015). Agricultural practices affect soil water; therefore, management of water uses in agriculture is critical for future sustainability. By using irrigation, artificial drainage, tillage and amendments, producers directly affect plant growth, water availability and losses of water to groundwater and streams. Infrastructure decisions, such as reservoir operations or water withdrawals for power plants, affect water resources for irrigation. Social and economic policies affect population growth, urban growth and pressures on agricultural land, and thus soil water at a regional scale.

Land management practices that affect soil water directly include irrigation, the selection of land cover and crop rotations and practices such as artificial drainage, which aim to prevent the accumulation of excess water. Alternatively, practices may focus on different aspects of management but still affect soil moisture. For example, tillage practices and terraces may target lower erosion and sediment transport. Similarly, crop rotations may seek to optimise crop synergies for fertilisation or pest and weed pressures. However, most land management practices have multiple effects, including some on agricultural production, soil water and the water balance. As such, agricultural practices affect water resources and water availability for other uses. Thus, at field scale, soil water management is linked to land management practices and changes in water balance and water quality. At regional scales, soil water management is linked to issues of water resource management and water availability. Although soil water management affects both water quantity and quality, this chapter focuses on water quantity and how climate change will affect the water balance.

Increased greenhouse gas (GHG) concentrations are altering the radiative balance of Earth's atmosphere, increasing mean temperatures and changing precipitation patterns (Solomon *et al.* 2007; O'Gorman and Schneider 2009). These changes will affect future soil water status, hampering agricultural productivity (Pokhrel *et al.* 2021). There is a pressing need to understand the future impact of climate change on hydrological processes in order to design sustainable mitigation and adaptation strategies. Simulation models are readily available tools for scientists and policy makers to explore trends in run-off, crop yields and erosion under future emission scenarios and identify what adaptation response may be warranted or necessary. The variability in both how climate change will affect local weather and the many variations in landscape make this a challenging task, but also highlights the value of modelling efforts.

Ecohydrological models are systems of equations that simulate the processes that occur on the landscape, including crop growth, water movement and soil temperature, in response to precipitation and air temperature inputs. Processes include surface run-off, percolation, lateral flow, transpiration associated with plant growth and evaporation. Soil water content throughout the soil profile is the nexus of all these processes. Ecohydrological models are important tools to simulate these processes under a range of scenarios that would neither be possible nor convenient to implement experimentally. Limitations to experimental testing include the availability of land resources to evaluate management scenarios and the time and expense it would take to quantify the impacts under the full spectrum of weather, soils and topographic conditions. Ecohydrological modelling started in the 1950s with the development of the Stanford Watershed Model, which became the Hydrological Simulation Program–Fortran (HSPF) model (Crawford and Lindsey 1966), and has since expanded with models becoming more complex in terms of spatial and temporal resolution and the number of processes simulated. A special collection of reviews of hydrological models used in agriculture and forestry (Moriasi *et al.* 2012) provides a starting point of available models at point, field and watershed scales.

The future climate of Earth is projected by Earth system models (ESM). Such models simulate Earth's climate as a function of atmosphere, ocean, land surface and sea ice (which are the physical components of the system), along with the biogeochemical processes that interact with these physical systems (Flato 2011). Climate fluctuates slowly: climate change is characterised by gradual changes in average conditions, as well as changes in the frequency and severity of extreme events (Rummukainen 2012). The detection of trends in average conditions and in extreme events, and the effects of these changes on soil water, require long-term observations (i.e. ≥20 years; Moran *et al.* 2008). Thus, long-term experimental studies are critical to improving our understanding of the effects of climatic fluctuations on soil water.

However, collecting the necessary meteorological and hydrological data to conduct long-term studies is costly and requires important commitments in terms of financial, human and infrastructure resources. In addition, long-term experimental studies only demonstrate past and current trends in soil water under the management scenarios that were implemented. Yet, land and water resource managers need to understand future trends under various management scenarios, how future climate will affect water resources and which mitigation and adaptation techniques can be effective. Ecohydrological models combined with ESMs are helpful in answering these questions. Climate forcing from ESMs is used as an input to ecohydrological models to simulate future trends in hydrological variables. The objectives of this chapter are to provide an understanding of the types of soil water management issues addressed with ESMs and ecohydrological models, describe the current state of the art regarding how ecohydrological modelling can help with soil water management decisions in a changing climate and describe opportunities for model improvement, which is necessary to address soil water management in a changing climate.

Current soil water management issues and challenges addressed by modelling

Resilience to climate extremes

Climate extremes such as droughts, floods and heat waves are projected to increase in both intensity and frequency in several regions worldwide due to climate change (Pachauri *et al.* 2014). Climate extremes are particularly worrisome for the agricultural sector because they can affect the potential for soil erosion, reduce soil health, decrease agricultural productivity and negatively affect food security and global sustainability (Lal *et al.* 2011). For example, drought can reduce soil water availability and contribute to reductions in crop yields or crop failure; temperature rises lead to increased water demand due to increases in crop potential evapotranspiration and changes in precipitation patterns can result in increased water shortages and an increased need for irrigation (Iglesias *et al.* 2011; Glotter and Elliott 2016). In contrast, increased precipitation may result in a reduction in the time available for field operations as well as increases in seasonal flooding.

Ecohydrological models are used to quantify the magnitude and frequency of the problems caused by climate extremes and determine which practices should be implemented, and where, to mitigate these problems. Although precipitation and temperature outputs of ESM help assess the potential for meteorological droughts or floods (Mishra *et al.* 2017), other models using these ESM outputs as inputs help assess the impacts of climate extremes on agriculture and water resources. For example, Grillakis (2019) estimated future soil moisture across Europe using the Joint UK Land Environment Simulator (JULES) land surface model combined with ESM projections through 2099 and found that drought conditions are expected to increase in occurrence, spatial extent and duration. Anache *et al.* (2018) combined a calibrated Water Erosion Prediction Project (WEPP) model with downscaled ESM projections to predict run-off and soil loss and found that run-off behaviour was distinct for different land uses (e.g. cropped land had greater run-off than pasture).

Ecohydrological models have also been used to identify the effects of climate change (increased temperature, increased or decreased precipitation volume or intensity, CO_2 concentration) that most affect outcomes such as run-off volume, soil erosion or crop yields. For example, Nearing *et al.* (2005) and Her *et al.* (2019) concluded that soil erosion is likely to be more sensitive to changes in precipitation and ground cover than run-off. This type of sensitivity analysis is useful for verifying that critical processes are implemented and correctly parameterised in a model. Understanding a model's capacity and limitations, as well as the data needed to construct the model, helps with model selection given the objectives of the study and keeps the interpretation of the results consistent with the processes that are simulated.

Among the most important data required for ecohydrological models are weather data. Although considerable progress has been made in generating future climate data, much of the work conducted to date is challenged by attempting to resolve weather data generated by climate models, which are often at a coarse spatial (and arguably temporal) scale rather than at the field scale, where decisions must be made. Efforts to integrate ESMs with crop yield models have improved over the past 10 years due to increased spatial resolution of ESM projections, downscaling techniques and coupling of climate, hydrological and agronomic models. The Agricultural Model Intercomparison and Improvement Project (AgMIP) is a hub for sharing research that connects ESMs to models that simulate processes critical to agriculture, such as hydrology, soil processes and crop yield (https://agmip.org/overview/). The 2018 *Agricultural Systems* special issue Modeling Cropping Systems Under Climate Variability and Change: Impacts, Risk and Adaptation provides updated insights (Nendel *et al.* 2018). Advances in computational capacity and efficiency, lessons learned from prior work and scientific research on key plant processes under a changing atmosphere (e.g. plant responses to elevated CO_2 (eCO_2); Ainsworth and Long 2005; Wang *et al.* 2020) all contribute to the ability to modify mathematical representation of key plant, atmosphere and weather interactions. Collectively, these contribute to advancing our ability to identify future risks with more certainty and to plan adaptation responses, as necessary.

Soil water management for agriculture intensification given available water resources

Increasing agricultural activity to support growing populations and resource-intensive consumption has accelerated the decline in available water resources (Chartres and Noble 2015). In many regions with limited water resources, soil water management practices that conserve water and increase crop productivity are needed to overcome water stress (Bodner *et al.* 2015). Soil water management can also be used to mitigate climate change because soils can sequester large quantities of atmospheric carbon (Lal *et al.* 2011). The sequestered carbon will help increase soil organic matter and water-holding capacity, increasing the ability of crops to tolerate drier conditions. Other examples of mitigation and adaptation to climate extremes include reducing soil evaporation by increasing soil cover (e.g. with crop residues; Bodner *et al.* 2015), nutrient management to promote crop growth and a denser canopy cover, a living canopy cover (cover cropping, relay intercropping) or non-crop mulch material, such as plastic foil and geotextiles, and reducing plant transpiration by fallowing (Verburg *et al.* 2012; Bodner *et al.* 2015). Organic fertilisation (Bodner *et al.* 2015) and crop species also influence soil hydrology, although these effects are weaker compared with the effects of soil cover and fallowing. Adjusting the cropping season or reducing its length by using shorter-duration varieties may help avoid overlap between the key crop growth periods and the drought period. Sowing methods such as dry direct seeding result in deeper, finer roots that are better able to access available soil water reserves (Haefele *et al.* 2016). However, for adaptation and mitigation efforts, it is necessary and challenging to accurately represent and project the hydrological response to these soil water management practices, more so under the uncertainty of future climate conditions.

Research studies that used ecohydrological models to study the impacts of soil water management on agricultural lands in regions with limited water supplies illustrate the potential of these models to address future water scarcity. Using AquaCrop, a crop-water productivity model, Mustafa *et al.* (2017) showed that deficit irrigation during the booting and flowering stages could increase grain yield up to 136% compared with rain-fed production, with significant water savings compared with full irrigation. Ahn *et al.* (2018) used a Soil and Water Assessment Tool (SWAT) model to study the impact of drought-adaptive agricultural water management in a semi-arid region. The results of that study suggested that greater groundwater recharge occurs in fields where corn and cotton are planted on silty clay loam soils compared with soil types such as sand or clay, because loam soils contain more nutrients, moisture and humus and have better drainage, infiltration and groundwater recharge (Ahn *et al.* 2018).

Certainly, there is great value to these modelling efforts, which can help with planning for soil water management. However, much remains to be understood to increase the certainty of modelling climate and soil water management in agricultural systems. Haacker *et al.* (2019) offer many insights into the challenges of combining process models to assess agricultural water management, and emphasise that to do so successfully requires three things: (1) sufficient calibration and validation data, including some that cover multiple management practices (Bhandari *et al.* 2017); (2) careful consideration of how to represent interdependence of hydrological, water quality and crop growth processes that have largely been studied individually; and (3) allowance for the conceptual combination of models that are designed to be useful to different stakeholders.

Irrigation management for production or environmental impacts

Proper irrigation management is key to successful crop production in arid regions where seasonal rainfall is historically inadequate to meet crop water demands. The frequency, timing and distribution of droughts under changing climate conditions have made irrigation management increasingly important (Rey *et al.* 2016). In more humid regions, where cumulative rainfall is sufficient, seasonal droughts may result in lower yields than would have been achieved had supplemental irrigation been used. Considering investment costs for mechanised irrigation systems (e.g. pumps, water distribution systems), designs that can provide full crop water needs are often preferred to ensure consistent yields from year to year.

In arid regions where annual evapotranspiration exceeds annual precipitation, proper irrigation management should include a means of maintaining soil salinity levels below the crop threshold limit (i.e. maximum salinity levels that do not limit crop yield). In locations where this may not be possible due to high saline water, irrigation schedules that result in a less-than-maximum yield may be more profitable. Transient-state salinity simulation models are preferred for estimating irrigation amounts with saline waters because such models demonstrate that less irrigation water is required to maintain acceptable crop yields than previously understood when using steady state models (Letey *et al.* 2011).

One of the greatest challenges to predicting irrigation needs and supporting proper irrigation management is obtaining timely and accurate field data (i.e. soil moisture, crop status, air temperature, humidity; Evett *et al.* 2020). Historically, data to support scheduling decisions have been collected through indirect means, such as evapotranspiration model estimates, or by taking in-field ground measurements. Current models available for irrigation management are either process based or statistical. Process-based crop models can be very useful, but require environmental, soil and plant and field data to calibrate and validate the models. Statistical models require an adequate history of crop and field data to correctly adjust algorithms to match actual crop water needs. Although limited in areal application, statistical models can be improved using machine learning methods (Goldstein *et al.* 2018). However, Zhang *et al.* (2021) have recently demonstrated that data–model fusion methods can be used to expand individual field measurements to larger, scalable applications. Simpler models may focus solely on soil moisture measurements and adjust for net water balance using evapotranspiration estimates and rainfall. Canopy temperature can be used to estimate crop water needs based on a crop water stress index (Idso *et al.* 1981). This method works best for arid climates because canopy temperatures may be elevated above threshold values in humid climates even though plants are not water stressed (Bockhold *et al.* 2011).

The selection of a model should be based on the specific application and level of expertise needed by the irrigator. In practice, a simpler model is more likely to be used, assuming that it saves time and provides reliable information. There is increasing incentive to automate the scheduling methods, but this typically increases cost and can limit the reliability due to the number and complexity of the sensors. Numerous irrigation scheduling models exist, ranging from freeware developed by university extension programs to sophisticated and comprehensive commercial systems. A fairly extensive list of scheduling software can be found in Zhang *et al.* (2021).

State-of-the-art process for climate change modelling studies

The overall process to assess soil water management strategies under future climate relies on using the ESM predictions as inputs to ecohydrological models. There are many ESMs available and many ecohydrological models from which one can choose. Limitations of ESMs in terms of spatial (~100 km) and temporal resolution imply that the predicted climate forcing should be corrected or benchmarked with local observations (Gautam *et al.* 2018). Here we explain the steps needed to assess climate change impacts using ESM outputs and process-based ecohydrological models, including ESM selection, ecohydrological model selection, downscaling of ESM outputs, synthesis of modelling results and reporting of uncertainties. A case study provides an example of these steps.

Selection of ESM

The selection of an ESM for impact assessment depends on the study objectives. The temporal and spatial scale of the ESM predictions, which processes are included and the accuracy of these process representations need to be consistent with the study objectives. As a guide, modellers may use ESM comparison studies. For example, Fasullo (2020) recently compared and ranked Coupled Model Intercomparison Project 6 (CMIP6) candidate ESMs based on their ability to represent the energy balance and individual atmospheric and land surface processes, as well as overall performance. In addition, one can select a model based on its performance to represent the historical climate. However, historical alignment of data does not necessarily guarantee future alignment. In all cases, the use of more than one set of climate data from ESMs is recommended to capture the range of projected future conditions (Fasullo 2020).

Selection of the ecohydrological model

Ecohydrological models simulate the hydrological cycle and plant growth as a function of climate, land use, topography, agricultural management and soil at different temporal and spatial scales. Crop growth models and biogeochemical models are also important for increasing our ability to model soil moisture and crop response to future climates. Selecting the most appropriate ecohydrological model for climate change assessment is not different from any other hydrological study, and should include criteria such as the intended use of the model in relation to the study objectives, the spatial and temporal scale, geographic location (e.g. temperate, tropical), land use type (e.g. agriculture, forest, urban), the processes simulated by the model and the amount of data available for parameterisation and calibration (Harmel *et al.* 2018). The decision making can be augmented with previous research work comparing different models and rankings based on their ability to represent different processes (Moriasi *et al.* 2012; Caldwell *et al.* 2015). For the assessment of climate change impacts, ecohydrological models should be sensitive to changes in temperature, precipitation and CO_2 concentration, and represent the effects of these perturbations on other processes.

Downscaling techniques

The coarse resolution of ESM climate data, typically a daily time step and a spatial resolution of hundreds of kilometres, frequently results in biased output temperatures and precipitation. When used as an input into a crop growth or ecohydrological model to project the future, these biased data lead to higher uncertainties of projections. Therefore, the ESM model output should be compared with historical observations from local weather stations to determine existing biases, a process referred to as 'downscaling'. Once biases are identified, they can be removed from future ESM outputs.

Downscaling involves matching ESM outputs with observed data over a past period to determine the correction needed based on historic trends to produce plausible future projections. Various downscaling techniques are used to correct climate model bias and these can be classified into three types: simple downscaling (Giorgi and Mearns 1991), statistical downscaling (Wilby *et al.* 1998) and dynamic downscaling (Dickinson *et al.* 1989).

Simple downscaling involves creating a fabricated weather dataset by adding or multiplying by a future change factor. Dynamic downscaling involves numerical meteorological modelling that relates large-scale patterns to local-scale variables. Dynamic downscaling requires more computational power and expertise and is not commonly used for impact assessment studies at finer field scales. Statistical downscaling is the most common approach used by environmental impact modellers. The statistical downscaling approach determines corrections that result in matching statistics between historical observations and model predicted data and applies those corrections to the future ESM data. The correction statistics include the mean, the mean and variance, or the full distribution described by its quantiles.

Synthesis of modelling results and reporting of uncertainties

Uncertainties in predicting climate change impacts on the hydrological response and crop growth have multiple sources. These sources include: (1) climate model emission scenarios that depend on global, political and economic conditions; (2) variation in ESM simulation methods; (3) bias correction or downscaling technique; and (4) the representation of interactions in the ecohydrological or crop growth model. The latter uncertainty has multiple causes: model structure, processes represented and parameterisation (Maurer 2007).

Studies have compared the uncertainty introduced by ESMs to that caused by the parameterisation of ecohydrological models. The results of these comparisons are likely to vary with geographic location, land use and management practices. However, researchers have reported that the ESM uncertainty is often an order of magnitude greater than that caused by the ecohydrological models for estimations of direct run-off (Shen *et al.* 2018; Her *et al.* 2019). Her *et al.* (2019) also demonstrated that for soil moisture or groundwater, the parameter uncertainty of the ecohydrological model has greater influence than the ESM projections. Soil moisture is a critical component of simulating crop growth; thus, improving the ability to simulate soil moisture can improve assessments of future risk to crop productivity. These recent studies highlight the need to continue to strive for higher-resolution climate and weather projections, downscaled using local weather data where possible, and to improve the representation of processes related to soil moisture in hydrological models.

One of the ways to represent the model structure, process and parameterisation uncertainties is to derive multimodel ensemble projections. The general approach involves running two or more related models and synthesising the results using statistical methods instead of using a single ESM and ecohydrological model. This strategy improves understanding of future trends with uncertainty bands and helps eliminate a potential bias that may be present when basing conclusions on only one model. Although the use of model ensembles for impact assessment research presents opportunities for understanding uncertainties, these ensembles represent a challenge when it comes to interpreting the results towards planning land and water resources management (Falloon *et al.* 2014). Efforts should be made to reduce these uncertainties.

Presently, a large group of researchers focusing on computational environmental engineering is using large sets of environmental observation data to refine the model parameters and process representation (Knüsel *et al.* 2019). With the availability of large historical datasets, advances in computational resources and new types of machine learning models, the ensemble of machine learning and process models could be a way to improve simulation models and better represent uncertainty in the future.

Case study: future drought projection for the Goodwater Creek Experimental Watershed

The Goodwater Creek Experimental Watershed (GCEW) is located in Boone and Audrain counties, Missouri, and is a 73-km^2 headwater watershed in the Salt River Basin (Plate 9). Land use in the GCEW includes row crops (72.8%), pasture (14.3%), forest (6.0%) and urban land (6.8%; Plate 9). The soils of the GCEW have a restrictive clay layer in the subsurface soil with lower hydraulic conductivity, restricting water movement and limiting the water storage capacity of soil to that of the topsoil layer, which ranges from 5 to 60 cm in depth (Sadler *et al.* 2015). This thin layer of topsoil leads to short-term drought whenever there is a lack of precipitation, even for a short period, and extremely high precipitation leads to flash floods due to fast saturation of the thin topsoil. The objectives of the case study were to project the frequency and severity of meteorological, hydrological and agricultural droughts in this small watershed and determine what adaptation measures may be needed.

This case study compares future occurrences of excessively dry and wet periods based on precipitation, temperature, stream flow, soil moisture and extreme events. The calibrated SWAT model of the GCEW was used to simulate the hydrology (Gautam *et al.* 2018) because it is a daily time step process-based distributed parameter model that enables representation of agricultural land cover and management practices. As such, it simulates how precipitation, temperature and CO_2 concentration affect plant growth and the water balance. Climate data (precipitation and temperature) from multiple ESMs with two different emission scenario representative concentration pathways (RCPs), namely RCP4.5 (moderate emission) and RCP8.5 (high emission), were obtained from CMIP5 for the near future (2016–45) and far future (2046–75). We selected the CMIP5 ESMs because the outputs are regionally downscaled and at the daily time step. Given the size of the GCEW, having regionally downscaled data was important. However, comparison of the outputs with historical data from 1980 to 2015 showed the need for further statistical downscaling to the GCEW. The outputs were then used as input for the SWAT model (Gautam *et al.* 2018). Three monthly drought indices were used. The meteorological and hydrological drought indices were calculated by fitting a gamma distribution to precipitation and streamflow data, and the resulting cumulative distribution functions were standardised to Gaussian values to classify different drought classes. Similarly, agricultural drought was calculated by fitting empirical cumulative probability functions to simulated soil water content, and drought classes were calculated based on a *Z*-score.

The meteorological drought results showed an increased frequency of dry months (i.e. lower than normal precipitation) and extremely dry months for both RCPs in winter, summer and autumn for future periods (Gautam *et al.* 2021). The agricultural (soil water-based) drought results showed an increased frequency of dry months for both RCPs for spring and summer. Results also showed an increase in the frequency of hydrological droughts (i.e. streamflow lower than normal throughout all four seasons). The results from the precipitation- and temperature-based extreme indices indicated an increase in the number of days with increased intensity and quantity of precipitation in the spring. In the summer, the results showed a decrease in precipitation intensity and an increase in the duration of warm spells.

The use of multiple drought indices, downscaled climate data and process model output facilitated drought prediction for different land surface processes and their comparison. This case study provides insights into the need for a multi-index approach and site-specific downscaled climate data for drought projection. The conclusion of the case study was that soils with low water-holding capacity make the watershed agriculture more challenging under projected futures with more extreme dry and wet conditions (Gautam *et al.* 2021). A shift to crops that have a longer planting window in the spring and are resilient to drought conditions (e.g. sorghum instead of corn) may be necessary. The difference in projected drought across multiple indices highlighted the benefits of using the outputs of a process-based model, which considers climate, management and landscape-scale processes.

The case study described above comes with some assumptions and limitations. For example, we used one set of parameters based on the calibration of the hydrological model for current conditions. However, multiple sets of parameters would produce more dynamic temporal and spatial results. We also assumed that land use, crop distribution and land management would not change in the future and that the model parameters are static and will not vary with changing climatic conditions.

The next section addresses challenges and opportunities for the modelling community to improve models and modelling techniques to improve our assessment of the need for adaptation measures.

Opportunities in ecohydrological modelling to address soil water management in the context of climate change

Success with addressing soil water management for agriculture in the context of climate change will depend on the ability of ecohydrological models to simulate the adaptation measures, such as technological improvements in agricultural practices and plant genetics and genomics that will allow the agricultural sector to adapt to changing climatic conditions. Success will also depend on the ability of ecohydrological models to simulate the processes affected by changing conditions, and will require models and modellers to acknowledge that model parameterisation is not static and may need to change as a function of climatic conditions. Finally, system modelling will need to address the trade-offs and the feedback that will exist as plants and people adapt to changing climatic conditions. This section addresses these points.

Including additional capabilities into ecohydrological models

Irrigation for improved crop production is an important tool for adjusting to changing weather patterns and more extreme weather events, particularly drought conditions. Various irrigation methods are practiced across the US, including gravity (surface), drip and sprinkler irrigation, but most new and existing irrigation systems are centre pivot sprinklers, accounting for more than 50% of all irrigated lands (United States Department of Agriculture, National Agricultural Statistics Service 2017). The actual percentage of each type of irrigation varies depending on location, available water supply and crop water needs.

Sources of irrigation waters include surface water, groundwater and man-made canal delivery systems. As competition for limited water supplies increases, there has been renewed interest in water reuse for irrigation, particularly in regions that have limited freshwater supplies but available waste water (Angelakis *et al.* 2018). Decentralised waste water treatment systems where waste waters are treated locally rather than directed to a single, large treatment location show promise for more practical reuse of waste water (Gikas and Tchobanoglous 2009).

Few of these options for sources and application methods of irrigation water are available in ecohydrological models. In many models, irrigation can only be specified by a specified amount at a specified time or as a function of soil water stress. This limitation originates from the initial objectives for model development, which were crop growth, stream discharge and water quality prediction, without any consideration for regional water availability. Yet, these options represent substantial differences in the management of water resources at regional scales.

Potential adaptation responses include precision agriculture and variable-rate irrigation. Precision agriculture for irrigation involves the application of high-speed sensors with near-real-time irrigation feedback and control using properly designed decision support systems (Evett *et al.* 2020). Variable-rate irrigation is the process of matching water application to specific temporal and areal field conditions (Evans *et al.* 2013). Simulating these management practices under a changing climate will give a better sense of the ability to adapt to these changes. Precision management is becoming more common, with nearly 20% of all USA corn and soybean growers using some form of global positioning system (GPS) mapping to determine crop selection and fertiliser needs (Schimmelpfennig 2016). Modelling for precision management is still in the developmental stage, but holds promise to take advantage of this technology (Bobryk *et al.* 2018).

The practice of agroforestry, where woody plants are integrated in the same field as crops and livestock, can improve the environmental quality of waters, soil health and carbon sequestration (Udawatta *et al.* 2011; Alagele *et al.* 2019). Agroforestry has long been practiced in low- and middle-income countries, but is now receiving interest in more high-income countries (Lovell *et al.* 2018). However, ecohydrological models do not simulate tree growth in a satisfactory manner, because this was not the original objective of these models. Specifically, tree root systems are deeper and grow in different ways than those of annual or perennial crops. Because of this, they may take up water from different aquifers (David *et al.* 2013), something not planned in most crop growth models.

Cover crops as part of a crop rotation of annual crops offer benefits in terms of soil health and GHG emissions (Poeplau and Don 2015) and reduced soil erosion (Daryanto *et al.* 2018), but present challenges due to the potential competition for soil water in areas with limited rainfall (Unger and Vigil 1998; Nielsen *et al.* 2015). Demonstration of the impact of cover crop mixes in cropping systems requires the ability to simulate, at the same time, multiple plants that will compete for light, water and nutrients. Different plant varieties may be successful during certain years, depending on the temperature and available water. Thus, it is important to simulate a mix as opposed to a single plant. Although certain models do have this ability (e.g. the Environmental Policy Integrated Climate (EPIC) model), not all do.

Continued improvements in plant genetics by developing crops that are more tolerant of limited water supplies, pests and diseases have the potential to increase water use efficiency. Because plant water use is regulated through root morphology, as well as through cuticles and stomata via guard cell responses to environmental stimuli, research that focuses on genes controlling these traits should be emphasised in breeding programs (Ruggiero *et al.* 2017). Although progress is being made in plant genetics to improve water use efficiency, and ultimately drought tolerance, real challenges exist in improving drought tolerance while simultaneously not decreasing yield potential (Ruggiero *et al.* 2017). The ability to account for water use efficiency improvements in plant models is equally challenging due to the numerous pathways that ultimately affect plant response to the environment, and the ability to include meaningful data that account for those responses.

Our understanding of CO_2 processes under changing climate scenarios continues to improve, but much uncertainty remains (Allen *et al.* 2020; Wang *et al.* 2020). Resolving molecular-scale representation of a process to a watershed modelling unit imposes another layer of complexity. However, plant scientists are gaining a better understanding of, and amassing data to better represent, detailed plant and molecular processes with the goal of improving the ability to model crop response to eCO_2 (Engineer *et al.* 2016; Dong *et al.* 2018).

Over the past 50 years, the annual growth rate of atmospheric CO_2 increased from 0.6 to 2.3 ppm year^{-1} (Lindsey 2020). The question becomes, what will the concentration of eCO_2 be in the future under a given emission scenario? Models based on representative concentration pathways have the advantage of using prescribed CO_2 concentrations, but, because individual models vary in predicted future temperature change, comparisons of model results are problematic (Schleussner *et al.* 2018). Rather than attempting to model changes in CO_2 concentrations as a function of climate change, the Intergovernment Panel on Climate Change (IPCC; 2013) recommended using a transient climate approach relating annual global mean temperature change with CO_2 doubling following a linear forcing of CO_2. This approach helps constrain probabilistic levels of CO_2 with the various RCP selections (IPCC 2014; Schleussner *et al.* 2018). Although uncertainty remains, this approach does provide a consistent framework within which to evaluate crop responses to future climate scenarios. In any case, the current overall understanding of the influence of eCO_2 on plant growth is sufficient, such that modelling of climate change scenarios should include the effects of eCO_2 (Toreti *et al.* 2020).

Dynamic change in model parameters

Model parameters are dependent on the weather characteristics during the calibration period (Vaze *et al.* 2010). This is particularly true for conceptual models (Osuch *et al.* 2015), but should not be ignored for process-based models because even these contain some conceptual components. One solution is to ensure that the calibration period is long enough to contain dry and wet periods with dry and wet extremes (Daggupati *et al.* 2015). Indeed, Vaze *et al.* (2010) concluded that a calibration period of 20 years is necessary for climate change studies provided that mean annual precipitation increases <20% or decreases <15%. Moreover, Vaze *et al.* (2010) concluded that model performance is more likely to remain satisfactory when going from drier to wetter conditions than from wetter to drier conditions. With further analyses, Coron *et al.* (2012) showed that model performance degradation is catchment dependent. In addition, Coron *et al.* (2012) showed that changes in temperature and evapotranspiration affected model performance to a lesser degree than precipitation. In some cases, the uncertainty of model predictions caused by the non-stationarity of model parameters can be greater than the uncertainty of model predictions caused by the uncertainty of final parameter values during calibration (Brigode *et al.* 2013). Additional strategies include adjusting model parameters as a function of weather characteristics (Westra *et al.* 2014).

Model parameters are also dependent on management. Management practices can alter soil properties (e.g. no-till practices and cover crops can increase soil organic content and soil hydraulic properties; Al-Kaisi *et al.* 2014; Nunes *et al.* 2018). However, current models poorly simulate the long-term changes in soil properties and soil health, and thus the impacts on hydrology. For example, the SWAT model uses a parameter that quantifies soil biological activity in the degradation of residues and cycling of nutrients, which affect plant growth and nutrient movement. However, this parameter differs for crop systems with different tillage and crop diversity. The current state of knowledge allows simulation of management scenarios with the appropriate soil properties (Baffaut *et al.* 2019). However, we are unaware of any study that has investigated the effect of a changing climate and the resulting altered plant growth effects on microbiological population and soil properties.

Integrating models to identify management trade-offs and feedback mechanisms

Integrating multiple models to simulate the different aspects of water management at different scales has been a successful strategy to study the effectiveness of soil water management on regional water availability (Cosgrove and Loucks 2015). For instance, Psomas *et al.* (2016) used the SWAT and Water Evaluation and Planning (WEAP) models to estimate water savings under various management scenarios. The SWAT model was used to simulate agricultural water management, whereas the WEAP model was used for regional allocation of available water. Psomas *et al.* (2016) found that using deficit irrigation, ~23% of water can be saved compared with the current management practices.

To highlight the trade-offs and feedback mechanisms of climate, land use, hydrological processes and soil water availability, we present the study by Phung *et al.* (2019), conducted for the Salt River Basin, an agricultural watershed located in north-east Missouri. This watershed has a drainage area of 6417 km^2. Its elevation varies from 305 to 158 m above mean sea level, with Mark Twain Lake's outlet at the lowest elevation. In its upper reaches, slopes vary from 0% to 7%, but become steeper near the major tributaries, with slopes of up to 20% on backslopes. The mean annual precipitation of the Salt River Basin is ~940 mm.

Future climate data from 19 ESMs were obtained from the bias correction and constructed analogues data at a resolution of 12 km × 12 km. Precipitation and temperature of the ESMs were further downscaled using statistical downscaling techniques such as the delta change method for temperature and quantile mapping for precipitation (Gautam *et al.* 2018). Two RCP scenarios, RCP4.5 (moderate emission) and RCP8.5 (high emission), were used to represent the moderate and high changes in CO_2 concentration projected to 2100.

A SWAT model of the basin was calibrated using historical meteorological data, the 2011 land cover map from the US Department of Agriculture National Land Cover Dataset 2011 (https://data.nal.usda.gov/dataset/national-land-cover-database-2011-nlcd-2011) and streamflow data from eight US Geological Survey streamflow stations (Phung *et al.* 2019). Evapotranspiration was calculated in SWAT using the modified Penman–Monteith method to incorporate variability in radiation-use efficiency, plant growth and transpiration induced by changes in atmospheric CO_2 concentration (Neitsch *et al.* 2011).

The Conversion of Land Use and its Effects at Small regional extent (CLUE-S) model (Verburg *et al.* 2002) was used to project different projections of forest, urban and agricultural land expansion or compression, including the best-case scenario (BCS) and the worst-case scenario (WCS; Plate 10). Low-meat diets were assumed for the BCS, resulting in reduced demand for crop and pasture land, which was replaced with forested land. The WCS assumes a doubling of urban land and more land conversion into crop and pasture land to maximise potential food production. Forest areas with slopes <15% were made available for conversion into crop and pastureland in CLUE-S because greater slopes would not be recommended for heavy machinery.

The ensemble results of Phung *et al.* (2019) projected an increase of 1.0°C in the mean annual temperature in 2020–39 and 2.0°C mid-century (2040–59). Precipitation was projected to increase between 4% and 7%. The results indicated that projected climate changes in the Salt River Basin would lead to seasonal changes (increase or decrease depending on the season) in evapotranspiration, base flow, surface run-off and total water yield in all scenarios during the early and mid-portions of this century. The results showed reduced precipitation in the summer, increased evapotranspiration in late spring and increased irrigation up to 38 mm (+11%) during the growing season (Phung *et al.* 2019). These findings indicate that agricultural water demand will likely increase in the future, which may exacerbate water scarcity issues. This potential increase raises questions about management approaches for long-term water planning.

Furthermore, Phung *et al.* (2019) demonstrated the feedback between land use and climate change (i.e. that land use change could mitigate or amplify the impact of climate change). Under the BCS, surface run-off for RCP4.5 was 4% lower than that under no land use change (Plate 11a). The changes in surface run-off suggested that converting ~25% of crop and pastureland to forest (BCS) would not only help mitigate the effects of climate change, but also bring the quantity of surface run-off below the historical level. Conversely, run-off would increase by 2% under the WCS, with statistically significant changes in the spring, autumn and winter. Base flow in both the BCS and WCS changed in all seasons compared with no land use change. Thus, changes in land use can substantially alter the water balance in the Salt River Basin watershed.

Summary

Globally, approximately 70% of the available fresh water is used in the agriculture sector; therefore, wisely managing water use in agriculture is critical to achieve sustainability. The projected increase in water demand across sectors (e.g. power generation, agriculture, households, industry) under higher population density and the complex interaction with the future climate change entails the need for intervention. Both ESM and ecohydrological models are necessary tools to predict how agricultural systems will respond to climate change and what soil water management practices producers and water resource managers can implement to mitigate the effects of climate change. Although there has been considerable progress, there is still a need to improve the ability of these models to accurately assess which management options can alleviate emerging risks across space and time and, moreover, with sufficient certainty as to warrant local, regional, national and even global decision making about investment in and the placement of infrastructure, biotechnology, crops and irrigation management options.

Experimental studies indicate that agricultural practices such as minimum tillage, terraces, precision fertiliser application, cover crops and other structural and non-structural best management practices have sustainable benefits for soil water. This wealth of knowledge can be used to benchmark or train ecohydrological models and design sustainable management systems that are resilient to future changes in climate. Simultaneously, there has been significant progress made in ESMs, which provide a range of ever-improving data on the future trajectory of Earth's climate. Although major challenges remain, and complete eradication of uncertainty may never be achievable, particularly as we aspire to represent more complex processes (e.g. effects of CO_2 on plant metabolism) the coupling of ESM with ecohydrological and crop models has improved. Ultimately, modelling efforts can improve our ability to understand the future trajectory of soil water status under different land use, crop and water management practices, as well as climate scenarios over time.

References

Ahn S, Abudu S, Sheng Z, Mirchi A (2018) Hydrologic impacts of drought-adaptive agricultural water management in a semi-arid river basin: case of Rincon Valley, New Mexico. *Agricultural Water Management* **209**, 206–218. doi:10.1016/j.agwat.2018.07.040

Ainsworth EA, Long SP (2005) What have we learned from 15 years of free-air CO_2 enrichment (FACE)? A meta-analytic review of the responses of photosynthesis, canopy properties and plant production to rising CO_2. *New Phytologist* **165**, 351–372. doi:10.1111/j.1469-8137.2004.01224.x

Al-Kaisi MM, Douelle A, Kwaw-Mensah D (2014) Soil microaggregate and macroaggregate decay over time and soil carbon change as influenced by different tillage systems. *Journal of Soil and Water Conservation* **69**, 574–580. doi:10.2489/jswc.69.6.574

Alagele SM, Anderson SH, Udawatta RP, Veum KS, Rankoth LM (2019) Effects of conservation practices on soil quality compared with a corn–soybean rotation on a claypan soil. *Journal of Environmental Quality* **48**, 1694–1702. doi:10.2134/jeq2019.03.0121

Allen LH, Kimball BA, Bunce JA, Yoshimoto M, Harazono Y, Baker JT, Boote KJ, White JW (2020) Fluctuations of CO_2 in free-air CO_2 enrichment (FACE) depress plant photosynthesis, growth, and yield. *Agricultural and Forest Meteorology* **284**, 107899. doi:10.1016/j.agrformet.2020.107899

Anache JAA, Flanagan DC, Srivastava A, Wendland EC (2018) Land use and climate change impacts on runoff and soil erosion at the hillslope scale in the Brazilian Cerrado. *The Science of the Total Environment* **622–623**, 140–151. doi:10.1016/j.scitotenv.2017.11.257

Angelakis AN, Asano T, Bahri A, Jimenez BE, Tchobanoglous G (2018) Water reuse: from ancient to modern times and the future. *Frontiers in Environmental Science* **6**. doi:10.3389/fenvs.2018.00026

Baffaut C, Ghidey F, Lerch RN, Kitchen NR, Sudduth KA, Sadler EJ (2019) Long-term simulated runoff and water quality from grain cropping systems on restrictive layer soils. *Agricultural Water Management* **213**, 36–48. doi:10.1016/j.agwat.2018.09.032

Bhandari AB, Nelson NO, Sweeney DW, Baffaut C, Lory JA, Senaviratne A, Pierzynski GM, Janssen KA, Barnes PL (2017) Calibration of the APEX model to simulate management practice effects on runoff, sediment, and phosphorus loss. *Journal of Environmental Quality* **46**, 1332–1340. doi:10.2134/jeq2016.07.0272

Bobryk CW, Yost MA, Kitchen NR (2018) Field variability and vulnerability index to identify regional precision agriculture opportunity. *Precision Agriculture* **19**, 589–605. doi:10.1007/s11119-017-9541-6

Bockhold DL, Thompson AL, Sudduth KA, Henggeler JC (2011) irrigation scheduling based on crop canopy temperature for humid environments. *Transactions of the ASABE* **54**, 2021–2028. doi:10.13031/2013.40654

Bodner G, Nakhforoosh A, Kaul H-P (2015) Management of crop water under drought: a review. *Agronomy for Sustainable Development* **35**, 401–442. doi:10.1007/s13593-015-0283-4

Brigode P, Oudin L, Perrin C (2013) Hydrological model parameter instability: a source of additional uncertainty in estimating the hydrological impacts of climate change? *Journal of Hydrology* **476**, 410–425. doi:10.1016/j.jhydrol.2012.11.012

Caldwell PV, Kennen JG, Sun G, Kiang JE, Butcher JB, Eddy MC, Hay LE, LaFontaine JH, Hain EF, Nelson SA (2015) A comparison of hydrologic models for ecological flows and water availability. *Ecohydrology* **8**, 1525–1546. doi:10.1002/eco.1602

Chartres CJ, Noble A (2015) Sustainable intensification: overcoming land and water constraints on food production. *Food Security* **7**, 235–245. doi:10.1007/s12571-015-0425-1

Coron L, Andreassian V, Perrin C, Lerat J, Vaze J, Bourqui M, Hendrickx F (2012) Crash testing hydrological models in contrasted climate conditions: an experiment on 216 Australian catchments. *Water Resources Research* **48**, W0552. doi:10.1029/2011WR011721

Cosgrove WJ, Loucks DP (2015) Water management: current and future challenges and research directions. *Water Resources Research* **51**, 4823–4839. doi:10.1002/2014WR016869

Crawford NH, Lindsey RK (1966) 'Digital simulation in hydrology: Stanford watershed model IV.' Technical Report No. 39, Stanford University, Stanford.

Daggupati P, Pai N, Ale S, Douglas-Mankin KR, Zeckoski RW, Jeong J, Parajuli PB, Saraswat D, Youssef MA (2015) A recommended calibration and validation strategy for hydrologic and water quality models. *Transactions of the ASABE* **58**, 1705–1719. doi:10.13031/trans.58.10712

Daryanto S, Fu B, Wang L, Jacinthe PA, Zhao W (2018) Quantitative synthesis on the ecosystem services on cover crops. *Earth-Science Reviews* **185**, 357–373. doi:10.1016/j.earscirev.2018.06.013

David TS, Pinto CA, Nadezhdina N, Kurz-Besson C, Henriques MO, Quilho T, Cermak J, Chaves MM, Pereira JS, David JS (2013) Root functioning, tree water use and hydraulic redistribution in Quercus suber trees: a modeling approach based on root sap flow. *Forest Ecology and Management* **307**, 136–146. doi:10.1016/j.foreco.2013.07.012

Dickinson RE, Errico RM, Giorgi F, Bates GT (1989) A regional climate model for the western United States. *Climatic Change* **15**, 383–422. doi:10.1007/BF00240465

Dong J, Gruda N, Lam SK, Li X, Euan Z (2018) Effects of elevated CO_2 on nutritional quality of vegetables: a review. *Frontiers in Plant Sciences* **9**, 924. doi:10.3389/fpls.2018.00924

Engineer CB, Hashimoto-Sugimoto M, Negi J, Israelsson-Nordström M, Azoulay-Shemer T, Rappel WJ, Iba K, Schroeder JI (2016) CO_2 sensing and CO_2 regulation of stomatal conductance: advances and open questions. *Trends in Plant Science* **21**, 16–30. doi:10.1016/j.tplants.2015.08.014

Evans RG, LaRue J, Stone KC, King BA (2013) Adoption of site-specific variable rate sprinkler irrigation systems. *Irrigation Science* **31**, 871–887. doi:10.1007/s00271-012-0365-x

Evett SR, O'Shaughnessy SA, Andrade MA, Kustas WP, Anderson MC, Schomberg HH, Thompson A (2020) Precision agriculture and irrigation: current U.S. perspectives. *Transactions of the ASABE* **63**, 57–67. doi:10.13031/trans.13355

Falloon P, Challinor A, Dessai S, Hoang L, Johnson J, Koehler A-K (2014) Ensembles and uncertainty in climate change impacts. *Frontiers in Environmental Science* **2**, 33. doi:10.3389/fenvs.2014.00033

Fasullo JT (2020) Evaluating simulated climate patterns from the CMIP archives using satellite and reanalysis datasets using the Climate Model Assessment Tool (CMATv1). *Geoscientific Model Development* **13**, 3627–3642. doi:10.5194/gmd-13-3627-2020

Flato GM (2011) Earth system models: an overview. *Wiley Interdisciplinary Reviews: Climate Change* **2**, 783–800. doi:10.1002/wcc.148

Gautam S, Costello C, Baffaut C, Thompson A, Svoma BM, Phung QA, Sadler EJ (2018) Assessing long-term hydrological impact of climate change using an ensemble approach and comparison with global gridded model – a case study on Goodwater Creek Experimental Watershed. *Water (Basel)* **10**, 564–588. doi:10.3390/w10050564

Gautam S, Costello C, Baffaut C, Thompson A, Sadler EJ (2021) Projection of future drought and extreme events occurrence in Goodwater Creek Experimental Watershed, Midwestern US. *Hydrological Sciences Journal* **66**, 1045–1058. doi:10.1080/02626667.2021.1906878

Gikas P, Tchobanoglous G (2009) The role of decentralized strategies in water resources management. *Journal of Environmental Management* **90**, 144–152. doi:10.1016/j.jenvman.2007.08.016

Giorgi F, Mearns LO (1991) Approaches to the simulation of regional climate change: a review. *Reviews of Geophysics* **29**, 191–216. doi:10.1029/90RG02636

Glotter M, Elliott J (2016) Simulating US agriculture in a modern Dust Bowl drought. *Nature Plants* **3**, 1–6.

Goldstein A, Fink L, Meitin A, Bohadana S, Lutenberg O, Ravid G (2018) Applying machine learning on sensor data for irrigation recommendations: revealing the agronomist's tacit knowledge. *Precision Agriculture* **19**, 421–444. doi:10.1007/s11119-017-9527-4

Grillakis MG (2019) Increase in severe and extreme soil moisture droughts for Europe under climate change. *The Science of the Total Environment* **660**, 1245–1255. doi:10.1016/j.scitotenv.2019.01.001

Haacker EMK, Sharda V, Cano AM, Hrozencik RA, Núñex A, Zambreski Z, Nozari S, Smith GEB, Moore L, Sharma S, *et al.* (2019) Transition pathways to sustainable agricultural water management: a review of integrated modeling approaches. *Journal of the American Water Resources Association* **55**, 6–23. doi:10.1111/1752-1688.12722

Haefele SM, Kato Y, Singh S (2016) Climate ready rice: augmenting drought tolerance with best management practices. *Field Crops Research* **190**, 60–69. doi:10.1016/j.fcr.2016.02.001

Harmel RD, Baffaut C, Douglas-Mankin K (2018) Review and development of ASABE Engineering Practice 621: guidelines for calibrating, validating, and evaluating hydrologic and water quality models. *Transactions of the ASABE* **61**, 1393–1401. doi:10.13031/trans.12806

Her Y, Yoo SH, Cho J, Hwang S, Jeong J, Seong C (2019) Uncertainty in hydrological analysis of climate change: multi-parameter vs. multi-GCM ensemble predictions. *Scientific Reports* **9**, 4974. doi:10.1038/s41598-019-41334-7

Idso S, Jackson R, Pinter JP, Reginato RJ, Hatfield JL (1981) Normalizing the stress-degree-day parameter for environmental variability. *Agricultural Meteorology* **24**, 45–55. doi:10.1016/0002-1571(81)90032-7

Iglesias A, Garrote L (2015) Adaptation strategies for agricultural water management under climate change in Europe. *Agricultural Water Management* **155**, 113–124. doi:10.1016/j.agwat.2015.03.014

Iglesias A, Mougou R, Moneo M, Quiroga S (2011) Towards adaptation of agriculture to climate change in the Mediterranean. *Regional Environmental Change* **11**, 159–166. doi:10.1007/s10113-010-0187-4

Intergovernment Panel on Climate Change (IPCC) (2013) *Climate Change 2013 – The Physical Science Basis. Contribution of Working Group I to the Fifth Assessment Report of the Intergovernmental Panel on Climate Change.* Cambridge University Press, Cambridge.

Intergovernment Panel on Climate Change (IPCC) (2014) *Climate Change 2014: Mitigation of Climate Change.* Cambridge University Press, Cambridge.

Knüsel B, Zumwald M, Baumberger C, Hadorn GH, Fischer EM, Bresch DN, Knutti R (2019) Applying big data beyond small problems in climate research. *Nature Climate Change* **9**, 196–202. doi:10.1038/s41558-019-0404-1

Lal R, Delgado J, Groffman P, Millar N, Dell C, Rotz A (2011) Management to mitigate and adapt to climate change. *Journal of Soil and Water Conservation* **66**, 276–285. doi:10.2489/jswc.66.4.276

Letey J, Hoffman GJ, Hopmans JW, Grattan SR, Suarez D, Corwin DL, Oster JD, Wu L, Amrhein C (2011) Evaluation of soil salinity leaching requirement guidelines. *Agricultural Water Management* **98**, 502–506. doi:10.1016/j.agwat.2010.08.009

Lindsey R (2020) *Climate Change: Atmospheric Carbon Dioxide.* National Oceanic and Atmospheric Administration (NOAA), Washington, DC, <https://www.climate.gov/news-features/understanding-climate/climate-change-atmospheric-carbon-dioxide>.

Lovell ST, Dupraz C, Gold M, Jose S, Revord R, Stanek E, Wolz KJ (2018) Temperate agroforestry research: considering multifunctional woody polycultures and the design of long-term field trials. *Agroforestry Systems* **92**, 1397–1415. doi:10.1007/s10457-017-0087-4

Maurer E (2007) Uncertainty in hydrologic impacts of climate change in the Sierra Nevada, California, under two emissions scenarios. *Climatic Change* **82**, 309–325. doi:10.1007/s10584-006-9180-9

Mishra N, Srivastava P, Singh S (2017) What do climate change projections say about future droughts in Alabama and Georgia? *Transactions of the ASABE* **60**, 1139–1151. doi:10.13031/trans.11854

Moran MS, Peters DPC, McClaran MP, Nichols MH, Adams MB (2008) Long-term data collection at USDA experimental sites for studies of ecohydrology. *Ecohydrology* **1**, 377–393. doi:10.1002/eco.24

Moriasi DN, Wilson BN, Douglas-Mankin KR, Arnold JG, Gowda PH (2012) Hydrologic and water quality models: use, calibration, and validation. *Transactions of the ASABE* **55**, 1241–1247. doi:10.13031/2013.42265

Mustafa S, Vanuytrecht E, Huysmans M (2017) Combined deficit irrigation and soil fertility management on different soil textures to improve wheat yield in drought-prone Bangladesh. *Agricultural Water Management* **191**, 124–137. doi:10.1016/j.agwat.2017.06.011

Nearing MA, Jetten V, Baffaut C, Cerdan O, Couturier A, Hernandez M, Le Bissonnais Y, Nichols MH, Nunes JP, Renschler CS, *et al.* (2005) Modeling response of soil erosion and runoff to changes in precipitation and cover. *Catena* **61**, 131–154. doi:10.1016/j.catena.2005.03.007

Neitsch SL, Williams J, Arnold J, Kiniry J (2011) *Soil and Water Assessment Tool Theoretical Documentation Version 2009.* Texas Water Resources Institute, College Station.

Nendel C, Rotter RP, Thorburn PJ, Boote KJ, Ewert F (Eds) (2018) Modelling cropping systems under climate variability and change: impacts, risk and adaptation. *Agricultural Systems* **159**, 139–143.

Nielsen DC, Lyon DJ, Hergert GW, Higgins RK, Holman JD (2015) Cover crop biomass production and water use in the Central Great Plains. *Agronomy Journal* **107**, 2047–2058. doi:10.2134/agronj15.0186

Nunes MR, van Es HM, Schindelbeck R, Ristow AJ, Ryan M (2018) No-till and cropping system diversification improve soil health and crop yield. *Geoderma* **328**, 30–43. doi:10.1016/j.geoderma.2018.04.031

O'Gorman PA, Schneider T (2009) The physical basis for increases in precipitation extremes in simulations of 21st-century climate change. *Proceedings of the National Academy of Sciences of the United States of America* **106**, 14773–14777. doi:10.1073/pnas.0907610106

Osuch M, Romanowicz RJ, Booij MJ (2015) The influence of parametric uncertainty on the relationships between HBV model parameters and climatic characteristics. *Hydrological Sciences Journal* **60**, 1299–1316. doi:10.1080/02626667.2014.967694

Pachauri RK, Allen MR, Barros VR, Broome J, Cramer W, Christ R, Church JA, Clarke L, Dahe Q, Dasgupta P (2014) *Climate Change 2014: Synthesis Report. Contribution of Working Groups I, II and III to the Fifth Assessment Report of the Intergovernmental Panel on Climate Change.* Intergovernment Panel on Climate Change (IPCC), Geneva.

Phung QA, Thompson AL, Baffaut C, Costello C, Sadler EJ, Svoma BM, Lupo A, Gautam S (2019) Climate and land use effects on hydrologic processes in a primarily rain-fed, agricultural watershed. *Journal of the American Water Resources Association* **55**, 1196–1215. doi:10.1111/1752-1688.12764

Poeplau C, Don A (2015) Carbon sequestration in agricultural soil via cultivation of cover crops – a meta-analysis. *Agriculture, Ecosystems & Environment* **200**, 33–41. doi:10.1016/j.agee.2014.10.024

Pokhrel Y, Felfelani F, Satoh Y, Boulange J, Burek P, Gädeke A, Gerten D, Gosling SN, Grillakis M, Gudmundsson L (2021) Global terrestrial water storage and drought severity under climate change. *Nature Climate Change* **11**, 226–233. doi:10.1038/s41558-020-00972-w

Psomas A, Panagopoulos Y, Konsta D, Mimikou M (2016) Designing water efficiency measures in a catchment in Greece using WEAP and SWAT models. *Procedia Engineering* **162**, 269–276. doi:10.1016/j.proeng.2016.11.058

Rey D, Holman IP, Daccache A, Morris J, Weatherhead EK, Knox JW (2016) Modelling and mapping the economic value of supplemental irrigation in a humid climate. *Agricultural Water Management* **173**, 13–22. doi:10.1016/j.agwat.2016.04.017

Ruggiero A, Punzo P, Landi S, Costa A, Van-Oosten MJ, Grilio S (2017) Improving plant water use efficiency through molecular genetics. *Horticulurae* **3**, 31. doi:10.3390/horticulturae3020031

Rummukainen M (2012) Changes in climate and weather extremes in the 21st century. *Wiley Interdisciplinary Reviews: Climate Change* **3**, 115–129. doi:10.1002/wcc.160

Sadler EJ, Lerch RN, Kitchen NR, Anderson SH, Baffaut C, Sudduth KA, Prato AA, Kremer RJ, Vories ED, Myers DB (2015) Long-term agroecosystem research in the Central Mississippi River Basin: Introduction, establishment, and overview. *Journal of Environmental Quality* **44**, 3–12. doi:10.2134/jeq2014.11.0481

Schimmelpfennig D (2016) 'Farm profits and adoption of precision agriculture.' Economic Research Report No. 2017, USDA Economic Research Service, Washington DC.

Schleussner CF, Deryng D, Müller C, Elliott J, Saeed F, Folberth C, Liu W, Wang X, Pugh TA, Thiery W, *et al.* (2018) Crop productivity changes in 1.5°C and 2°C worlds under climate sensitivity uncertainty. *Environmental Research Letters* **13**, 064007. doi:10.1088/1748-9326/aab63b

Shen M, Chen J, Zhuan M, Chen H, Xu CY, Xiong L (2018) Estimating uncertainty and its temporal variation related to global climate models in quantifying climate change impacts on hydrology. *Journal of Hydrology* **556**, 10–24. doi:10.1016/j.jhydrol.2017.11.004

Solomon S, Manning M, Marquis M, Qin D (2007) *Climate Change 2007 – The Physical Science Basis: Working Group I Contribution to the Fourth Assessment Report of the IPCC.* Cambridge University Press, Cambridge.

Toreti A, Deryng D, Tubiello FN, Muller C, Kimball BA, Moser G, Boote K, Asseng S, Pugh TAM, Vanuytrecht E, *et al.* (2020) Narrowing uncertainties in the effects of elevated CO2 on crops. *Nature Food* **1**, 775–782. doi:10.1038/s43016-020-00195-4

Udawatta RP, Garrett HE, Kallenbach R (2011) Agroforestry buffers for nonpoint source pollution reductions from agricultural watersheds. *Journal of Environmental Quality* **40**, 800–806. doi:10.2134/jeq2010.0168

Unger PW, Vigil MF (1998) Cover crop effects on soil water relationships. *Journal of Soil and Water Conservation* **53**, 200–207.

United States Department of Agriculture (USDA), National Agricultural Statistics Service (2017) *2017 Census of Agriculture.* USDA, Washington DC, <https://www.nass.usda.gov/Publications/AgCensus/2017/>.

Vaze J, Post DA, Chiew FHS, Perraud JM, Viney NR, Teng J (2010) Climate non-stationarity – validity of calibrated rainfall-runoff models for use in climate change studies. *Journal of Hydrology* **394**, 447–457. doi:10.1016/j.jhydrol.2010.09.018

Verburg PH, Soepboer W, Veldkamp A, Limpiada R, Espaldon V, Mastura SS (2002) Modeling the spatial dynamics of regional land use: the CLUE-S model. *Environmental Management* **30**, 391–405. doi:10.1007/s00267-002-2630-x

Verburg K, Bond WJ, Hunt JR (2012) Fallow management in dryland agriculture: explaining soil water accumulation using a pulse paradigm. *Field Crops Research* **130**, 68–79. doi:10.1016/j.fcr.2012.02.016

Wang S, Zhang Y, Ju W, Chen JM, Ciais P, Cescatti A, Sardans J, Janssens IA, Wu M, Berry JA, *et al.* (2020) Recent global decline of CO_2 fertilization effects on vegetation photosynthesis. *Science* **370**, 1295–1300. doi:10.1126/science.abb7772

Westra S, Fowler HJ, Evans JP, Alexander LV, Berg P, Johnson F, Kendon EJ, Lenderink G, Roberts NM (2014) Future changes to the intensity and frequency of short-duration extreme rainfall. *Reviews of Geophysics* **52**, 522–555. doi:10.1002/2014RG000464

Wilby RL, Wigley T, Conway D, Jones P, Hewitson B, Main J, Wilks D (1998) Statistical downscaling of general circulation model output: a comparison of methods. *Water Resources Research* **34**, 2995–3008. doi:10.1029/98WR02577

Zhang J, Guan K, Peng B, Jiang C, Zhou W, Yang Y, Pan M, Franz TE, Heeeren DM, Rudnick DR, *et al.* (2021) Challenges and opportunities in precision irrigation decision-support systems for center pivots. *Environmental Research Letters* **16**, 053003. doi:10.1088/1748-9326/abe436

Index